This book is to be returned on
or before the date stamped below

Post Processing Treatment of Composites

Mel M. Schwartz

Copies of all SAMPE® publications may be obtained from

SAMPE International Business Office
P.O. Box 2459
Covina, California 91722

ISBN 0-938994-75-1

PREFACE

Within the past three decades, the special requirements for military aircraft and space vehicles have brought composites from research oddity to acceptable materials for end products. All this has been accomplished with a surprisingly high level of success. What has been an especially troubling area is what I choose to call "post processing treatment." By which, I mean those finishing activities which take place after the entire part or the subassemblies which compose the part have been manufactured. These include joining, machining, drilling, etc. What I have found both in my own efforts and those reported in the literature is that, regardless of the type of composite (polymer, metal, or ceramic matrix), these systems do not behave like metals during post processing treatment. As might be expected, much of this was learned by sad experience. Therefore, it is my intention (and SAMPE's) in publishing this book, to provide information as to how such composites can be finished and the precaution to be observed. Although my personal experience is primarily with aircraft applications, that of others as well as my own, can serve as a guide to products which are finished in a proper and satisfactory manner.

DEDICATION

To Perry P. Klein, whose courage
and strength of spirit inspires all
who know him.

TABLE OF CONTENTS

CHAPTER I

1.1. INTRODUCTION

In its simplest context, a composite is composed of a reinforcement and a matrix material. This chapter is devoted to a discussion of the fiber types and resins which are employed to produce the particular composite.

1.2 FIBERS

Reinforcements come in a variety of forms including continuous and discontinuous fibers, whisker particulates, platelets, flakes, etc.

1.2.1 Glass

Glass fibers are the most widely used and least expensive of all fibers. A glass-fiber reinforced plastic (GFRP) may contain between 30% and 60% glass fibers by volume. Glass fibers are made by drawing molten glass through small openings in a platinum die. There are two principal types of glass fibers; 1) E type, a borosilicate glass, which is used most; and 2) the S type, a magnesia alumina-silicate glass, which has higher strength and stiffness and is more expensive.[1-2]

Glass fiber accounts for almost 90% of the reinforcement in thermosetting resins. Forms of glass-fiber materials are roving (continuous strand), chopped strand, woven fabrics, continuous-strand mat, chopped-strand mat, and milled fibers. The longer fibers provide the greatest strength; continuous fibers are the strongest.

1.2.2 Graphite

Although they are more expensive than glass fibers, these fibers have a unique combination of low density, high strength, and high stiffness.

Most graphite fibers are made by pyrolysis of organic precursors, commonly polyacrylonitrile (PAN) because of its lower cost. Rayon and pitch (the residue from catalytic crackers in petroleum refining) can also be used as precursors. Pyrolysis is the term for inducing chemical changes by heat, such as burning a length of yarn, which becomes graphitic and black in color. The temperatures for carbonizing range up to about 3,000°C.

The difference between carbon and graphite, although the words are often used interchangeably, depends on the temperature of pyrolysis and the purity of the material. Carbon fibers are generally 93-95% carbon, and graphite fibers are usually more than 99% carbon.[1]

1.2.3 Aramid

Aramids are the toughest fibers available and have the highest specific strength of any fiber. They can undergo some plastic deformation before fracture and, thus, have higher toughness than brittle fibers.

Aramid is an aromatic organic compound of carbon, hydrogen, oxygen, and nitrogen. Aramid yarns have been converted into broad goods, woven fabrics, and tapes, as well as continuous-filament yarns, rovings, and chopped fiber. When marketed by DuPont, aramid fiber is identified as Kevlar®.

1.2.5 Boron

Boron fiber (B_f) used as a reinforcement for polymeric and metallic materials is available in many forms, several diameters, and on substrates of tungsten and carbon.

Boron-Tungsten

The most common method for producing continuous boron filaments is by a chemical vapor deposition (CVD) plating process in which the reduction of boron trichloride by hydrogen gas takes place on a moving incandescent tungsten filament.

Boron-Carbon

Substituting a graphite monofilament for tungsten is both practical and economical. Manufacturers have successfully produced boron on a graphite monofilament. The graphite monofilament has a diameter of 0.03 mm. The mechanical properties of boron fiber made from it are somewhat less than those of boron on tungsten, but, superior to those of regular graphite fiber.

1.2.5 Silicon Carbide and Alumina (SiC and Al_2O_3)

SiC and Al_2O_3-based fibers have received much attention but are prone to oxidation and thermal degradation and are, therefore, restricted to temperatures below ~ 1,200°C. SiC fiber also requires more cost-effective fiber coating systems to limit fiber/matrix interaction and to provide the correctly engineered balance of mechanical, thermal, and physical interfacial properties. Successful composites having excellent oxidation stability and electrically insulating properties have been achieved with the use of Al_2O_3 fiber in both alumino-silicate glass and LAS (lithium alumino-silicate) glass matrices. Al_2O_3 fiber-reinforced glass was unaffected by exposure to temperature up to 1,000°C.[3]

Wetting is rather poor between different forms of Al_2O_3 and molten aluminum, with contact angles ranging from 180° (no wetting) at the melting point to greater than 60° at 1,527°C. In particular, α-Al_2O_3 (FP) based systems may involve the formation of spinels at the fiber/matrix interface, and they enhance the chemical bond at the interface.

Fiber FP has been demonstrated as a reinforcement with emphasis on Al and Mg matrices. One advantage of the Al_2O_3 fiber is its compatibility with vacuum infiltration casting, a manufacturing technique that could improve the economics for producing many MMC components.

PRD-166 fiber is a continuous fiber with the composition of α-Al_2O_3 with about 15-20 wt.% of ZrO_2.

Whiskers of SiC, Si_3N_4, and Al_2O_3 have been used but, only the SiC whisker have been widely developed and tested. Many different sizes are available with relatively high aspect ratios up to average values of 30-50. Small whiskers of diameters of about 0.3 μm have however, been mostly used.

Particles and platelets have been processed successfully in composite components. Many particulates have been investigated, these include:

- reinforcement of Al_2O_3 by SiC, TiC, BN, and TiN;

- reinforcement of SiC by TiB_2, TiC, and AlN; and

- reinforcement of Si_3N_4 by SiC and TiC.

Different types of platelets have been developed, but, most work has been done on SiC and Al_2O_3 platelets.

1.3 MATRICES

The matrix has three functions in a composite:

- Support and transfer the stresses to the fibers, which carry most of the load.

- Protect the fibers against physical damage and the environment.

- Prevent propagation of cracks in the composite by virtue of the ductility and toughness of the plastic matrix.

1.3.1 Thermosetting Resins

Thermosetting resins react at elevated temperatures to produce three dimensional crosslinked networks. This results in a change in the physical state of the resin, improved heat and chemical resistance, as well as mechanical behavior. Because these reactions are irreversible, the ability to reuse thermoset scrap (resin or composite) is very difficult.

1.3.2 Polyesters

The most commonly used thermosetting matrix (especially in the commercial market) is the combination of an unsaturated polyester dissolved in a liquid reactive monomer (usually styrene). Where somewhat higher temperature capability is required, vinyl esters replace the polyesters in such systems.

1.3.3 Epoxy Resins

Epoxies are the resin used in the largest volume for high performance composites. The combination of 1) ease of processing (low pressure, low temperature, autoclave); 2) excellent mechanical properties in composites; 3) high hot- and wet-strength properties (150°C); and 4) cost, dictate their use.[4-8]

1.3.4 Phenolic Resins

Phenolic resins, developed over 80 years ago, are beginning to find many nonstructural applications in the automotive industry, especially for highly stressed, elevated-temperature (150-200°C) applications and exhibit excellent fire-resistant properties.

1.3.5 Bismaleimides

BMIs have an upper use temperature of 232°C, compared to the epoxy upper use temperature of 175°C and are of interest for higher temperature applications.

3

1.3.6 Polyimides

Polyimides have been developed which perform in the temperature range (275-300°C) considerably beyond the capability of either epoxy or BMIs.

1.4 THERMOPLASTICS

Thermoplastics are the second general type of RMC. Thermoplastics have the potential for much higher damage tolerance (toughness), improved microcrack resistance, negligible moisture absorption, and high flame resistance. Additionally, since thermoplastics require no chemical reactions, they offer an unlimited shelf life, no necessity for cold storage, and are meltable and reformable.

1.4.1 Engineering Thermoplastics

Thermoplastic-matrix composites can be used over a temperature range from 100 to 300°C, depending on the particular thermoplastic matrix. Like thermoset resins, thermoplastic matrices are used with both continuous and chopped glass and graphite fibers.

1.4.2 Other Thermoplastics[9-10]

An example of one of the new thermoplastics is polyethersulfone (PES) which combines outstanding high-temperature characteristics with dimensional stability and surface resistivity. Other amorphous thermoplastics include:[11]

- LARC-TPI
- Polyetherimide (PEI)
- PI 2080
- PMR-15
- Polyetheretherketone (PEEK)

1.5 METAL MATRICES

Most MMC components use either aluminum, magnesium, or titanium as the matrix material.

1.5.1 Aluminum Alloys

Aluminum alloys, because of their low density and excellent strength, toughness, and resistance to corrosion, find important applications in the aerospace field. Of special interest are the Al-Cu-Mg and Al-Zn-Mg-Cu, precipitation hardenable alloys. The latest, and by far, most important development in the precipitation hardenable aluminum alloys is a series of aluminum-lithium alloys. Lithium, when added to aluminum as a primary alloying element, has the unique characteristics of increasing the elastic modulus and decreasing the density of the alloy. Understandably, the aerospace industry has been the major target of this development.

Aluminum matrices have been produced with the following fibers, whiskers, and particulates:

- Graphite (Gr)
- Silicon Carbide (SiC)

4

- Alumina (Al_2O_3)

- Boron (B)

Gr/Al composites have been fabricated from several aluminum alloys including Al-13Si, Al-10Mg, and Al-1Mg-0.6Si (6061). The graphite fiber yields a composite material that can be bent or otherwise formed without breaking the fibers.

SiC/Al composites have successfully been made which include the 2000, 6000, and 7000 series of aluminum alloys.

The B/Al MMC system has many potential aerospace applications, e.g., jet-engine fan blades and tubular struts in the mid-fuselage section of the Space Shuttle. B/Al composite has the weight of aluminum and the strength and stiffness of steel. The densities of aluminum and B/Al are about the same, and the modulus of B/Al along the direction of the fibers is about 221 GPa compared with about 207 GPa along any direction for steel.

1.5.2 Titanium Alloys

Titanium is one of the important aerospace materials. It has a density of 4.5 g cm^{-3} and a Young's modulus of 115 MPa. For titanium alloys, the density can vary between 4.3 and 5.1 g cm^{-3} while the modulus can have a range of 80-130 GPa. High strength/weight and modulus/weight ratios are important. Titanium has a relatively high melting point (1,672°C) and retains strength to high temperatures with good oxidation and corrosion resistance. All these factors make it an ideal material for MMC aerospace applications.

Titanium matrix materials, Ti-3Al-2.5 and Ti-13V-10Mo-5Zr-2.5Al have been successfully used with boron fibers and the composite materials exhibited higher strengths at 538 and 649°C exposure temperatures than the base titanium alloys themselves.

Titanium matrix composites (TMC) have been examined for applications as turbine engine fans and compressor blade materials operating at 427°C temperature and above. Tests on TMCs have exhibited higher room temperature moduli of elasticity than either stainless steel or titanium up to 599°C temperature.

1.5.3 Magnesium Alloys

Magnesium and its alloys form another group of very light materials. Magnesium's density is 1.74 g cm^{-3}. Magnesium and its alloys adhere well to Fiber FP. Magnesium matrix composites containing 70 V_f fibers have been produced. Alloys of magnesium have been successfully vacuum-infiltrated by reinforcements such as ZK60A and ZE41A (Mg-4.25Zn-1.25 rare earths-0.50 Zr). Casting of boron fibers in magnesium composites has been successfully accomplished. Preform shapes include: rods; tubes; I-beams; Z-stiffeners; D-shapes; and tapes. The process is adaptable to any system and has become a model MMC system exhibiting outstanding interfacial bond and load-redistribution characteristics.

AZ91 or AZ61 magnesium alloys have been successfully used with graphite fibers for Gr/Mg composite materials. Longitudinal tensile strength for the composite has been shown to be three times that of monolithic magnesium and the modulus value for the composite has been double that of the monolithic material.

5

1.5.4 Copper Alloys

The degradation of pure copper's strength properties, even at moderately elevated temperatures, requires either special cooling systems or a metal alloy such as beryllium copper (BeCu). At higher temperatures, BeCu is actually six times as strong as pure copper compared with three to four times for Cu/Gr. This difference, however, is more than compensated for in electrical conductivity: Cu/Gr has about 65% the conductivity of pure copper, while that of BeCu is only 25% that of pure metal.

1.5.5 Lead Alloys

Applications for Gr/Pb composite could enhance lead's potential in the chemical, battery, construction, and bearing industries. Where pure lead has a strength of 14 MPa and a typical bearing alloy (72 MPa), composites show values of 345 to 1,103 MPa. The modulus of elasticity of the composites is also four to eight times higher than that of the above mentioned bearing alloy.

1.6 CERAMIC MATRICES (CMC)

The main goal of ceramic composites is to increase the toughness. For this purpose, two types of reinforcement can be used:

• Particles, either with or without phase transformation

• Fibers or whiskers

Several mechanisms have been identified:

• Phase transformation

• Microcracking

• Crack deflection

• Crack bridging

• Fiber reinforcement

CMCs have the potential and, therefore, have received a great deal of attention for use in high-temperature structural applications. Adding a reinforcing or toughening phase should improve the functional reliability of ceramics. A reinforcing phase having the proper mechanical, physical, and chemical properties can introduce internal deformation and fracture mechanisms that would decrease the sensitivity of the matrix to flaws and increase the fracture toughness of the ceramic. As with other composite systems, the reinforcing ceramic second phase may be in a variety of forms, ranging from nearly spherical particles through whiskers and chopped fibers to essentially continuous fibers. Compact particles and whiskers lend themselves to conventional ceramic processing methods. In addition to toughening parts in all directions, this approach is relatively inexpensive, and is well suited for mass production.[12-22]

The fundamental reinforcement mechanisms of fiber composites are dominated by the interface between the fibers and the matrix. These mechanisms are governed mainly by the relative fiber/matrix interface debond toughness, the residual stresses which are mostly of thermal origin, the friction coefficient, the fiber strength distribution, the matrix properties (such as toughness and statistical nature of matrix cracking) and the fiber volume fraction (V_f).

CMCs can be processed via either liquid or gas phase routes. Liquid routes for fibers include:

- Slurry infiltration

- Melt infiltration

- Sol-gel and polymer pyrolysis

- Reaction bonding

while gas phase routes include chemical vapor deposition (CVD) and chemical vapor infusion (CVI).

The main characteristic of particle, whisker, and platelet CMCs is their ability to be processed by a route similar to that for monolithic ceramics, i.e., powder preparation, compaction, and sintering. This is possible because of the small size of the reinforcement and its stability at sintering temperatures.

The principal reinforcements for CMCs are:

- ZrO_2-Al_2O_3 (PRD-166, FP)-SiC (Nicalon)-Si_3N_4 -TiB_2-TiC-Mullite (Nextel 440)-BN

Using deposition, infiltration, and the various liquid methods, the following have been utilized as matrices in CMCs:

- SiC-Al_2O_3-Si_3N_4-B_4C-BN-TiB_2-Mullite

1.7 GLASS-CERAMIC COMPOSITES

From a fabrication point of view, glass matrix composites, compared to other ceramic candidates, probably offer the greatest commercial potential for ease of densification, low cost, and attainment of high performance aspects. They have the following important attributes:

- Glass can be produced with a wide range of thermal expansion coefficients to essentially match those of the reinforcing fibers.

- The low elastic modulus of glasses (50-90 GPa) permits high elastic modulus fiber reinforcement.

The composite densification process can be rapid since glass matrix flow is all that is required. Moreover, with the low viscosity, a high density can be achieved without damage to the fibers.

Carbon fiber reinforced glass ceramics with strengths above 700 MPa were demonstrated in the 1970s. The key to the successful development of glass matrix composites lies in being carried out as a direct extension of MMC and PMC efforts. Indeed, the elastic moduli of glass and glass ceramic allow reinforcement by load transfer similar to what occurs in fiber-polymer composites.

The type of fibers and their properties most frequently used in industry to reinforce glasses and glass-ceramics is shown in Table I-1. The variety of the fiber forms and the compositions makes available different levels of tensile strength, elastic modulus, chemical reactivity, electrical conductivity, and density. Of all the reinforcements listed, the carbon yarns offer the greatest range of mechanical properties and also potential for the lowest cost.

While fibers, whiskers, and particles used for improving the properties of the composites are important, the matrix glass and glass-ceramic are equally important to tailor a successful material. Glass-ceramic matrices can be easily fabricated by the following processes:

- Hot pressing of tapes and fabric lay-up.

- Hot matrix transfer into woven preforms
- Hot injection molding of chopped fiber compounds or preforms
- Slurry infiltration plus HIPing
- Pyrolysis

Many of the matrices used to date are indicated in Table I-2.

Table I-1 Materials Used to Reinforce Glass and Glass-Ceramic Matrices

Material	Daimeter, μm	Density, g/cm^3	E Modulus, GPa	Ultimate Tensile Strength, GPa	Thermal Expansion Coeeficient, $10^{-6}/°C$
Boron Monofilament	100 - 200	2.5	400	2.75	4.7
Silicon Carbide Monofilament	140	3.3	425	3.45	4.4
Carbon Yarn	7 - 10	1.7 - 2.0	200 - 700	1.4 - 5.5	-0.4 to -1.8
Silicon Carbide Yarn*	10 - 15	2.55	190	2.4	3.1
FP Alumina Yarn[†]	20	3.9	380	1.4	5.7
Alumino-Borosilicate Yarn[‡]	10	2.5	150	1.7	—
VLS-SiC Whisker[§]	6	3.3	580	8.4	—

*Nicalon, Nippon Carbon Co., Tokyo, Japan
†E.I. DuPont de Nemours & Co., Inc., Wilmington, DE
‡Nextel 312, 3M Co., St. Paul, MN
§Los Alamos National Laboratory, Los Alamos, NM
Source: United Technologies Research Center, East Hartford, CT

Table I-2 Glass-Ceramic Matrices

Matrix	Type	Major Constituents	Minor Constitutents	Major Crystalline Phase	T$_{max}$ °C	E GPa
Glasses						
7740	Borosilicate	B_2O_3, SiO_2	Na_2O, Al_2O_3		600	65
1723	Aluminosilicate	Al_2O_3, MgO, CaO, SiO_2	B_2O_3, BaO		700	90
7933	Silica	SiO_2	B_2O_3		1150	65
Glass-Ceramics						
LAS-I		Li_2O, Al_2O_3, MgO, SiO_2	ZnO, ZrO_2, BaO	β-Spodumene	1100	90
LAS-II		Li_2O, Al_2O_3, MgO, SiO_2, Nb_2O_5	ZnO, ZrO2, BaO	β-Spodumene	1100	90
LAS-III		Li_2O, Al_2O_3, MgO, SiO_2, Nb_2O_5	ZrO2	β-Spodumene	1200	90
MAS		MgO, Al_2O_3, SiO_2	BaO	Cordierite	1200	
BMAS		BaO, MgO, Al_2O_3, SiO_2		Barium Osumilite	1250	105
CAS		CaO, Al_2O_3, SiO_2		Anorthite	1250	90
MLAS		MgO, Li_2O Al_2O_3, SiO_2		β-Spodumene χ-Cordierite	1250	

1.8 CARBON-CARBON COMPOSITES

Carbon/carbon composites, sometimes denoted as C-C or G/G or C/C composites, offer superior potential as high performance engineering materials at elevated temperatures for aeronautic or aerospace applications where resistance to thermal shock, coupled with high strength, is of importance. No other engineering materials can compete when specific strength (maximum strength to density ratio) is considered the prime factor at extremely high temperatures (above 1,500°C). However, two limiting factors restrict the use of C-C composites in certain areas in the aerospace industry. The first is their susceptibility to oxidative degradation at high temperatures for a long period of service. The second limitation is the uncertainty of obtaining controllable quality from suitable design methodology and processing. In spite of these limitations, and because of their unique high temperature properties, C-C composites are considered the most promising of tomorrow's materials for future hypersonic flight and reentry vehicles.

C-C composites have a carbon or graphite matrix reinforced with graphite fibers or yarns. For three-dimensional structures, graphite fibers are woven or compacted into a dense, multidirectional shape. The fiber preform is infiltrated with precursor or carbon vapor onto the filament, and then graphitized to rigid composites. The quality of the 3-D C-C composites is influenced by composition, processing, and structure.

1.8.1 Processes

There are two processes used to fabricate C-C composites. The first group consists of woven or prefer-oriented structure, which is intended to be used at stress bearing hot spots for high speed vehicles such as nose-cones, throats of nozzles, and leading edges of structural components. The second consists of random distributed filaments or fibers, intended for heat shield areas where thermal shock and extended high temperature exposure created problems in, for example, the liner of ramjet engines, the exit cone of nozzles, or in brake disks.

The main processes for both groups include 1) preform formation; 2) infiltration; 3) carbonization; 4) pyrolization/graphitization; and 5) densification.

1.8.2 Properties

Too many factors, such as preform design, fiber composition and types, variety, and conditions of the matrix precursor, methodologies and conditions of infiltration, carbonization and graphitization, influence the properties of C-C composites. Therefore, the uncertainty of obtaining controllable quality is the major limitation of 3-D C-C composite. Moreover, there is a lack of analytical methodology to make a precise estimate of failure mode and strengthening mechanism. Consequently, it is necessary to be more familiar with the properties of C-C composites. Some mechanical properties of C-C composites are listed in Table I-3.

Highly densified C-C composites reinforced in two directions (2-D) with high-performance fibers can have strengths in the principal directions of 250 to 340 MPa and elastic moduli of 90 to 120 GPa. When considering flight-related applications, the most pertinent way of viewing strength is to consider the ratio of strength to density. Clearly, C-C composites are unparalleled in terms of strength retention at extremely high temperatures.

Carbon fibers, when arranged unidirectionally in any carbon matrix, produce a solid that exhibits anisotopic mechanical properties. In this case, since the fibers take most of the load,

9

Table I-3 Mechanical Behavior of C-C Composites

Property	3D Orthogonal C-C Composites, HIP 5 Cycles	C-C Composites HIP 5 Cycles + CVD	C-C Composites from Felt Perform	G-90 (Axial)
Density (g/cm)	1.88	1.88	1.9	1.9
Tensile				
Strength (MPa)	37	45	24	19
Modulus (GPa)	13	24	14	7
Strain to Failure (%)	0.34	0.54	0.35	0.22
Compression Strength (MPa)	66.1	—	56	60
MOR (MPa)	105	112	34	30

both the stiffness and strength of the composite are large when measured parallel to the fibers, but, are small when measured perpendicular to them. In addition, the composite becomes tough because a pseudo-plasticity is exhibited if debonding of the fibers and matrix occurs during failure. The frictional resistance of pulling the fibers out of the matrix behind the crack front then contributes to the work of fracture.

Fibers can be aligned unidirectionally, multidirectionally, or they can be present as various fiber weaves; these include braided yarns, stacked 2-D fabrics, orthogonal fabrics, or multidimensional weaves.[23] An important variable that influences the properties of C-C composites is the volume fraction (V_f) of fibers relative to the carbon matrix. Generally, the higher the V_f of strong/stiff fibers in a matrix, the greater the strength and stiffness of the composite. The type of weave can vary, with those more commonly used consisting of either a plain weave or a harness satin weave.[30] The harness satin weaves allow a higher V_f of fibers and usually produce higher strengths due to the floating yarns.[24-26] C-C composites are generally considered among the most competitive materials in structures designed for use at high temperature. Unfortunately, although their stability is extremely good in nonaggressive mediums, their performance quickly degrades if oxidation occurs.[27-29]

Regarding multidimensional properties, ordinary 2-D laminates have low interlaminar shear strength or peel resistance. Many demanding structures such as airframes and rocket nozzles require enhanced out-of-plane or isotropic properties – hence, the interest in more three-dimensional (3-D) fabrications.[30] SEP of France offers 2.5-D C-C composites with obliquely interwoven wraps of carbon fiber that are easier to form than 3-D C-C composites. These are used in rocket nozzles and future Space Shuttle aircraft. Engineers can preform thin woven or braided 3-D structures (such as rocket nozzles) to near net shape (NNS), then add a matrix rather than machine out solid C-C composite blocks.[30-32]

Oxidation above 600°C limits the application of this material. But, protection against oxidation can be obtained by multilayer coatings. A coating consisting of pack-cemented SiC and silica glass offers good protection up to 1,600°C, due to microcrack healing. Beyond 1,600°C, chemical reaction takes place between SiC and SiO_2 coating and a new protection concept with aluminum nitride and alumina successive coatings over SiC has been designed for providing oxidation resistance at higher temperature.

Other efforts have been made to create a protective layer to inhibit the diffusion of oxygen to an exposed graphite surface. Bowman developed the concept of the addition of a "protective gas" to react with graphite-based components to form a protective carbide coating. Wallace developed a CVD process to deposit refractory carbides such as ZrC, TiC, and NbC which

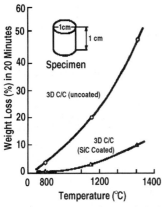

Figure I-1. Comparison of C-C composites with SiC-coated and uncoated material.

serve as protective coatings. Carnahan et.al., developed two techniques on the surface of composites of particulate and CVD SiC coatings.[24-25,30,33-34]

Figure I-1 reflects the oxidation resistance of C-C composites between the SiC-coated and -uncoated material and, as can be seen, the SiC coated composites exhibited improved resistance to oxidation.

1.8.3 Advantages and Disadvantages of C-C Composites

C-C composites' main advantage is that they enhance the toughness and strength of brittle monolithic carbon with a carbon-fiber reinforced phase, thus making the material usable in applications that demand greater structural service. The designer can tailor the properties to each application by fine-tuning production-process variables.

After processing, the finished composite offers the following advantages:

- Low weight (for example, 1/4 the density of steel).

- Good strength or specific strength (strength divided by weight) at temperatures over 1,000°C, surpassing other high-temperature materials in the 1,000 to 2,000°C range.

1.8.4 Other Matrices

A new family of ceramic fiber-reinforced ceramic composite materials referred to as 2-D C-C-TiC has recently been developed to improve the properties of the already known 2-D C-C composite. The stack of carbon fiber fabrics remains the same, but, part of the carbon matrix is replaced by TiC. Compared with 2-D C-C-SiC composite, 2-D C-C-TiC materials appears to have a higher thermal stability and wear resistance, but, a somewhat lower oxidation resistance. The processing technique consists of two CVI steps. The first one is to make a 2-D C-C composite by infiltration of carbon in a preform made of a stack of carbon fiber fabrics. In the second step, TiC is deposited within the remaining pores at 950°C. This material have been reported to have a brittle behavior (although the composite is tougher than the related monolithic ceramic).

1.8.5 Applications

Very specialized C-C composites have been used for aircraft brakes and seals while prototype integrally loaded gas turbine discs have been evaluated with some oxidation resistance being conferred by converting the surface, via a diffusion treatment, to a SiC coating. Further potential for protected C-C composite materials are gas turbine components as well as exhaust nozzle flaps and combustor panels.

Nose cones or caps and rocket nozzles can require materials able to withstand 2,200 to 2,760°C service temperatures, and C-C composites have, therefore, proved useful. However, C-C composites' most important use today is in aircraft brakes, in which C-C offers two main

11

benefits: lighter weight and more landings (two times those of steel brakes) between over-hauls.[35] Most of the heavier civil jets and virtually all military jets in production today have carbon brakes. Analysts also expect carbon brakes to penetrate high-performance ground vehicle markets, beginning with racing cars and motorcycles.

Many non-aerospace uses show promise but need further technical development and cost reduction. Monolithic carbon is less expensive than C-C composites and may work for furnace liners or parts, but, C-C composites' structural strength allows its use as load-bearing fasteners (for example, liner attachments) in these furnaces. Volume industrial applications must await reduced processing costs; then, even in some lower-temperature applications, users will benefit from the unique combination of C-C composite properties, such as light weight and high strength or stiffness. One driver for the industrial market for C-C composites is the environmental movement. C-C composite brakes and clutch materials can substitute for asbestos in high-temperature service.

In recent years, two commercial aerospace applications – brake discs for transport planes and rocket propulsion components for satellite launch vehicles – have also become prominent. Also notable is the coated C-C composite used for the nose cap and wing leading edges on the Space Orbiter vehicles.[36-37] A wing leading edge cooling system has been proposed for the NASP aircraft containing heat pipes embedded in C-C composites or C/SiC composite materials. The NASA-designed system, which would use a refractory metal such as molybdenum, or possibly lithium, to dissipate the heat, would save more than 3,429 kg compared with an actively cooled leading edge; a 900 mm test pipe has been fabricated for test.[38]

A recent announcement by LTV covered an LTV-designed and fabricated turbine rotor made of C-C composite which was successfully tested in a turbojet engine operating in excess of 1,649°C.

The test represented the first successful demonstration of the use of C-C composite for rotating components in a turbojet engine in the U.S. It achieved a new milestone in uncooled turbine engine-operating temperatures, exceeding previous temperatures by more than 1,000°C.

Conventional engine technology uses internal cooling to maintain metal component temperatures of less than 1,900°C. This cooling system, with its inherent performance penalties, is not needed with C-C, due to its increased material strength retention at very high temperatures.

A component of a different sort was designed and fabricated by W.R. Grace Co. The part, a valve for automotive engines, was conceived as a possible alternative to ceramic versions. Using C-C composites could reduce fuel consumption and improve engine power density. The C-C composite prototype valve has demonstrated the toughness, hardness, heat resistance, and strength needed for the application in internal combustion engines. Unidirectional tests indicate the valve stem could deliver good performance: 276 MPa tensile strength; 1.4×10^{-5} MPa modulus.

The present commercial industrial applications for C-C composites, hot glass handling tools and hot press dies and pistons, are minor compared to the uses in defense and aerospace.

Finally, C-C composites use in glass container forming machines as an asbestos replacement for hot-end glass contact applications illustrates its potential growth. 2-D C-C composites are used as pushout pads, stacker bars, ware transfer pads, and machine conveyor ware guides. The C-C composite material has shown wear characteristics from 100 to 300 times greater than asbestos for these applications, and because it does not get wet by molten glass and does not require external cooling or frequent replacement, it is a cost-effective replacement of asbestos.[39]

1.8.6 Future

The potential for C-C composites is no doubt substantial. They exhibit the desirable characteristic of high refractory – high resistance to chemical attack, high specific strength, high thermal conductivity as well as high thermal shock resistance. These unique properties make C-C composites possibly the most versatile material system for today and the future. However, the complexity of manufacturing process to obtain their expected quality causes C-C composites to have high cost and, thus, restricts their use only to certain special critical applications.

Preliminary tests indicate that 3-D C-C composites possess excellent mechanical, thermal, and ablative properties. After improvement, the unique nature of the 3-D C-C composites should make it possible to prepare structural components, such as leading edges with properties adjusted to meet the requirements in vehicles for future hypersonic flight.

1.9 INTERMETALLIC MATRIX COMPOSITES

Compared to conventional aerospace materials and despite some recent successes, development of intermetallic alloys into useful practical structural materials remains a major scientific and engineering challenge.

The combination of good high-temperature strength and creep capability, improved high-temperature environmental resistance, and relatively low density make these materials good candidates to replace the more conventional titanium alloys and, in some instances, nickel-base superalloys. Moreover, titanium aluminide matrix composites appear to have the potential to surpass the monolithic titanium aluminides in a number of important property areas. Fabrication into composite form may be a partial solution to some of the current shortcomings attributed to monolithic titanium aluminides.

The search for advanced materials to surpass the high-temperature capabilities of the superalloys has concentrated attention on composite materials. The effort to develop new material systems which are more chemically stable and, therefore, able to meet the demands of long-term exposure to elevated temperatures than the traditional MMCs has shifted the focus to intermetallic-fiber-reinforced (particularly aluminide) matrix composites (IMCs).[40-44]

It is expected that intermetallic systems, used both monolithically and as components in composite systems, will leapfrog the superalloys and redefine the absolute use temperature ceiling. Consequently, a dedicated effort towards the basic research and development of these materials has been expanded and is being championed throughout the high temperature materials community.

By going the composite route, the low density of the intermetallic compounds can be utilized to good advantage and, if low density, high strength fibers are available, the low strength of intermetallic matrices becomes less of an issue. Thus, the matrix can be optimized for other properties (ductility, oxidation resistance, and density). the influence of the fiber on the strength properties of a composite has been discussed by McDanels and Stephens[45] for the use of SiC_f in aluminide matrices where the predicted strength-to-density for SiC reinforced aluminide composites was shown to be essentially independent of matrix strength. An example of this concept is illustrated by the results of Brindley.[46] The tensile behavior was shown to achieve the predicted rule of mixture (ROM) behavior. On a density corrected basis, the composite was shown to have superior tensile properties compared to wrought nickel base and cobalt base alloys and a single crystal superalloy, NASAIR 100.

Table I-4 The Future for IMCs in Aircraft

Vehicle	Enabling Technologies
Subsonic Transport*	Metal/intermetallic composites for compressor, HP turbine, LP turbine. Ceramic composites for combustor and turbine. Polymer composite for fan, fame, and duct.
Commuter Rotocraft[†]	Metal/intermetallic composites for compressors. Uncolled ceramic composites for turbine. Unique eccentric design.
High-Speed Transport[‡]	Intermetallic composites for compressor. Ceramic turbine and combustor components. Intermetallic and ceramics for nozzle.
* GE study contract. †Allison study cotnract – core only. ‡Pratt & Whitney Aircraft study contract.	

Table I-5 IMCs and Other Composites and
Potential Application to Generic Components

Component Materials	Fans			Compressors					Turbines			
	Blades	Cases	Fairings	Ducts	Cases	Blades	Rings	Vanes	Blades	Vanes	Ducts	Cases
PMC	X	X	X	X	X							
IMC/MMC						X	X	X	X	X	X	X
CMC									X	X	X	

One of the major problems facing the successful development of IMCs is the compatibility between fiber and matrix both from a chemical viewpoint and the mismatch in coefficient of thermal expansion (CTE). An example of the extreme complexity of fiber matrix interaction is the SiC_f which has a two-layer carbon zone on the surface which further contributes to the chemical reactions in this system.

The problems associated with the IMCs provide challenges and opportunities for materials and structures researchers. To overcome the compatibility problems of the fibers and the matrices in IMCs, studies in kinetics of reaction types and methodology have been undertaken to identify the most appropriate reinforcements for the intermetallic matrices.

The strength requirements for engine applications may not be attainable with the aluminides by alloying without compromising their density and melting point advantages. However, the aluminides can be used as matrices for a range of IMCs because of their light weight, oxidation resistance, and/or high melting temperatures and thus achieve the required strength and resistance to fracture via fiber strengthening. It is anticipated that the aluminides meet engine structural requirements imposed by future turbines and compressors in the 815 to 1,260°C temperature range, Tables I-4 and 5.

The ideal fiber for reinforcement of IMCs would be chemically compatible with the matrix, have a CTE that matches that of the matrix, and have a low density and a high modulus and strength at elevated temperatures. No commercially available fibers meet all the above criteria, and therefore, developments in several government agencies are underway to meet these goals.[47-49] The most important goal is that the fiber should have a CTE close to that of the

14

matrix, so that the thermal expansion mismatch between the fiber and the matrix can be minimized. A fiber with a CTE of about 10 ppm/°C would be needed to match Ti-Al, while a CTE of up to 16 ppm/°C would be needed for matching NiAl matrices. The fiber must be chemically compatible with potential matrix materials, so that the properties of the composite are not severely degraded during fabrication or service. For convenient fabrication with IMC matrices, the fiber should be a monofilament in the 2-10 mm diameter range. For example, carbon-coated SiC_f has successfully reinforced α-Ti_3Al which has a room temperature ductility of 1.5-3% and Ti_3Al is ductile enough to adjust to the different CTE of it and SiC_f.[50]

Al_2O_3 is the most common ceramic oxide fiber. Conventional Al_2O_3 fiber is weaker than SiC or TiB_2, and a poorer thermal conductor and heavier than SiC. But, it stands out in other ways. Heat resistance is exceptional, and fully oxidized Al_2O_3 will not degrade in oxygen at high temperatures. Its expansion coefficient closely matches TiAls, and it does not react with TiAl at high temperatures.

Al_2O_3 may also be a candidate to reinforce NiAl, which has been under consideration for turbines. NiAl does not ordinarily wet out Al_2O_3 fibers, therefore, composites are not very tough. By adding two strong oxidizers, titanium, and yttrium, to the intermetallic matrix, wettability has been improved.[51].

TiB_2 works well with γ-TiAl and good mechanical properties and good ROM properties have been developed. Additionally, the CTE of TiB_2 matches more closely that of TiAl, and the material is less reactive at 982°C than SiC.[61]

1.9.1 Processing

Development of IMCs is, to a certain extent, following a path parallel to the continuing development of MMCs.[51] Two broad families of materials are emerging: those having particulate or short fiber reinforcement, where emphasis is on moderate property benefits at acceptable cost; and those having continuous fiber reinforcement, where potential property benefits are greater, as are potential costs.[52-54]

A number of processing approaches for making IMCs have been pursued, certainly more than the total number of viable IMC systems. In fact, manufacturing-oriented comparisons of processing techniques that have been made by systematic and objective evaluation still do not point to a single manufacturing process.[55] Perhaps the most widely recognized approach is the foil-fiber-foil technique, and its variations, which have been used to produce Ti-based MMCs and IMCs.[56-57] This technique's primary advantage is that the very thin foils of Ti alloys can be produced rather easily, and with minimal interstitial pick-up. However, foil techniques are labor intensive and limited to relatively ductile matrices. Common problems can include incomplete diffusion bonding of the intermetallic foils, fiber swimming, fiber breakage and cross weave/matrix alloy reaction. Powder techniques, such as powder tape casting, may be a tenuous processing route for interstitially sensitive materials such as the titanium aluminides. Excessive interstitial pick-up is unacceptable because it can result in embrittlement of the matrix. A significant effort in plasma spraying has been performed using titanium-matrix composites.[58] Problems have included thermal shock to the fibers and breakage, interstitial pick-up and loss of microstructural control. Despite these issues, both powder processes and plasma spraying techniques offer excellent control of fiber distribution and are amenable to scale-up and automation. Another method is PVD, although the slow rate of deposition is a concern, this technique can offer good control of fiber spacing.[59] In terms of production rate, melt infiltration is an attractive approach. Unfortunately, there are serious problems in retaining fiber spacing and severe chemical reactivity between the matrix and the fiber at these very elevated temperatures. All of the processing methods exhibit some degree of difficulty in manipulating the matrix microstructure.

A critical barrier to IMC application is the absence of commercial suppliers who are able to support production part manufacturing. The first step to building the required supplier base for IMC manufacturing is the establishment of manufacturing capability for conventional titanium-based composites.

The α_2-TiAl-based materials, matrix alloys with high Nb contents resulting in $\alpha_2 + \beta$ microstructures have received the most attention for composite applications within aircraft engines and the NASP, since the $TiAl_3$-based composites have a potentially higher payoff.

Many issues must be resolved before using $TiAl_3$-based composites in aircraft engines and NASP related applications.

These issues are

- Insufficient environmental resistance which occurs with Ti-based composites.

- Reaction between the SiC_f and the $TiAl_3$-based matrix.

- A large CTE mismatch between the SiC reinforcing fibers and the $TiAl_3$-based matrix.

Coating the fibers to retard the reaction between the fiber and matrix and to provide a compliant layer between the fiber and matrix is one way of getting around the incompatibilities between the SiC_f and the $TiAl_3$-based matrix. As is the case for titanium-based alloys, Al_2O_{3f} are chemically incompatible with $TiAl_3$-based alloys. Environmentally resistant coatings for the entire composite are also being developed. However, most work to resolve the major issues related to $SiC/TiAl_3$-based composites has had only limited success. Substantial improvements need to be made before $TiAl_3$-based composites can be seriously considered for aircraft engine applications.

However, phase stability and environmental resistance are key issues that need to be addressed. Composites having a Ti_2AlNb-based matrix are being considered to augment work on Ti_3Al-based composites.

1.9.2 Prealloyed Powder Mixing

The PM route most frequently applied to Al-based composites is to blend prealloyed matrix powders with reinforcement particles and then to consolidate and form. This does not appear to have been widely applied to IMCs possibly reflecting the absence of fine ($\sim 10 \mu m$) prealloyed powders for these materials, together with concerns regarding agglomeration of fine particles or the use of coarse particles, which will act as defects in these defect sensitive materials.

However, the powder route involves blending of chopped fibers with either prealloyed powders, elemental powders, or a mixture of the two, followed by either hot pressing, hot isostatic pressing or extrusion.[60-69]

Powder metallurgical processes have successfully produced Ti_3Al/SiC[70] and Al_2Ta/Al_2O_3[71] systems.

Brindley[70] prepared TiAl powder cloths from prealloyed powders; these cloths were interspersed with layers of continuous SiC fibers, and consolidated by hot pressing. Anton[71] infiltrated a Al_2O_3 fibrous preform with elemental Al and Ta powders. Consolidation of the composites or of the Al_3Ta matrix was performed by reactive hot pressing.

1.9.3 Reactive Consolidation

The process of reactive sintering for monolithic intermetallics can be readily extended (at least in concept) to production of IMCs. The addition of ceramic fibers or particles will resist densification and so the process has been undertaken simultaneously with hot isostatic pressing (reactive hipping or RHIP).[72-73] The reinforcements result in increases in strength, but also a substantial decrease in elongation.

Talia, Mickle, Frazier, et.al.[74] used a novel approach in developing IMCs. They cold isostatically pressed elemental aluminum and titanium powders or wires at 344 MPa. The compacts were reactively sintered at 680°C in air. It was observed that reactive sintering produced Al_3Ti. In addition, an aligned intermetallic reinforced composite was produced when aluminum powders were reactively sintered with titanium wires. It was also observed that the processing parameters could be used to tailor the interface and control the mechanical properties of the composite.

1.9.4 Injection Molding

Reactive consolidation of mixed elemental powders and fibers does not give fiber alignment. One approach to producing aligned fiber composites by this route is to mix the fibers and powders with an organic binder and produce a preform by injection molding. Control of the material flow can align the fibers in the preform. Work to date has encountered difficulties with preform debinding and contamination of the powders with preform residues, thus inhibiting reactive sintering.[73,75]

1.9.5 XD Processing

The XD process developed by Martin Marietta Corp.* has as its central feature the *in situ* formation of second phases by reaction of the ceramic forming constituents in a metal solvent medium to provide very finely dispersed ceramic particles in the metal matrix.[76] The advantages claimed for this process include the fine size of the dispersed reinforcement produced (e.g., about 1 μm TiB_2) and the clean reinforcement/matrix interfaces.[77-78]

XD titanium aluminides can be converted to powders by attrition or atomization routes for further processing giving a range of reinforcement sizes.[78]

The XD process affords the flexibility of a PM route in addition to the incorporation of a large V_f of the reinforcing phase (as high as 60 vol.%) in the matrix material. Alternatively, the IM route can be used to produce XD composites containing up to 10 vol.% of the reinforcement; for larger V_f, melt viscosity becomes a limitation.

Several variants of the XD process exist. The process relies on the exothermic heat of formation of intermetallic compounds. Control of reinforcement size is dependent on the magnitude of the exotherm and the thermal conductivity of the immediate environment. Thus, particle size and distribution are extremely system specific. Similarly, appropriate matrix-reinforcement combinations allow some flexibility in reinforcement shape. In general, the starting materials

---

Now Lockheed Martin Corp.

Table I-6 Room Temperature Mechanical Properties of Investment-Cast Ti-48Al-2V-2Mn Alloy With and Without 7 Vol% TiB$_2$ Reinforcement

Yield Strength MN m^{-2}	Ultimate Tensile Strength MN m^{-2}	Plastic Elongation %	Modulus GN m-2
Ti-48Al-2V-2Mn 400 - 428	538 - 573	1.1 - 1.2	159
XD Ti-48Al-2V-2Mn + 7 Vol % TiB$_2$ 573 - 587	704 - 731	0.9 - 1.1	180

can be either elemental or prealloyed powders; these are blended appropriately and heated to the reaction temperature.

The patented XD (exothermic dispersion) technology is based on sustained high temperature synthesis. In the process, elemental boron, titanium, and aluminum powders are blended and compacted, then ignited. As the flame front passes through, titanium diboride and titanium aluminide form spontaneously. Because of the *in situ* development of reinforcement, the process eliminates oxide formation that could weaken the interface between the reinforcement and the matrix. The resultant master alloy has a high (20 to 75%) concentration of TiB$_2$, which can be added to a matrix material to achieve required levels of reinforcement.

Particulate reinforcement in γ TiAl not only enhances mechanical properties, but also signifi-cantly influences the kinetics of phase transformations that are critical to mechanical properties. Detailed investigations have been carried out on near γ TiAl-TiB$_2$ composites, obtained by the addition of an XD produced TiB$_2$ containing master alloy to TiAl, followed by ingot casting.[79]

Investment casting, which reduces machining requirements significantly, appears to be an attractive process for producing near-net-shape (NNS) components. Additionally, the proce-dure for investment casting of titanium aluminides is similar to that for conventional titanium alloys.[80] Room temperature mechanical properties of investment cast, HIPed, and heat treated materials with and without additions of TiB$_2$ are given in Table I-6.[80] Improvements in strength and modulus with no loss in ductility are attributed to grain refinement and particulate strengthening. This approach has been successfully used to investment cast low pressure turbine blades to NNS.[80-81]

In rotating components, the XD process can produce a low density in TiAl/TiB$_2$ which in turn reduces strength requirements of the whole system. For example, reducing the weight of low-pressure jet-engine turbine blades by 30 to 50% could allow weight reductions in the turbine disk, turbine shaft, and bearing support housings. Other examples of rotating applica-tions include compressor vanes, rotating seals, diffusers, and turbocharger wheels.

In structural components, TiAl composites reduce both stress and running clearances because of their low density, high modulus, and low CTE. Replacing steel with the composite in automotive exhaust valves lowers the spring tension required to retain the valves during operation. This results in lower friction in the valve train, allowing fuel economy to be increased by as much as 0.3 kpl. In a development program at Ford Motor Co., investment-cast TiAl/TiB$_2$ NNS blanks were machined to shape in just two minutes. Other structural applications include compressor cases, compressor vanes, diffuser housings, missile fins, and structural brackets.

Howmet Corp. has adapted the XD technology by combining the XD master alloy with other alloying elements to form an electrode which is then triple vacuum-arc remelted. The resulting ingots are well suited to investment casting because the reinforcements are stable in the liquid phase, retain their fine size through the investment casting process, and generate a fine, homogeneous structure. TiB_2 reinforcement levels for the ingots range from 0.8 to 7 vol.%.

1.9.6 CoSpray Forming

Cospray deposition of Al and reinforcing particles has been developed as a route that gives material properties approaching those of PM materials at a potentially much lower cost. To date, application of this technique to production of IMCs has not been reported, although spray forming has been used. Although the cospray route offers the potential advantages of low interaction times over other melt routes for composite fabrication, practical problems associated with melt contaminant and contamination must be overcome if it is to be used for reactive matrix composite systems.[82]

1.9.7 Liquid Metal Infiltration

This is one of the first techniques discussed for production of continuous fiber reinforced IMCs. As with MMCs, composite properties will be dominated by fiber/matrix interface related effects, controlling key factors such as the retention of fiber integrity, the avoidance of stress raising interaction products and matrix contamination, and the accommodation of differential thermal expansion stresses.

The pressure assisted liquid metal infiltration technique has been applied by Nourbakhsh and co-workers to NiAl, Ni_3Al, and TiAl.[83-85]

A series of Ni_3Al, TiAl and Fe_3Al composites, unidirectionally reinforced with either DuPont's fiber FP (Al_2O_3) or PRD-166 (Al_2O_3-20 wt% ZrO_2) continuous fibers which were 20 μm in diameter were pressure cast by Nourbakhsh and Margolin.[86] In the case of nickel and iron aluminides, it was found necessary to add a small amount (0.5-1 at%) of a strong oxide former such as Ti, Y, and Zr to improve wetting and therefore to achieve complete infiltration. Mechanical testing indicated that reinforcing Ni_3Al with PRD-166 fibers resulted in a twofold increase in specific modulus and a fivefold increase in strength.[80]

A modified TiAl alloy employed in the investigation had a composition Ti-48.4 at% Al-1 at.% Mn and was prepared by arc melting of the constituents. Manganese has been shown to be beneficial in enhancing the ductility of TiAl[87] and was therefore incorporated in the alloy. Diffusion of titanium from the molten metal into the outer perimeter of the PRD-166 fiber enhances the rate of Al_2O_3 grain growth in that region. The ZrO_2 particles near the fiber surface were brought to the fiber surface by the migrating Al_2O_3 grain boundaries. Since the CTE of TiAl and Al_2O_3 are almost identical, the dislocation content of the matrix near the fiber/matrix interface was rather low.

Fukunaga, Wang, and Aramaki[88] reported on a series of reaction squeeze casting tests applied to IMCs. It is difficult to form Ti_3Al by reaction squeeze casting for a large amount of molten aluminum. The calculation of the complete reaction leads to a fiber V_f of 47 to 48% for Al_2O_3 + TiAl and of 32 to 33% for Al_2O_3 + $TiAl_3$.

One reaction mechanism and the formation process for a certain reaction product by squeeze casting was demonstrated by Fukunaga and associates. Applying this process to another system, they showed the possibility of fabricating hard, heat-resistant matrix composites with a NNS: for instance, mixing the substances of easy-to-react TiO_2, iron, nickel, and so on, with the reinforcements such as SiC, Si_3N_4 and Al_2O_3, which may make IMCs of the future.[89]

1.9.8 Powder Cloth

Chandra, Ananth, and Garmestani[90] evaluated the micromechanical modeling of process-in-duced residual stresses in Ti-24Al-11Nb/SCS-6 composites by fabricating consolidated fiber mats over matrix available as powder cloth or cast tape.

Draper, Brindley, and Nathal investigated the effect of fiber strength variability on RT composite strength using the powder cloth technique to fabricate composite SCS-6 plates of SiC_f/Ti-24Al-11Nb (at.%).[91] They found that:

- The strength of the composite was directly related to the strength of the as-received fiber with higher strength fiber resulting in higher strength and larger strain to failure of the composite.

- The strength of the composite plates containing mixed fiber strengths was dominated by the lower strength fiber.

- The powder cloth fabrication process resulted in only a slight degradation of SCS-6 SiC fiber strength.

- With the variation in fiber strength accounted for, the composite UTS could be correlated with V_f of fiber.

The properties of these first-generation SiC/Ti-24Al-11Nb (Ti-24-11) composites compare favorably with those of current nickel-base superalloys on a strength/density basis. However, the SiC_f is too reactive with the matrix material above 815°C, and also with the other candidate matrix materials. Therefore, researchers are focusing on using Al_2O_3 as the reinforcing fiber for these materials. There is a need for new fibers, however, and new compositions and fiber-processing techniques, such as the laser floating-zone process, have been identified.

1.9.9 Foil/Fiber

Stacking and hot pressing of alternate layers of monofilament fibers and matrix foils is a standard route for MMCs. Conceptually this is easily transferred to IMCs. However, the difficulties in producing the foil should not be underestimated and alternate foil production routes to thermomechanical processing (TMP) should be considered, e.g. by consolidation of chopped rapidly solidified ribbon.[103-104]

1.9.10 Spray Deposition

Both arc spray and plasma spray processes have been used by NASA to produce IMC monotapes by spraying matrix powder over a collimated array of monofilament wound onto a mandrel.[105] Consolidation is again by hot pressing.

Chandra and his associates also used plasma spray as an alternate method by plasma spraying the matrix material onto a rotating drum wound with the reinforcing fiber, which produced a monolayer of fiber and matrix; monolayers were then assembled to a NNS by the simultaneous application of temperature and pressure over time.[101]

1.9.11 SHS

Using elemental powders, combustion reaction has been carried out to form intermetallic-ceramic composites in the Ti-Al-N system. Ti and Al powders reacted exothermically in gaseous nitrogen and formed a mixture product which had a fine distribution of the Ti_2AlN particles in the matrix TiAl with a small amount of Ti$_3$Al. Subsequently, these reacted products were arc-melted to obtain fully dense button ingots. The resulting composites had about 30 V$_f$ Ti$_2$AlN, and the Ti$_2$AlN particles were ellipsoidal or columnar in shape with sizes of 2-10 μm and appeared to be homogeneously distributed and well bonded to the matrix TiAl. It was found that such composite materials have a high strength at both room and elevated temperatures and some intrinsic compressive ductility at room temperature. Therefore, this processing technique is of interest as a new combustion reaction process to make intermetallic-based composite materials.[105]

Metin and Inal[106] studied the mechanical properties, microstructure and crystallographic structure of TiAl + ZrO$_{2p}$ composites as well as TiAl$_3$ without ZrO$_2$, consolidated by explosive-induced pressure. They prepared compacts by the explosive compaction technique which were crack-free, highly dense and structurally sound. Slight improvements in fracture toughness values of compacts were obtained with ZrO$_2$ addition and attributed to the presence of the tetragonal form of ZrO$_2$. Post-shock annealing at 850°C for four hours retained approximately 12% of the ZrO$_2$ in the tetragonal form.

Eliezer and associates[96-97] formed IMCs containing discontinuous ceramic or intermetallic particulate reinforcements by high-energy high-rate powder consolidation. BCC-stabilized Ti$_3$Al powder matrices was consolidated with second phases including AlN, TiB$_2$, and TiAl powders. Processing was accomplished by discharge of a single high electrical energy pulse through a powder blend under pressure. Control of energy input produced solid-state and liquid-phase-assisted consolidation and the high-temperature stability of the matrix/reinforcement interfaces were compared by annealing treatments of up to 100 hours at 800-1,000°C.

1.9.12 Nickel Aluminides

Processing techniques for nickel and iron aluminide composites have been reviewed in general by Stoloff and Alman,[98-99] Feest and Tweed,[100-101] Bowman and Noebe,[102] and Yang and his co-workers.[103]

Seybolt[104] mechanically alloyed NiAl with Al$_2$O$_3$, Y$_2$O$_3$ and ThO$_2$ particles. These particles improved the high temperature tensile properties of NiAl. The effects of fine TiB$_2$ (1 to 3 μm) particles on NiAl have also been studied.[78,105-106] The proprietary XD process has been employed to fabricate NiAl dispersed with 2.7 to 30 v/o TiB$_{2p}$.[78,105-106] These particles significantly improved the compressive strength of stoichometric NiAl; the addition of 30 v/o TiB$_2$ improved the compressive flow stress of NiAl from 25 MPa to 95 MPa at 1,027°C and a strain rate of 2 x 10^{-6} sec^{-1}.[78,105] The XD process also has been used to disperse fine TiB$_{2p}$ in Ni-38.5 a/o Al,[107] NiAl/Ni$_2$AlTi alloys,[108] CoAl[55], and Ti-45 a/o Al.[109] Similarly, NiAl/2.7 v/o TiB$_2$,[110] fabricated by rapid solidification techniques, had a tensile yield stress of 208 MPa as compared to 85 MPa for monolithic NiAl at 760°C. Not all efforts have been aimed at fabrication of particulate composites based on NiAl. Kumar, Whittenberger, et.al., also fabricated NiAl and NiAl/TiB$_2$ alloys reinforced with random Al$_2$O$_{3w}$. With 15 v/o whiskers the fracture toughness improved to 9 MPa m$^{1/2}$ compared to 6 MPa m$^{1/2}$ for monolithic NiAl. However, the scatter associated with the fracture toughness of the whisker-reinforced composite was significant.

1.9.13 Prealloyed Powder Mixing

Povirk et.al.,[111] did consolidate a composite from prealloyed Ni$_3$Al powders with 20 V$_f$ DuPont FP alumina fibers which were dispersed by high shear mixing in hexane. After consolidation via HIPing, some agglomeration of fibers and fiber breakage was noted. A binary alloy matrix gave a composite with 10% elongation at room temperature, whereas a more complex alloy matrix gave poor ductility, owing to the formation of coarse particles at the fiber/matrix interface.

1.9.14 Mechanical Alloying

Y_2O_3 was successfully incorporated in prealloyed Fe-40Al-0.1Zr (at.%) by attritor milling. On extrusion, this material had a uniform distribution of Y_2O_3 and a grain size of 3-5 μm, giving high room temperature ductility. Other systems have been discovered by serendipity. Whittenberger et.al.,[112] produced a composite of NiAl with a dispersion of AlN by trying to mill NiAl with Y_2O_3 in liquid nitrogen. Comparative tests at 1,025°C showed this composite to be twice as strong as NiAl-10% TiB_2 over a range of strain rates up to 2 x 10^{-4} sec^{-1}.

1.9.15 Reactive Consolidation

Reactive sintering, also termed self-propagating high temperature synthesis or gasless combustion has been described earlier.[72-73,113]

Alman and Stoloff[98] fabricated NiAl/TiB_2 composites by RHIP at 750 and 1,200°C. The composite alloy fabricated at 750°C was inhomogeneous. The application of pressure during RHIP caused the transient liquid phase to become nonuniformly dispersed, which lead to the inhomogenity.

A rule of mixtures (ROM) calculation provides an upper limit for the strength of NiAl reinforced with TiB_2, assuming that there is no interaction (i.e. dislocation motion impedance) between the matrix and particles. It is important to remember that the properties of NiAl are sensitive to grain size, impurity content, deviations from stoichiometry, and strain rate effects. However, it is important to recognize that fine TiB_{2p} are needed to significantly strengthen NiAl at elevated temperatures. Also, NiAl fabricated by reactive synthesis has equivalent mechanical properties to materials produced by other techniques.

Mei, Yuan, and Duan[114] examined the possibilities of preparing TiC-reinforced NiAl-matrix composites by self-propagating high temperature synthesis and melting process (SM).* Two kinds of composites, namely, commercial TiC-reinforcement and synthesized TiC-reinforcement Ni_3Al-matrix composites, were fabricated and the effects of particle size of the commercial TiC on the mechanical properties of the Ni_3Al-matrix composites were studied. Their results showed that the mechanical properties of the composites decrease with increasing particle size of the commercial TiC. The microstructures of 35 wt.% TiC + 65 wt.% Ni_3Al composites produced by SM technology from the four elements Ti, C, Ni, and Al were examined and the results showed that in these composites, the particle size of TiC synthesized *in situ* is fine and that the materials have considerable high-temperature bending strength and fracture toughness.

Misiolek and German[116] studied the processing (reactive sintering) of NiAl matrix composites reinforced with NiAl/$15V_f$ TiB_2 continuous fibers. The whole composite material i.e., both the reinforcing fibers and the matrix material were produced by extrusion of powder binder mixtures, CIPed together and then HIPed. The authors claim that this fabrication method enables one to tailor the properties of the manufactured composite material. This can be done

---

*Similar to SHS.

22

by controlling the fiber composition, fiber fraction and their layout in the matrix. Extrusion provides the flexibility in terms of fiber diameter, while the HIP process allows for the usage of reliable process parameters such as temperature, pressure and time.

Reactive consolidation also may be applied to injection molding, a process increasingly being used for powder processing of both monolithic and composite alloys.[116] In injection molding, a mixture of powders, short fibers and a binder is extruded through a tapered die to achieve fiber alignment. Extrusion must be performed above the softening temperature of the binder. After extrusion, the binder is removed (thermally or by wicking action) and the compact is consolidated to approximately full density by HIP. Apart from the alignment of fibers, which is achieved only when particles and fibers are very small in diameter (10 μm), this process offers the possibility of producing complex P/M parts. However, the principal disadvantages are the difficulty of complete binder removal and the inability to produce continuous-fiber-reinforced composites. Nevertheless, fully dense Al_2O_3-reinforced composites of NiAl have been produced successfully by this method.

1.9.16 XD Process

Incorporation of up to 30 vol.% TiB_2 of 1 μm size in NiAl has led to substantial increases in strength under slow compressive loading at 1,025°C. The ductility of NiAl may be improved by using a Ni-38.5 at.% Al composition. However, incorporation of 20 vol.%: TiB_2 by XD has led to a high temperature strength little better than unreinforced NiAl.[117]

TiB_2 , as a 1 μm in diameter particle, has been incorporated into NiAl-Ni_2AlTi and CoAl matrices in the form of a weakly bonded compact. However, other ceramic reinforcements (carbides, nitrides, and silicides) have been incorporated into Nb-Ti-Al matrices.[118]

Interfaces of TiB_2-NiAl and α-Al_2O_3-NiAl in TiB_2/NiAl composites produced by the XD process were investigated by analytical electron microscopy by Wang and Arsenault.[119] Although no consistent crystallographic orientation relationships were found between NiAl and TiB_2 or Al_2O_3, semicoherent interfaces between α-Al_2O_3 and NiAl were observed by high-resolution electron microscopy (HREM) in areas where the low indexed crystallographic planes of α-Al_2O_3 aligned with that of NiAl.

Kumar and Whittenberger[78,120-123] compiled an extensive review of recent results obtained for discontinuously reinforced IMCs produced using the XD process. Their review covered work on the properties of intermetallics (single phase and multiphase) reinforced with 0-30 V_f $(TiB_2)_p$ using the XD approach The versatility of the XD method was evident: both IM and PM routes can be used to process the particulate composites; single crystal reinforcements of different types, morphologies, volume fractions, sizes, and distributions were obtainable in a range of matrices; the method relies on the exothermic heat of formation of intermetallic compounds and is therefore energy efficient; the IM route produces grain refinement and microstructural homogeneity, both of which are central to improved workability and permit casting of NNS parts.

When judiciously incorporated into an intermetallic, particulate reinforcements can enhance physical and mechanical properties such as density, stiffness, CTE, ambient and intermediate temperature strength, and, to some extent, creep resistance. By refining grain size, they can also enhance phase transformation kinetics, thereby encouraging certain desirable microstructural transformations.

23

There are also some drawbacks associated with particulate reinforcement of intermetallic matrices. Improvements in toughness require a non-uniform dispersion with a small mismatch in modulus and CTE, but these criteria counteract effects that are beneficial to other properties. Thus, the usefulness of large V_f of discontinuous reinforcements in intermetallic matrices is not yet well established.

1.9.17 Powder Cloth Process

Brindley and Barlotta[123-125] were the first who reported on the use of the powder cloth method to produce Fe-40 at.% Al-W and NiAl-W composites. The technique they described, mixed the intermetallic alloy powder (which may be produced by atomization or by pulverization of melt-spun ribbon) with a powdered Teflon binder (the most successful results were obtained in the range 4-11V_f) and Stoddard solution (high grade kerosene). The mixture, after most of the Stoddard solution has been evaporated, has a dough-like consistency. The dough is rolled into thin sheets with a stainless steel rolling pin. The balance of the Stoddard solution is evaporated from the rolled sheet, which is then trimmed to size for hot pressing.

The fiber arrangement is prepared by winding a continuous fiber on a drum mounted in a lathe. When the desired fiber spacing and mat width have been achieved, the lathe is stopped and the mat is coated with a polymer-base binder such as polymethylmethacralate in a solvent base. After evaporation of the solvent, the fiber mat is removed from the drum and cut to size in preparation for hot pressing. The desired number of fiber plies is obtained by stacking alternate layers of powder cloth and fiber mat. Vacuum hot pressing is used for final consolidation; a dynamic vacuum and liquid-nitrogen-cooled trap to retain organics during binder removal are necessary for best results. As in injection molding, complete binder removal is required to avoid contamination of the composite with carbon or other constituents such as fluorine in the binder.

Bowman, Noebe, Doychak et.al.,[97] reported the fabrication of NiAl reinforced with continuous fibers of Al_2O_3 or Mo using the powder cloth method. The authors found and concluded the following:

- Matrix/fiber bond strength of NiAl/Al_2O_3 composites fabricated by standard powder-cloth techniques was entirely frictional in nature (~ 20-80 MPa). Use of a compliant layer was recommended.

- A strong bond was present between matrix and fiber in directionally solidified NiAl containing Al_2O_{3f} (> 250 MPa).

- Composites of NiAl/Al_2O_3 fabricated without the use of binders had bond strengths of > 250 MPa.

- NiAl/Mo composites had bond strengths of > 250 MPa and higher toughness than monolithic NiAl.

- A modification of the powder-cloth technique which eliminates the use of binders was desirable.

1.9.18 Spray Deposition

Preliminary results were reported by Liang, Kim, Earthman, et.al.,[126] on the microstructure and elevated temperature creep rupture behavior of a spray-atomized and co-deposited Ni_3Al composite reinforced with SiC and TiB_2 particulates. Microstructural studies suggest that SiC reacts extensively with Ni_3Al, leading to the formation of other carbide phases, e.g., Cr_3C_2.

Regarding the elevated temperature behavior of the hot-extruded materials, the monolithic Ni_3Al exhibited longer creep rupture lifetimes compared with that for the $Ni_3Al/SiC/TiB_2$ IMC. This behavior was attributed to the formation of cavities around small carbide particles, resulting from the decomposition of SiC_p which may account for the shorter lifetimes of the $Ni_3Al/SiC/TiB_2$ IMC.

1.9.19 Liquid Infiltration Techniques

Glushko, Mileiko, and Kondrashova[127] fabricated a $Ni_3Al/Al_2O_3 + ZrO_2 + Y_2O_3$ IMC using a liquid infiltration technique under low pressure. The fibers were placed in an alumina and/or quartz mold and the N_3Al placed above the fiber bundle, heated to 1,480°C in vacuum chamber and argon gas under a pressure of 0.1 MPa. Initial test results were very promising and they obtained matrix strength equal to about 2,000 MPa, whereas the pure matrix strength was only about 1,200 MPa.

They also found that the values of composite strength up to temperature 1,200°C are higher than those reported recently.[128-129] Evaluation of effective fiber strength based on one set of experimental data cannot be conclusive, but the fiber strength was reduced at temperatures above 900°C only and, at a temperature of 1,200°C, it was about 500 MPa.

So, assuming matrix stress to relax completely at temperature 1,200°C and neglecting fiber creep, the author's concluded that creep strength on the corresponding time base for the composites with fiber V_f 20% should be about 100 MPa, and, for those with 30% fibers, it should go up to 150 MPa.

1.9.20 Pressure Casting

Rhee et.al.,[130] examined a pressure-cast NiAl composite with polycrystalline alumina (PRD-166) fibers containing 0.2 wt. fraction of partially stabilized ZrO_2. Titanium was added to facilitate the infiltration of the fiber preform by the molten NiAl. The fibers in the preform used for casting were forced into contact, and fiber bonding occurred in a number of instances. Fiber V_f was increased from an initial value of 0.4 to 0.6 as a consequence of the applied pressure. The theory on the increase in final V_f of the fibers in the composite is offered: the interaction of applied pressure, wetting angle, and the rigidity of the fiber preform. At the fiber/matrix interface, the Al_2O_3 was free of ZrO_{2p}.

It was proposed that Al_2O_3 grain growth forced the $ZrO2$ into the molten NiAl, where it dissolved. As solidification took place, the concentration of ZrO_2 in the molten NiAl increased to a point where ZrO_2 reacted with Al_2O_3 to form $ZrO2$ again.

1.9.21 Plasma Spray Deposition

Tiwari, Herman, Sampath, et.al.,[131] used plasma spray processing to produce dense, rapidly solidified IMCs. This technique combines, melting, quenching, blending and consolidation into a single step. Dense Ni_3Al-based composites with TiB_{2p} were obtained by vacuum plasma spraying (VPS) processing and showed a relatively uniform distribution of the TiB_{2p}. The room temperature mechanical properties indicated significant strengthening due to the inclusion of the diboride particles. A strong particle-matrix bond was also obtained by this technique. The degree of strengthening in VPS-processed IMCs is greater than that observed in P/M-processed composites. Failure in the VPS-processed IMCs occurs by matrix separation and matrix particle decohesion.

1.9.22 Vacuum Hot Press/Electroplating

Kesapradist, Ono and Fukawa[132] produced an Ni_3Al matrix/Al_2O_3 IMC using a consolidation temperature 460°C lower than is typical for other aluminide composites. Aluminum sheets were electroless nickel plated to obtain sheets with 3Ni:Al plus 0.26 wt.% B. Alternating layers of these sheets and Al_2O_{3f} were consolidated by vacuum hot pressing at 800°C for one hour under 55 MPa. The consolidated material was then heat treated at 1,200°C for three hours to homogenize the matrix. Phase equilibrium caused ordered Ni_3Al to form. The result was an IMC with Al_2O_{3f} reinforcing a matrix of Ni_3Al.

The process appears to be applicable to other IMC systems where the matrix material has reduced properties prior to reactive heat treatment. Prior to the reaction, the IMC can be consolidated and formed at lower temperature and pressure, reducing tooling cost and complexity. The material properties would be finalized by heat treatment of the NNS product.

1.9.23 Cryomilling

Arsenault reported an investigation to determine if, by cryomilling NiAl and TiB_2 powders, ultra fine subgrain or grain size material could be obtained. Cryomilling decreased the grain or subgrain size but the strength of the 20 vol% TiB_2/NiAl composite was the same as 20 vol% TiB_2 made by the XD^{TM} process. Also, a 0 vol% TiB_2/NiAl composite produced from cryomilled NiAl was about the same strength as the 20 vol% TiB_2/NiAl composites. The above mentioned composites were made by HIPing and extrusion. If the 0 Vol% TiB_2/NiAl cryomilled composite is tested in the as HIPed condition, the strength is much greater.

1.10 IRON AND NIOBIUM ALUMINIDES

1.10.1 Pressure Infiltration

Sakamoto, Akiyarna, Kitahara, et.al.,[134] manufactured an FeAl alloy matrix in a high-frequency induction furnace in an Ar atmosphere, using electrolytic Fe (99.9%) and pure Al (99.99%). Subsequently, they made a preform using Saffil-RG (δ Al_2O_3, diameter 3 µm, length 300 µm) by pressurized infiltration, sintered in vacuum at 1,000°C.

Their results indicated the ability to produce composite materials with good fiber dispersion and no voids or other defects. Using a preform with a V_f of about 8%, the V_f systematically changed within the range of 10-38% due to the preform manufacturing method and a preform crush at the time of melting liquid injection. Grains of the matrix were fine, compared with those of the noncomposite section. Grains became finer from the center of the composite materials toward the perimeter, apparently reflecting differences in cooling speeds. The compression strength of the composite materials rose remarkably when compared with the noncomposite section, showing the effects of compounding.

Nourbakhsh, Margolin and Liang[135] pressure cast a composite of Fe-28Al-2Cr-lTi (at.%) reinforced with 20 µm diameter ZrO_2-toughened Al_2O_{3f}, PRD-166 (Al_2O_{3f} containing 20 wt.% ZrO_2).

During infiltration of the molten alloy, grain growth took place at the surface of the fiber and some ZrO_2 from the fiber dissolved into the matrix. A second phase, Fe_2AlZr, is formed and found at the fiber/matrix interface. The matrix contained a high density of dislocations resulting from a difference in the CTE between the matrix and fiber.

26

Suganuma[136] fabricated Fe_3Al matrix composites reinforced with 0-13 vol.% fine Al_2O_{3p} by hot pressing powder mixtures of Fe and Al. The microstructure and mechanical and several other properties of the Al_2O_3-reinforced composites were examined. As a result of the dispersion of Al_2O_3 powder, the grain size of the matrix decreased. No third phase was observed between the matrix and the Al_2O_3 by TEM. The matrix alloy showed quite good tensile strength (above 800 MPa) as well as good elongation (up to 5%) at room temperature. On increasing the V_f of Al_2O_3, the tensile strength increased slightly. The 0.2% proof stress was improved considerably more than the tensile strength and reached its maximum value at an Al_2O_3 V_f of 2.5%. At 500°C the strength of the composite was higher by 100 MPa than that of the monolithic alloy. The oxidation resistance of the Fe_3Al and its composites up to 1,000°C proved to be quite good.

Thus, the best improvement obtained with the dispersion of Al_2O_{3p} was the increase in strength at room temperature to elevated temperature. However, accompanying the increment in strength, the elongation decreased, which is always the case in composite materials. Therefore, the optimum choice for the Fe_3Al matrix composite will be in the range 2.5-5 vol.% Al_2O_{3p} addition.

1.10.2 Pressure Casting

Nourbakhsh, Sahin, Rhee et.al.,[137] produced a $NbAl_3 + Nb_2Al$ composite reinforced with continuous PRD-166 fiber, a 20 µm diameter ZrO_2-toughened Al_2O_{3f} by pressure casting. The micro- structure of the composite was characterized by optical and TEM and energy dispersive spectroscopy (EDS).

They found the following:

- An external pressure was needed to induce infiltration. The external pressure caused an increase in fiber V_f from 0.4, in the preform, to 0.5, in the composite.

- Exposure of the fiber to the molten metal at high temperatures led to Al_2O_3 and ZrO_2 grain growth, transformation of ZrO_2 and formation of a thin layer of an amorphous phase on the grain boundaries of Al_2O_3.

- Preferential Al_2O_3 grain growth, near the fiber surface, forced the ZrO_{2p} present in the region out of the fiber, into the molten metal, where it dissolved.

- A very small amount of a third phase, tentatively identified as Nb_3Al_3Zr, was observed at only one location at the fiber/matrix interface.

- Three types of dislocations were found in the Nb_3Al matrix.

1.11 FUTURE OF IMCS

Currently, the practical choice of fibers (monofilaments and tows) for IMCs is limited to SiC, Al_2O_3, YAG/Al_2O_3 eutectic, and Al_2O_3-mullite nanocomposite, all of which have low CTEs compared to that of many intermetallics. Among all fibers, monofilament CVD SiC fibers have the best mechanical properties at temperatures of interest for IMC application. Thus, the potential of monofilament SiC fibers for various intermetallic matrices needs to be fully explored.

With the currently available fibers, it becomes apparent that the problem of CTE mismatch is going to be there for most IMCs except for Al_2O_3 or YAG/Al_2O_3 fiber reinforced $MoSi_2$ and titanium aluminide composites. The effect of CTE mismatch can be manageable for composites with ductile matrices. The main challenge here is to preserve the matrix ductility during composite processing. For composites with brittle matrices, alternative solutions must be devised to accommodate the effect of CTE mismatch. One potential solution is the use of small diameter tow fibers to eliminate matrix cracking resulting from CTE mismatch induced residual stresses. The recent development of Al_2O_3-mullite nanocomposite and stoichiometric SiC (HiNicalon and Dow Corning) tow fibers with improved high temperature properties offers some new opportunities. Clearly the potential of using small diameter fibers in IMCs needs to be explored further. Interfacial reaction barrier coatings will be needed for many composites using Al_2O_3-mullite nanocomposite and SiC tow fibers.

Single crystal Al_2O_3 and YAG/Al_2O_3 eutectic fibers can be potential reinforcements for titanium aluminides if suitable interfacial reaction barriers can be found and for $MoSi_2$. However, they will not find use in high-CTE and brittle matrices like NiAl. Based on stress rupture properties, YAG/Al_2O_3 eutectic is the preferred fiber; however the lack of availability of this fiber in large quantities may preclude its use for near-term applications. The successful use of single crystal Al_2O_3 fibers in composites requires that the problem of fiber damage during composite processing be solved. Whether similar fiber damage occurs in Al_2O_3/YAG eutectic fiber reinforced composites needs to be determined.

Fiber coatings will be needed to accommodate various fiber-matrix incompatibilities. Because of the problem of fiber-matrix reaction in many IMCs, one major function of a fiber coating would be to act as an interfacial reaction barrier.

Although the interfacial reaction barrier coatings have been proved to be effective in preventing fiber-matrix reaction during composite processing, their effectiveness over long time periods is yet to be demonstrated.

Another clear function of a fiber coating is to optimize the interfacial bond strength. With regard to reducing the CTE mismatch induced residual stresses, fiber coatings may not offer practical solutions. Thus, the primary role of fiber coatings in IMCs is to prevent fiber-matrix reaction and to optimize the interfacial bond strength.

The morphology and microstructure of the coating are important variables in determining the long-term effectiveness of coatings as reaction barrier and need to be optimized for any composite. Furthermore, failure mechanisms within the coating itself need to be examined. For example, fracture may originate within the coating which could adversely affect composite properties.

28

REFERENCES – CHAPTER I

1. M.M. Schwartz, Composite Materials Handbook, 2nd ed., McGraw-Hill, New York, 1992, pp 2.53-2.69.

2 C.J. Wang, B.Z. Jang, J. Panus, et.al., *J. of Reinf. Plast. and Comp.,* **110** (4) 356 (1991).

3. R.J. Diefendorf, S.J. Grisaffe, W.S. Hillig, ECS 8902528, Loyola College, Balto. MD, March 1991, 139p.

4. G.R. Almen, et.al., SAMPE Int'l Symposium/Exhibition, **33**, pp 979-89 (1988).

5. G. Browning, 28th SPI Reinforced Plastics/Composites, Sec 15A, 1973.

6. N. Odigiri, T. Muraki, K. Tobukuro, *SAMPE Int'l Symposium/Exhibition,* **33,** 272 (1988).

7. Ciba-Geigy Product data sheet, R-6376, High Impact Resin System, 1985.

8. Fiberite/ICI data sheet HY-E 1377-25.

9. S.G. Hill, E.E. House, J.T. Hoggatt, Advanced thermoplastic composite development, Contr. N00019-17-C-0561, for Naval Air Systems Command, Washington, DC, May 1979.

10. J.T. Hoggatt, S.G. Hill, E.E. House, Environmental exposure of thermoplastic adhesives, D180-24744-1, Seattle, WA, Dec 1978.

11. S. Kalpakjian, Manufacturing processes for engineering materials, 2nd ed., Addison-Wesley Pub. Co. Inc., 920p, 1991.

12. F.S. Galasso, Advanced fibers and composites, Gordon and Breach, New York, 1988, 145 pp.

13. R.M Laine, Silicon nitride ceramic fibers from preceramic polymers, TR 8 (SRI), N0014-84C0392, Jun 1987.

14. M.U. Islam, Artificial composites for high temperature applications, Review, Natl. Research Council of Canada, DME007, Jan 1987, 84 pp.

15. W.H. Atwell et.al., Polymer Processing, AFWAL-TR-85-4099, F33615-83-C-5006, Dec. 1987, vol. 1: "Fiber Technology," 394 pp, vol. 2: "Composites Technology," 622 pp.

16. R.D. Veltri, F.S. Galasso, *J. Am. Ceram. Soc.,* **73,** 2,137, (1990).

17. N.H. Tai, T.W. Chou, Modeling of chemical vapor infiltration (CVI) in Al_2O_3/SiC Composites Processing, in J.D. Buckley (ed.), Proc. 12th Ann. Mtg., Metal Matrix, Carbon, and Ceramic Matrix Composites 1988, NASA Conf., Cocoa Beach, FL, Jan 20-22, 1988, pp 237-46.

18. P. Reagan, F.N. Huffman, Chemical vapor composite deposition, ibid, pp 247-58.

19. M.I. Mendelson, Characterization of Nextel™ 3-D woven fiber structures, ibid, pp 259-70.

20. C.V. Burkland, J.M. Yang, Development of 3-D braided Nicalon/Silicon Carbide composite by chemical vapor infiltration, ibid, pp 271-80.

21. L. Leonard, Ceramic-matrix composites; Mettle for the nasty jobs, *Adv. Composites,* pp. 37-43, Jul/Aug 1990.

22. M.J. Koczak et.al., Inorganic composite materials in Japan; Status and Trends, ONRFEM 7, Nov 1989, 53p.

23. G.W. Meetham, *J. Mater. Sci.,* **26** (4) 853 (1991).

24. Onrasia Scientific Information Bulletin, SIB- **17** (3) 173 (1992).

25. A.L. Klein, *AM&P,* **3** (3) 40 (1986).

26. L.E. McAllister, W.L. Lachman, Multidirectional carbon-carbon composites, Fabrication of Composites, eds., A. Kelly, S.T. Mileiko, Elsevier Sci. Publ. Co. Inc., 1983, pp 109-75.

27. J.X. Zhao, R.C. Bradt, P.L. Walker, Jr., *Carbon,* **23** (1) 9 (1985).

28. I.M. Pickup, B. McEnaney, R.G. Cooke, *Carbon,* **24** (5) 535 (1986).

29. L. Leonard, *Adv. Composites*, May 1989, p 27.

30. J.D. Buckley, D.D. Edie, Carbon-carbon materials and composites, NASA RP 1254, 286p, 1992.

31. L.M. Rinek, Prospects for carbon-carbon composites, SRI International, SRI BIP D 91-1597, Dec 1991, 14p.

32. N A. Alfutov, P A. Zinov'ev, O.S. Selyanin, *Comp. Sci. and Technol.,* **45,** 189 (1992).

33. J.E. Sheehan, Ceramic coatings for carbon materials, *Proc. of the Fourth Ann Matls. Conf.,* M. Genisio, ed., S. Illinois University at Carbondale, IL, May 1987, pp 56-8.

34. J. Strife, J. Sheehan, *Ceram. Bull.,* **67** (2) 369 (1988).

35. S. Kimura, Y. Tanabe, E. Yasuda, *Proc. of the Fourth Japan-U.S. Conf. on Comp. Matls.,* 1990, 867 pp.

36. E. Fitzer, B. Terwiesch, *Carbon,* **11,** 750 (1973).

37. C.F. Lewis, *Matl. Engr.,* Jan 1989, p 27.

38. J.W. Sawyer, T.M. Rothgeb, Carbon-carbon joint and fastener test results at room and elevated temperatures, NASA TM-107638, Jul 1992, 32p.

39. Report on the economics achieved by replacing asbestos with K-Karb carbon-carbon, *Kaiser Aerotech Tech. Bull.,* 1980.

40. C A. Ginty, H.R. Gray, Overview of NASA's advanced temperature engine materials technology program, 24th SAMPE Tech. Conf., vol 24, editors, T.S. Reinhart, M. Rosenow, R A. Cull et.al., Oct. 20-22, 1992, Toronto, Canada, pp T1029-1045, 1991.

41. J.R. Stephens, Intermetallic and ceramic matrix composites for 815 to 1,370°C gas turbine engine applications, Int. Conf. on Metal and Ceramic Matrix Composites: Processing, Modeling and Mechanical Behavior, Anaheim, CA, 1990 Mtg., Feb 1922, 1990, ed. by R.B. Bhagat, A.H. Clauer, P. Kumar, A.M. Ritter, pp 3-12.

42. C.K. Elliot, G.R. Odette, G.E. Lucas et.al., Toughening mechanisms in intermetallic gamma-TiAl alloys containing ductile phases, *Mat. Res. Soc. Symp. Proc.,* **120,** (120) 95 (1988).

43. E.A. Feest, *Met. Mater.,* **4,** 273 (1988).

44. Synthesis of metallic materials for demanding aerospace applications using powder metallurgy techniques, *Proc. 1991 P/M in Aerospace and Defense Technologies Symp.,* Tampa, FL, 4-6 Mar 1991, MPIF., pp 5-36.

45. D.L. McDanels J.R. Stephens, High temperature engine materials technology-intermetallics and metal matrix composites, NASA TM-100844, 1988.

46. P.K. Brindley, P.A Bartolotta, S.J. Klima, Investigation of a SiC/Ti-24Al-11Nb composite, NASA TM-100956, 1988.

47. N.S. Stoloff, D.E. Alman, *MRS Bull,* **15,** 47 (1990).

48. R. Bowman, R. Noebe, *Adv. Mater. Process.,* **8,** 35 (1989).

49. J.-M. Yang, W.H. Kao, C.T. Liu, *Mater. Sci. and Eng.,* **A107,** 81 (1989).

50. K.S. Kumar, Nickel aluminide/titanium diboride composites via XD synthesis, MML TR 89-102(C), Cont # N00014-85C0639, Nov 1989, 120p.

51. R.J. Arsenault, Strengthening of NiAl matrix composites, Univ. of MD, Metall. Matls. Lab., Oct 1993, 32p, NML-1993.

52. R.S. Rastogi, K. Pourrezaei, *J. Mater. Proc. Technol.,* **43** (2-4) 89 (1994).

53. Y.-W. Kim, S. Krishnamurthy, G. Das et.al., Microstructure/processing/property interactions of metallic aerospace alloys and composites, WL-TR-92-4078, Final Rept. Aug 1989-Mar 1992, Oct 1992, 195p.

54. D.L. Anton, D.M. Shah, D.N. Duhl et.al., *JOM,* **41** (9) 12 (1989).

55. *Advanced Materials and Manufacutirng Processes,* **3** (1) (1988) 151p.

56. C.H. Ward, A.S. Culbertson, "Issues in potential IMC application for aerospace structures," MRS Symposium Proceedings, **350,** 3-11 (1994).

57. R.L. Fleischer, A.L. Taub, *JOM,* **41** (9) 8 (1989).

58. A. Rose, R.M. German, *AM&P,* **3,** 37 (1988).

59. Intermetallic Matrix Composites, Vo. 194, 1990 MRS Spring Mtg., San Francisco, CA, ed. D.L. Anton, P.L. Martin, D.B. Miracle, R. McMeeking, 440 p.

60. L.J. Westfall, D.L. McDanels, D.W. Petrasek, "Advanced fiber development for intermetallic matrix composites," NASA TM-102109, E 4878, June 1989.

61. C.J. Smith, J.E. Johnson, "Advanced materials for 21st century UBE," ASME, 34th Int. Gas Turbine Conf., Toronto, Canada, June 7, 1989.

62. D.W. Petrasek, "High temperature intermetallic matrix composites,: HITEMP Review, 1988, Advanced High Temperature Materials Technology Prog., NASA CP-10025, Nov. 1988, p 67-82.

63. Y.W. Kim, *J. Met.,* **41,** (7) 24 (1989).

64. S. Nourbakhsh, H. Margolin, Fabrication of high temperature fiber reinforced intermetallic matrix composites, Int. Conf. Metal and Ceramic Matrix Composites: Processing, Modeling and Mechanical Behavior, Anaheim, CA, 19-22 Feb 1990, R.B. Bhagat, A.H. Clauer, et.al., ed., TMS, pp 75-90.

65. J.D. Whittenberger, D.J. Gaydosh, K.S. Kumar, *J. Mater. Sci.,* **25,** 2771 (1990).

66. J.D. Whittenberger, Solid State Powder Processing, A H. Clauer and J.J. deBarbadillo, ed., TMS, 1990, p 137.

67. N.S. Stoloff, D.E. Alman, Intermetallic Matrix Composites, D.L. Anton et.al., ed., **194,** MRS, 1990, p 31.

68. J.D. Whittenberger, E. Arzt, M.J. Luton, Intermetallic Matrix Composites, D.L. Anton et.al., ed., **194,** MRS, 1990, p 221.

69. *Advanced Materials,* ed. P. West, **14** (8) 2 (1992).

70. P.K. Brindley, High Temperature Ordered Intermetallic Alloys II, 1987, editors, N.S. Stoloff, C.C. Koch, C.T. Liu et al, MRS Symp. Proc., **81,** MRS, Pittsburgh, PA, p 419.

71. D.L. Anton, High temperature/High Performance Composites, 1988, editors, F.P. Lemkey, S.G. Fishman, A.G. Evans et al, MRS Symp. Proc., **120,** MRS, Pittsburgh, PA, p 57.

72. A. Bose, B. Moore, R.M. German et.al, *J. Met.,* **40** (9) 14 (1988).

73. R. German, A. Bose, *Mater. Sci. and Eng.,* **A107,** 107 (1989)

74. J.E. Talia, T.T. Mickle, W.E. Frazier et.al, Interface strengthened, reactive sintered Al-Ti composites, editor, K. Upadhya, Developments in Ceramic and Metal-Matrix Composites, TMS Ann. Mtg. Mar 1-5, 1992, pp 136-70, 1991.

75. K. Kato, A. Matsumoto, Y. Nozaki, Powder metallurgy of Ti-Al intermetallic compound by injection molding, Proc. of 4th Symp. on High-Performance Materials for Severe Environments, Jun 1-2, 1993, Nagoya, Japan, pp 83-91.

76. J.M. Brupbacher, L. Christodoulou, D.C. Nagle, Process for forming metal-ceramic composites, US Patent 4,710, 348, Martin-Marietta, 1987.

77. .L. Adams, S.L. Kampe, L. Christodoulou, *Int. J. Powder Metall.,* **26,** 105 (1990).

78. J.D. Whittenberger, R.K. Viswanadham, S.K. Mannan et.al, *J. Mater. Sci.,* **25,** 35 (1990).

79. D.E. Larsen, S.L. Kampe, L. Christodoulou, Intermetallic Matrix Composites, D.L. Anton et.al., ed., **194,** MRS, 1990, p 285.

80. D.E. Larsen, L. Christodoulou, S.L. Kampe et.al, *Proc. Conf. Innovative Inorganic Composites,* ASM Int. Ann. Conf. Detroit, MI, Oct 1990.

81. S.K. Mannar, K.S. Kumar, J.D. Whittenberger, *Metall. Trans.,* **21A,** 2179 (1990)

82. E.A. Feest, Co-spray route to fine particle reinforced MMC, UK Patent GB 2172825, AEA Technology, 1988.

83. S. Nourbakhsh, F.L. Liang, H. Margolin, An apparatus for pressure casting of fibre-reinforced high-temperature metal-matrix composites, *J. Phys. E:,* Sci. Instrum., **21,** 898 (1988).

84. S. Nourbakhsh, H. Margolin, F.L. Liang, Metall. Trans., **20A,** 2159 (1989).

85. S. Nourbakhsh, F.L. Liang, H. Margolin, Interaction of Al_2O_3-ZrO_2 fibers with a Ti-Al matrix during pressure casting, Metall. Trans., **21,** 213 (1989).

86. S. Nourbakhsh, H. Margolin, *Mater. Sci. and Eng.,* **144,** 133 (1991).

87. T. Tsujimoto, K. Hashimoto, editors, C.T. Liu, A.I. Taub, N.S. Stoloff et.al, High-temperature Ordered Intermetallic Alloys III, MRS Symp. Proc., **13,** MRS, pp 391-96.

88. H. Fukunaga, X. Wang, *J. Mater. Sci. Lettrs.,* **10** (1) 23 (1991).

89. H. Fukunaga, Cast reinforced metal composites, editors, S.G. Fishman, A.K. Dhingra, Conf. Proc., ASM, Chicago, 1988, p 101.

90. N. Chandra, C.R. Ananth, H. Garmestani, *J. Composites Technol. & Res.,* JCTRER, **16** (1) 37 (1994).

91. S.L. Draper, P.K. Brindley, M.V. Nathal, Effect of fiber strength on the room temperature tensile properties of SiC/Ti-24Al-11Nb, ed., K Upadhya, Developments in Ceramic and Metal-Matrix Composites, Proc. of TMS, 1992 TMS Ann. Mtg, San Diego, CA, Mar 1-5, 1992, pp 189-202.

92. N.S. Stoloff, D.E. Alman, *Mater. Sci. and Eng.,* **144,** 51 (1991).

93. J.W. Pickens, R.D. Noebe, G.K. Watson et al, NASA TM-102060, 1989.

94. H. Mabuchi, H. Tsuda, Y. Nakayarna, *J. Mater. Res.,* **7** (4) 894 (1992).

95. E.S. Metin, O.T. Inal, *Mater. Sci. and Eng.,* **148** (1) 115 (1991).

96. Z. Eliezer, B.-H. Lee, C.J. Hou et.al., Matrix-reinforcement interface characteristics of (Ti_3Al+Nb)-based powder composites, consolidated by high-energy high-rate processing, Int. Conf. on Metal and Ceramic Matrix Composites: Processing, Modeling and Mechanical Behavior, Anaheim, CA, 1990 Mtg., Feb 19-22, 1990, ed. by R.B. Bhagat, A.H. Clauer, P. Kumar, A.M. Ritter, pp 401-12.

97. R.R. Bowman, R.D. Noebe, J. Doychak et.al., Effect of interfacial properties on the mechanical behavior of NiAl based composites, 4th HITEMP Review, NASA CP 10082, pp 43-1 to 43-13, NASA-Lewis Res. Ctr., Cleveland, OH, Oct 29-30, 1991.

98. D.E. Alman, N.S. Stoloff, *Int. J. of P/M,* **27** (1) 29 (1991).

99. T.G. Nieh, K.R. Forbes, T.C. Chou et.al., Microstructures and deformation properties of an Al_2O_3-Ni_3Al composites from room temperature to 1,400°C, ibid, pp 85-98.

100. .A. Feest, J.H. Tweed, *Mater. Sci. and Technol.,* **8** (4) 308 (1992).

101. K. Nishiyarna, M. Mohri, S. Umekawa, Fabrication and mechanical properties of C_f/ NiAl and SiCw/NiAl composites, *Metal Matrix Composites I,* pp 417-24.

102. R.D. Noebe, R.R. Bowman, J.I. Eldridge, Initial evaluation of continuous fiber reinforced NiAl composites, editors, D.L. Anton, P.L. Martin, D.B. Miracle et.al., Intermetallic Matrix Composites, MRS Symp. Proc., **195**, Pittsburgh, PA, 1990, pp 323-31.

103. J.-M. Yang, W.H. Kao, C.T. Liu, *Mater. Sci. and Eng.,* **A107,** 81 (1989).

104. A.U. Seybolt, Trans. ASM, **59,** 860 (1966).

105. J.D. Whittenberger, K.S. Kumar, S.K. Mannan, *Mater. at High Temperatures,* **9** (1) 3 (1991).

106. J.D. Whittenberger, K.S. Kumar, S.K. Mannan, et.al., *J. Mater. Sci. Lettrs.,* **9,** 326 (1990).

107. J.D. Whittenberger, S.K. Mannan, KS. Kumar, *Scr. Metall.,* **23,** 2055 (1989).

108. J.D. Whittenbergcr, R.K. Viswanadham, S.K. Mannan et.al., *J. Mater. Res.,* **4,** 1164 (1989).

109. L. Christodoulou, P.A. Parrish, C.R. Crowe, High temperature/high performance composites, 1988, editors, F.P. Lemkey, S.G. Fishman, A.G. Evans et.al., *MRS Symp. Proc.,* vol 120, MRS, Pittsburgh, PA, p 29.

110. S.C. Jha, R. Ray, *J. Mater. Sci. Lettrs.,* **7,** 285 (1988).

111. G.L. Povirk, J.A. Horton, C.G. McKamey et.al., *J. Mater. Sci.,* **23,** 3945 (1988).

112. J. Whittenberger, E, Arzt, M.J. Luton, *J. Mater. Sci.,* **5,** 271 (1990).

113. D.M. Sims, A. Bose, R.D. German, Progress in Powder Metallurgy, MPIF, Princeton, NJ, **43,** 575 (1987).

114. B. Mei, R. Yuan, X. Duan, *J. Mater. Res.,* **8** (11) 2,830 (1993).

115. W.Z. Misiolek, R.M. German, Processing of continuous fiber reinforced NiAl matrix composite, *Proc of Advances in Powder Metallurgy-1991,* MPIF and APMI, Volume 6, Jun 9-12,1991, Chicago, IL, compiled by L.F. Pease III and R.J. Sansoucy, pp 167-76.

116. R.M. German, Powder Injection Molding, Metal Industries Federation, Princeton, NJ, 1990.

117. F.H. Froes, *JOM,* **41** (9) 6 (1989).

118. RM. Aikin, P.E. McCubbin, L. Christodoulou, editors, D.L. Anton, P.L. Martin, D.B. Miracle et.al., Intermetallic Matrix Composites, *MRS Symp. Proc.,* MRS, Pittsburgh, PA, **194,** 307 (1990).

119. L. Wang, R.J. Arsenault, *Metall. Trans. A,* **22** (12) 3,013 (1991).

120. P.K. Brindley, editors, N.S. Stoloff, C.C. Koch, C.T. Liu, et.al., High Temperature Ordered Intermetallic Alloys II, *MRS Symp. Proc.,* MRS, Pittsburgh, PA, **81,** 419 (1987).

121. J. Bowling, G.W. Groves, *J. Mater. Sci.,* **14,** 443 (1979).

122. L.S. Sigl, et.al., *Acta Metall.,* **36,** 945 (1988).

123. C.K. Elliot, et.al., High temperature, *MRS Symp. Proc. on High Temperature, High Performance Composites,* editors F. Lemkey et.al., **120,** 95 (1988).

124. P.K. Brindley, P A. Bartolotta, High temperature review, NASA Pub. 10025, 1988, p 225.

125. D.W. Petrasek, Intermetallic matrix composites, *HITEMP Review,* 1989, p 8.1, Advanced High Temperature Engine Materials Technol. Prog., Clev., OH, NASA CP 10039, 1989.

126. X. Liang, H.K. Kim, J.C. Earthman, et.al., *Mater. Sci. and Eng.,* **153,** 646 (1992).

127. V.I. Glushko, S.T. Mileiko, *J. Mater. Sci. Ltrs.,* **12** (12) 915 (1993).

128. Y. Le Peticorps, F.D. Martina, J.M. Queinisset, Metal matrix composites- processing, microstructure and properties, *Proc of the 12th RISO Symp.,* editors, N. Hansen, D.J. Jensen, T. Lefters, et.al., RISO, Roskilde, Denmark, 1991, p 461.

129. J.H. Schneibel, E.P. George, C.G. McKamey, et.al., *J. Mater. Res.,* **6,** 1673 (1991).

130. S. Nourbakhsh, O. Sahin, W.H. Rhee, et.al., *Metall. Trans. A,* **22** (12) 3,059 (1991).

131. R. Tiwari, H. Herman, S. Sampath, et.al, *Mater. Sci. and Eng.,* **144,** 127 (1991).

132. J. Kesapradist, K Ono, K. Fukaura, *Mater. Sci. and Eng.,* **153** (1) 641 (1992).

133. R.J. Arsenault, Strengthening of NiAl matrix composites, Ann. Rept. Oct 1991-Sep 1992, Univ. of MD, Cont # N00014-91-J-1353, ONR, Oct 1992, 6p.

134. M. Sakamoto, S. Akiyarna, et.al., Strengthening Al_2O_3 short fibers of B2-type FeAl intermetallic compounds, JPRS-JST-92-058-L, Jul 23, 1992, Japan Insti. of Metals, 109th Conf., p33.

135. S. Nourbakhsh, H. Margolin, F.L. Liang, *Metall. Trans.,* **21,** 2881 (1990).

136. K. Suganuma, *J. Alloys and Compounds,* **197,** 29 (1993), JALCOM 602.

137. S. Nourbakhsh, O. Sahin, W.H. Rhee, et.al., *Acta. Metall. Mater.,* **40** (2) 285 (1992).

BIBLIOGRAPHY – CHAPTER I

B.A. Lerch, M.E. Melis, M. Tong, Experimental and analytical analysis of the stress-strain behavior in a [90/0] 2_s SiC/Ti-15-3 laminate, NASA TM-104470, 1991.

G.N. Morscher, J A. DiCarlo, A simple test for thermomechanical evaluation of ceramic fibers, NASA TM103767, 1991.

B. Budiansky, J.C. Amazigo, A.G. Evans, Small-scale crack bridging and the fracture toughness of particulate-reinforced ceramics, *J. Mech. Phys. Solids*, **36** (2) 167 (1988).

J. Eberhart, M.F. Ashby, Cambridge Univ. Eng. Dept., Rept. CVED/C-MATS/TR 140, 1987.

G.R. Odette, H.E. Deve, C.K. Elliot et.al., The influence of the reaction layer structure and properties on ductile phase toughening in titanium aluminide-niobium composites, *Interfaces in Metal Ceramics Composites,* editors, R.Y. Lin, R.J. Arsenault, G.P. Martins and S.G. Fishman, TMS Ann. Mtg., Feb 18-22, 1990, Anaheim, CA, pp 443-64, l990.

A.J. Pysic, I.A. Aksay, M. Sarikaya, Ceramic microstructures, MSR, editors, J.A. Pask, A.G. Evans, **21,** 45 (1986).

K. Vedula, *Mater. Manuf. Process,* **4,** 39 (1989).

S. Nourbakhsh, F.L. Liang, H. Margolin, Characterization of a zirconia toughened alumina fiber, PRD-166, *J. Mater. Sci. Lett.,* **8,** 1252 (1989).

S.L. Draper, P.K. Brindley, M.V. Nathal, Effect of fiber strength on the room temperature tensile properties, editor, K. Upadhya, Developments in Ceramic and Metal-Matrix Composites, TMS Ann. Mtg. Mar 1-5, 1992, pp 156-70, 1991.

J. Cook, E.A. Feest, Control of fibre matrix interactions in SiC/Ti MMC, FR BRITE Project P-1204, Pub. No. EUR 13614, CEC, Luxembourg, 1991.

C. Fujiwara, Fabrication of SiC (CVD)/TiAl composites, JPRS-JST -93-055-L, 22 Jul 1993, *Int. Symp. on Adv. Materials,* pp 188-99.

J.L. Walters, H.E. Cline, *Metall. Trans.,* **4,** 33 (1973).

M.D. Skibo, D.M. Schuster, Cast reinforced composite material, US Patent 4,759,995, Dural Aluminum Composites Corp., San Diego, CA, 1986.

Proceedings of 1989 Symposium on High Temperature Aluminides & Intermetallics, editors, S.H. Whang, C.T. Liu, D.P. Pope, J.O. Stiegler, TMS Fall Mtg., 1990, sponsored by TMS and ASM, Indianapolis, IN, Oct 1-5, 1989, 593 p.

J.C. Rawers, W.R. Wrzesinski, Reaction-sintered hot-pressed TiAl, *J. Mater. Sci.,* **27,** (11) 2,877 (1992).

Characterization of Fibre Reinforced Titanium Matrix Composites, AGARD Rept. 796, Papers of 77th Mtg. of the AGARD Structures and Materials Panel, Bordeaux, France, 27-28 Sep 1993, 125 p.

M. Khobaib, Damage evolution in creep of SCS-6/Ti-24Al-11Nb metal-matrix composites, *Proc. of the Am. Soc. for Composites, 6th Tech. Conf.,* Composite Materials, Mechanics and Processing, Oct 7-9, 1991, Albany, NY, pp 638-41.

D.B. Marshall, M.C. Shaw, M.R. James, et.al., Fatigue resistant Ti$_3$Al composites, FR Sep 1990-Sep 1992, WL-TR-93-4034, SC-71034-FRD, Jan 1993, 67p.

35

P.A. Bartolotta, P.K. Brindley, High-temperature fatigue behavior of a SiC/Ti-24Al-11Nb composite, Composite Materials: Testing and Design, 10th vol., ASTM STP 1120, G.C. Grimes, editor, ASTM, Phila, PA, 1992, pp 192-203.

Intermetallic Matrix Composites III, editors, J A. Graves, R.R. Bowman, J.J. Lewandowski, *MRS Sym. Proc.,* San Francisco, CA, **350,** 312 (1994).

P.K. Brindley, SiC reinforced aluminide composites. High-temperature ordered intermetallic alloys II, MRS Symp. Proc., N.S. Stoloff, et.al., editors, MRS, Pittsburgh, PA, **81,** 419 (1987).

S.F. Baumann, P.K. Brindley, S.D. Smith, Reaction zone microstructure in a Ti$_3$Al+Nb/SiC composite, *Metall. Trans.,* **21A,** 1559 (1990).

C.G. Rhodes, R.A. Amato, R A. Spurling, Fiber-matrix interactions in SiC reinforced titanium aluminides, *Symp. on High Temperature Composites,* Am. Soc. for Composites, Dayton, OH, Jun 1989.

P.K. Brindley, S.L. Draper, M.V. Nathal, et.al., Factors which influence the tensile strength of a SiC/Ti-24Al-11Nb (at%) composite. Fundamental relationships between microstructures and mechanical properties of metal matrix composites, *TMS Fall Mtg. Proc.,* P.K. Liaw, M.N. Gungor, editors, TMS, Indianapolis, IN, 1989, pp 387-401.

P.K. Brindley, P.A. Bartolotta, R A. Mackay, Thermal and mechanical fatigue of SiC/Ti$_3$Al+Nb, NASA CR-10039, 1989.

P.A. Bartolotta, M.A. McGaw, A high temperature fatigue and structures testing facility, NASA TM-100151, 1987.

R.W. Hayes, The creep behavior of the Ti$_3$Al alloy Ti-24Al-11Nb, *Scr. Metall.,* **23,** 1931 (1989).

J.-M. Yang, S.M. Jeng, Interfacial reactions in titanium-matrix composites, *JOM,* **41** (11) 56 (1989).

S.M. Jeng, J.-M. Yang, C.J. Yang, Fracture mechanisms of fiber-reinforced titanium alloy matrix composites Part II: Tensile behavior, *Mater. Sci. and Eng.,* **138** (2) 169 (1991).

S.M. Jeng, J.-M. Yang, C.J. Yang, Fracture mechanisms of fiber-reinforced titanium alloy matrix composites Part III: Toughening behavior, *Mater. Sci. and Eng.,* **138** (2) 181 (1991).

A.J. Misra, Modification of the fiber/matrix interface in aluminide-based intermetallic-matrix composites, *Composites Sci. and Technol.,* **50** (1) 37 (1994).

P A. Bartolotta, Predicting fatigue lives of metal-matrix/ fiber composites, NASA Tech Briefs, Apr 1994, pp 28-30.

S.L. Draper, P.K. Brindley, M.V. Nathal, Effect of fiber strength on the room temperature tensile properties of SiC/Ti-24Al-11Nb, *Metall. Trans.,* **23A,** 2541 (1992).

D.R Pank, A.M. Ritter, R A. Amato et.al., Titanium aluminide composites, editors, P.R. Smith, S.J. Balsone, T. Nicholas, WL-TR-91-4020, Wright-Patterson AFB, OH, 1991, pp 382-98.

J. Gayda, T.P. Gabb, A.D. Freed, Fundamental relationships between microstructures and mechanical properties of metal matrix composites, *TMS Fall Mtg. Proc.,* P.K. Liaw, M.N. Gungor, editors, TMS, Warrendale, PA, 1990, pp 497-514.

H. Gigerenzer, P.K. Wright, Titanium aluminide composites, editors, P.R. Smith, S.J. Balsone, T. Nicholas, WL-TR-91-4020, Wright-Patterson AFB, OH, 1991, pp 251-64.

A.M. Ritter, F. Clark, P. Dupree, *Proc. Symp. on Light Weight Alloys for Aerospace Applications II*, editors, E.W. Lee, N.J. Kim, TMS, Warrendale, PA, 1991, pp 403-12.

P.K. Brindley, S.L. Draper, J.I. Eldridge, et.al., *Metall. Trans. A*, **23**, 2527 (1992).

T. Nicholas, S.M. Russ, Elevated temperature fatigue behavior of SCS-6/ Ti-24Al-11Nb, *Mater. Sci. and Eng.*, **153** (1) 514 (1992).

S.M. Yeng, J.-M. Yang, D.G. Rosenthal, et.al., Mechanical behaviour of SiC-fibre-reinforced titanium/titanium aluminide hybrid composites, *J. Mater. Sci.*, **27**, 5357 (1992).

P.B. Aswath, S. Suresh, *Metall. Trans.*, **22A,** 817 (1991).

K.S. Chan, Fatigue fracture, *Engineering Material Structures*, **13,** 171 (1990).

D.R Schuyler, M.M. Sohi, R.L. Hollars, et.al., Feasibility of titanium aluminide metal matrix composites (MMC) for 1,400°F applications, FR Sep 1986-Jan 1991, Cont # N62269-86-C-0248, 30 Apr 1991, Rept. No. NADC-91066-60, 340p.

A.K. Misra, Reaction of Ti Ti-Al alloys with alumina, *Metall Trans.*, **22** (3) 715 (1991).

C. Jones, C.J. Kiely, S.S. Wang, *J. Mater. Sci.*, **4,** 327 (1989).

J.-M. Yang, S.M. Jeng, *Scr. Metall.*, **23,** 1559 (1989).

S. Krishnamurthy, Interfaces in Metal-Ceramic Composites, R.Y. Lin, R.J. Arsenault, G.P. Martins, et.al., TMS, Warrendale, PA, 1990, pp 75-84.

R. Tressler, T.L. Moore, R.L. Crane, *J. Mater. Sci.*, **8,** 151 (1973).

H.F. Merrick, M.L. Labib, Advanced reinforcement systems for intermetallic applications, NASA CR-4488, E-7557, Mar 1993, 51 p.

W.O. Soboyejo, K.T. Venkateswara Rao, S.M.L. Sastry, et.al., Strength, fracture, and fatigue behavior of advanced temperature intermetallics reinforced with ductile phases, *Metall. Trans.*, **24** (3) 585 (1993).

C.H. Weber, J.Y. Yang, J.P.A. Lofvander, et.al., The creep and fracture resistance of reinforced with Al_2O_3 fibers, *Acta Metall. Mater.*, **41** (9) 2,681 (1993).

E.U. Lee, T. Kircher, J. Waldman, Oxidation of titanium aluminide and its XD composite, FR 1988-89, NADC-90122-60, Dec 1990, 19 p.

K. Sadananda, C.R. Feng, The creep of intermetallics and their composites, *JOM,* **45** (5) 45.

M.J. Donachie, editor, Superalloys-Source Book, ASM, Metals Park, OH, 1984.

M.L. Gambone, Fatigue and fracture of titanium aluminides, Volume 1, FR 1 Jul 85- 31 Jul 89, Rept No. Allison-EDR-14249-Vol 1, 70 p, TR-89-4145-Vol 1, Cont # F33615-85-C-5111, Feb 1990.

M.L. Gambone, Fatigue and fracture of titanium aluminides, Volume 2, FR 1 Jul 85-31 Jul 89, Rept No. Allison-EDR-14249-Vol 2, 235 p, TR-89-4145-Vol 2, Cont # F33615-85-C-5111, Feb 1990.

P.R. Smith, W.C. Revelos, Environmental aspects of thermal fatigue of a SiC/Ti-24Al-11Nb (at%) composite, *Proc. of FATIGUE 90,* Honolulu, Hawaii, 15-20 Jul 1990.

P.K. Brindley, R A. Mackay, P.A. Bartolotta, The effects of thermal cycling in vacuum on tensile properties of a SiC/Ti-24Al-11Nb (at%) composite, *WESTEC 89,* Los Angeles, CA, 21-23 Mar 1989.

T.G. Nieh, K.R. Forbes, T.C. Chou, et.al., Microstructures and deformation properties of an Al_2O_3-Ni_3Al composite from room temperature to 1,400°C, editor, K. Upadhya, Developments in Ceramic and Metal-Matrix Composites, TMS Ann. Mtg., Mar 1-5, 1992, pp 115-27, 1991.

J. Doychak, J.A. Nesbitt, R.D. Noebe, et.al., Oxidation of Al_2O_3 continuous fiber reinforced/ NiAl composites, 15th Conf. on Metal Matrix, Carbon, and Ceramic Matrix Composites, editor, J.D. Buckley, NASA CP-3133, Part 1, Dec 1991, pp 251-61.

R.R. Petrich, C.A. Moore, J.R. Hellman, et.al., Interfacial shear behavior of sapphire-reinforced niobium and nickel aluminide matrix composites, ibid, pp 233-39.

C.A. Moose, D.A. Koss, J.R. Hellman, Interfacial shear behavior of sapphire-reinforced nickel aluminide matrix composites, editors, D.L. Anton, P.L. Martin, D.B. Miracle, et.al., Intermetallic Matrix Composites, *MRS Symp. Proc.,* Pittsburgh, PA, **195,** 293 (1990).

A. Kelly, K.N. Street, Creep of discontinuous fibre composites II. Theory for the steady-state, *Proc. Roy. Soc.* London, **A328,** 283 (1972).

V.C. Nardone, J.R. Strife, K.M. Prewo, Processing of particulate reinforced metals and intermetallics for improved damage tolerance, *Mater. Sci. and Eng.,* **144,** 267 (1991).

V.C. Nardone, J.R. Strife, NiAl-based microstructurally toughened composites, *Metall. Trans.,* **22** (1) 183 (1991).

A.L. DeNardo, The national aerospace plane: technology transfer, *Aerospace Engineering,* Oct 1994, pp 15-19.

M. Nobuki, D. Vanderschueren, M. Nakamura, High-temperature mechanical properties of vanadium alloyed gamma-base titanium-aluminides, *Acta Metall. Mater.,* **42** (8) 2,623 (1994).

W. jie Fan, R A. Varin, Z. Wronski, Roll-bonding of a laminate aluminum alloy/Ni_3Al-type alloy composite, Zeitschrift fur Metallkunde, **85** (7) 522 (1994).

D. Kouris, D. Marshall, Damage mechanisms in Ti_3Al matrix composites, *J. Eng. Mater. & Technol.,* Trans. of ASME, **11** (3) 319 (1994).

M.G. Benz, M.R. Jackson, J.R. Hughes, Composite strengthening of Nb-Ti alloys, L. Murugesh, K.T. Venkateswara Rao, L.C. DeJonghe, et.al., Fabrication of Nb_3Al intermetallic *in situ* composite microstructures, ibid, pp 65-84.

K. Natesan, W.D. Cho, High-temperature corrosion of iron aluminides, ANL/ET/CP-82282, CONF-9405143-1, 8th Ann. Conf. on fossil energy materials, Oak Ridge, TN, 10-12 May 1994, 12 p.

M. Guermazi, Temperature-dependence of interfacial shear-strength in SCS-6 fiber-reinforced Ti-24Al-11Nb metal-matrix composites, *Mater. Sci. & Technol.,* **10** (5) 399 (1994).

V.K. Sikka, C.T. Liu, Iron-aluminide alloys for structural use, *Matls. Technol.,* **9** (7-8) 159 (1994).

E. Hellum, M. Luton, A. Olsen, Structure and properties of dispersion strengthened aluminium-10% titanium, AIAA-93-5035, AIAA/DGLR 5th Int. Aerospace Planes and Hypersonics Technologies Conf, 30 Nov-3 Dec 1993, Munich, Germany, 6 p.

K. Sadananda, C.R. Feng, Creep of intermetallic composites, *Mater. Sci. & Eng. A,* **179** (1-2) 199 (1993).

H.C. Cao, J.P.A. Lofvander, A.G. Evans, Mechanical properties of an in-situ synthesized Nb/Nb$_3$Al layered composite, *Mater. Sci. & Eng. A,* **185** (12) 177 (1994).

R.D. Field, R. Darolia, D.F. Lahrman, Precipitation in NiAl/ Ni$_2$AlTi alloys, *Scripta Metall.,* **23** (9) 1,469 (1989).

J. Doychak, T. Grobstein, The oxidation of high-temperature intermetallics, *JOM,* **41** (10) 30 (1989).

T.G. Nieh, C.M. McNally, J. Wadsworth, Superplasticity in intermetallic alloys and ceramics, *JOM,* **41** (9) 31 (1989).

L. Zhao, J. Beddoes, D. Morphy, et.al., Effect of HIP conditions on the microstructure of a near gamma-TiAl + W powder alloy, *Mater. and Manufact. Proc.,* **9** (4) 695 (1994).

D.B. Marshall, W.L. Morris, B.N. Cox, et.al., Transverse strengths and failure mechanisms in Ti$_3$Al matrix composites, *Acta Metall. Mater.,* **42** (8) 2,657 (1994).

R.J. Arsenault, Interfaces in metal- and intermetallic-matrix composites, *Composites,* **25** (7) 540 (1994).

P.I. Ferreira, R.M.L. Neto, Reactive sintering of NbAl$_3$, *Int. J. of Powder Metall.,* **30** (3) 313 (1994).

J.D. Whittenberger, R. Ray, S.C. Farmer, Elevated-temperature deformation properties of in-situ carbide particle strengthened Ti-48Al materials, *Intermetallics,* **2** (3) 167 (1994).

J.R. Stephens, M.V. Nathal, Status and prognosis for alternative engine materials, TMS, Superalloys 1988, editors, S. Reichman, D.N. Duhl, G. Maurer, et.al., 1988, pp 183-92.

A.K. Misra, S.M. Arnold, Compliant layer for the Ti$_3$Al+Nb/SCS-6 composite system, editor, J.D. Buckley, 15th Conf. on the Metal Matrix, Carbon, and Ceramic Matrix Composites, NASA CP-3133, Part 1, 416 p, Cocoa Beach, FL, Jan 16-18, 1991, Report L-17016, Dec 1991.

P.R. Smith, C.G. Rhodes, W.C. Revelos, Interfacial evaluation in a Ti-25Al-17Nb/SCS-6 composite, *Interfaces in Metal-Ceramics Composites,* editors, R.Y. Lin, R.J. Arsenault, G.P. Martins and S.G. Fishman, TMS Ann. Mtg. Feb 18-22, 1990, Anaheim, CA, pp 35-57, 1990.

A.K. Misra, Fibers and fiber coatings for IMCs, *MRS Symp. Proc.,* **350,** 73 (1994).

D.R. Baker, P.J. Doorbar, Fibre-matrix reaction zones in model silicon carbide-titanium aluminide metal-matrix composites, Rolls-Royce Ltd., Derby, U.K., PNR 90713, Aug 1989, 7 p.

J.K. Tien, M.W. Koop, Interfacial reactions in high temperature metallic and intermetallic matrix composites: A status review, Metal & Ceramic Matrix Composites: Processing, Modeling & Mechanical Behavior, editors, R.B. Bhagat, A.H. Clauer, P. Kumar and A.M. Ritter, TMS, 1990, pp 443-56.

M. Takeyarna, Microstructural evolution and tensile properties of titanium-rich TiAl alloy, *Mater. Sci. and Eng.,* **152,** 269 (1992).

D.D. Himbeault, J.R. Cahoon, Creep regimes for directionally solidified Al-Al$_3$Ni eutectic composite, *Metall. Trans. A,* **24A,** 2721 (1993).

H.C. Cao, E. Bischoff, O. Sbaizero, et.al., Effect of interfaces on the properties of fiber-reinforced ceramics, *J. Am. Ceram. Soc.,* **73** (6) 1,691 (1990).

D.B. Marshall, Fiber matrix interface effects in failure of ceramic matrix fiber composites, FR 7/85-7/89, Rockwell Int. Corp., SC 5432. FR, Cont # N00014-85C0416, Sep 1990.

Intermetallic Matrix Composites II, vol 273, 1992 MRS Spring Mtg., San Francisco, CA, editors, D.B. Miracle, D.L. Anton, J A. Graves, 450 p.

P.K. Brindley, P.A. Bartolotta, Failure mechanisms during isothermal fatigue of a SiC/Ti-24Al-11Nb composite, Session I, ASM Int. Ann. Mtg., Oct 2-6, 1994, Rosemont, IL.

P. Smith, J. Graves, Tensile and creep behavior of high temperature "neat" titanium matrices, Session V, ibid.

S. Kraus, A. Kumnick, P. Nagy, Fabrication and evaluation of Ti-22Al-23Nb "Orthorhombic" and Ti-6Al-4V composites utilizing high strength silicon carbide fiber, Session V, ibid.

C. McCullough, Manufacture of orthorhombic titanium aluminide composites by PVD methods, Session V, ibid.

O. Ivasishin, A. Demidik, V. Prozorov, Synthesis of TiAl-SiC composites by reactive sintering using TiH$_2$ powder, Session V, ibid.

D. Upadhya, F. Froes, M. Smith, et.al., Effect of processing on the frequency of thermal cracks in a continuous silicon carbide fiber-reinforced TiAl metal matrix composite, Session V, ibid.

C. McCullough, R. Kieschke, Protective interphase barriers for Al$_2$O$_3$/orthorhombic titanium aluminide, Session V, ibid.

A. Vassel, F. Brisset, Relationships between interfacial phenomena and fibre properties in Al$_2$O$_3$-reinforced titanium aluminide, Session V, ibid.

H. Chiu, J. Yang, Modification of interfacial properties and mechanical behaviors through fiber coating in a SCS-6/TiAl composite, Session V, ibid.

I. Gotman, M. Koczak, E. Gutmanas, et.al., Processing, synthesis and thermal stability of TiAl-SiC composites cold sintered from ultrafine powders, Session V, ibid.

Y. Chen, D.D.L. Chung, Aluminum-matrix Al$_3$Ti particle in-situ composites prepared by stir casting a liquid aluminum slurry containing TiO$_2$ and Na$_3$AlF$_6$ particles, Session V, ibid.

P. Ramasundaram, W.O. Soboyejo, R. Bowman, Ductile phase toughening of NiAl, Session VIII, ibid.

K.T. Venkateswara Rao, R.O. Ritchie, High-temperature fatigue and fracture properties of ductile phase toughened γ-TiAl intermetallic composites, Session VIII, ibid.

P. Krishnan, M.J. Kaufman, Development and characterization of NiAl/Al$_2$O$_3$ composites produced by *in situ* displacement reactions, Session II, ibid.

L. Christodoulou, *in situ* composites by exothermic dispersion, Session II, ibid.

D.E. Alman, C.P. Dogan, Processing, structure and properties of *in situ* TiAl-based composites by reaction synthesis techniques, Session II, ibid.

H.J. Feng, J.J. Moore, *in situ* exothermic forming of advanced composite materials, Session II, ibid.

O. Sahin, S. Nourbakhsh, W.H. Rhee et al, Oxidation of a zirconia-toughened alumina fiber-reinforced Ni₃Al composite, *Metall. Trans. A,* **23** (11) 3,151 (1992).

A. Joshi, T.C. Chou, J. Wadsworth, High temperature interactions of metallic matrices with ceramic reinforcements, FR Dec 1987 Aug 1991, LMSC-P010621, Cont # F49620-88-C-0021, Aug 1991, 200 p.

M.P. Metelnick, R.A. Varin, Z. Wronski, As-cast structure and the effects of temperature on an Al-Si metal matrix composite reinforced with Ni₃Al type ribbons, *Zeitschrift fur Metallkunde,* **83** (4) 227 (1992).

M.P. Metelnick, R.A. Varin, Effects of elevated temperature on an aluminum metal-matrix composite containing Ni₃Al-type ribbons, *Zeitschrift fur Metallkunde,* **82** (5) 346 (1991).

H.-P. Chiu, S.M. Jeng, J.-M. Yang, Interface control and design for SiC fiber-reinforced titanium aluminide composites, *J. Mater. Res.,* **8** (8) 2,040 (1993).

G.L. Povirk, J.A. Horton, C.G. McKamey, et.al., Interfaces in Nickel Aluminide/Alumina Fibre Composites, *J. Mater. Sci.,* **23,** 3,945 (1988).

C.G. McKamey, G.L. Povirk, J A. Horton, et.al., Fabrication and Mechanical Properties of Ni₃Al-Al₂O₃ Composites, High-Temperature Ordered Intermetallic Alloys, C.T. Liu, A.I. Taub, N.S. Stoloff, et.al., Ed., MRS, **133,** 609 (1989).

S. Nourbakhsh, W.H. Rhee, O. Sahin, et.al., Mechanical behavior of a fiber reinforced Ni₃Al matrix composite, *Mater. Sci. and Eng.,* **153** (1-2) 619 (1992).

J.R. Porter, M.C. Shaw, Fiber-matrix compatibility studies in ceramic-metal systems, 15th Conf. on Metal Matrix, Carbon, and Ceramic Matrix Composites, editor, J.D. Buckley, NASA CP-3133, Part 1, Dec 1991, pp 181-97.

Penn State investigates composite, *Ceramic Industry,* May 1994, p 14.

S. Guruswamy, M.K. McCarter, M.E. Wadsworth, Explosive compaction of metal-matrix composites and deuterides, editors, L.F. Pease III, R.J. Sansoucy, Advances in P/M-1991, vol 6, Aerospace, Refractory and Advanced Materials, MPIF, Jun 9-12, 1991, Chicago, IL, pp 251-66.

J.R. Stephens, Composites boost 21st century aircraft engines, *AM&P,* **137,** (4) 35 (1990).

P. Rangaswamy, N. Jayaraman, Residual stresses in SCS-6/Ti-24Al-11Nb composite: Part II-finite element modeling, *J. Composites Technol. & Res.,* JCTRER, **16** (1) 54 (1994).

D.G. Konitzer, M.H. Loretto, Microstructural assessment of interaction zone in titanium aluminide/TiC metal matrix composite, *Mater. Sci. and Technol.,* Jul 1989, pp 627-31.

JPRS-JST-92-019, 14 Jul 1992, *2nd Int. SAMPE Symp. and Exhibition on Advanced Materials Technol.,* 90 p.

S. yarnamoto, I. Barron, *ONRASIA Scientific Information Bull.,* NAVSO P-3580, ONRASIA **18** (2) 184 p (1993).

D.A. Clarke, Fabrication aspects of glass matrix composites for gas turbine applications, PNR 90752, Warwick Univ., IOP, 10-12 Apr 1990.

Advanced Materials, **14** (13) 1 (1992)

SAMPE Journal, July-Aug 1992.

Advanced Materials, **14** (6) 4 (1992).

V. Wigofsky, Thermosets expand role in design innovation, *Plastics Engineering,* July 1988.

D. Horton, D. Admas, *21st SPI,* sec 10A, 1967.

L.J. Blankenship, M.N. White, P.M. Pucket, **34,** 234 (1989).

D.A. Scola, R.H. Pater, *J. SAMPE,* **13,** 487 (1981).

Monsanto Resinox SC-1008 data sheet no. 2849D.

D.A. Shimp, *J. SAMPE,* **32,** 1063 (1987).

D A. Jarvie, *J. SAMPE,* **33,** 1405 (1988).

D.A. Scola, Polyimide Resins, in Engineered Materials Handbook, *Composites,* ASM International, 1977, pp 78-89.

H.D. Stenzenberger, et.al., *J. SAMPE,* **18,** 500 (1986).

H.D. Stenzenberger, et.al., *J. SAMPE,* **33,** 1546 (1988).

D.A. Scola, United Technologies Research Center (UTRC) Corporate Report R76-412819-1.

S. Clemans, T. Hartness, Advanced thermoplastic preforms, *SME Fabricating Composites, 1988 Conf,* Sept 1988, pp 103-09.

G.L. Brown, Jr., Pultrusion-flexibility for current and future automotive applications, COG-SME AUTOCOM'87 Conf, June 1987, pp 220-34.

J.H. Campbell, J.L. Kittelson, Winding the tape, *Aerospace Comp. and Materials,* **3** (6) 21 (1991).

D. Stover, Filament winding and fiber placement: stretching the bounds of an automated process, *Advanced Composites,* **5** (6) 20 (1990).

H.B. Soebroto, V.T. White, Braiding of complex shape beam composite preform, Session on Processing 1: Prefoming 1, *7th Annual ASM/ESD Adv. Comp. Conf and Expo,* Detroit, MI, Sept 30-Oct 3, 1991.

R.A. Florentine, The designer of 3-D braided preforms and the automotive design engineer: communications for innovation and profit, ibid.

G.C. Sharpless, Advancement of braiding/resin transfer molding from commercial to aerospace applications, ibid.

M. Jander, Industrial RTM-new developments in molding and preforming technologies, ibid.

B.E. Spencer, Designing and manufacturing filament-wound composites: the basics, ibid.

R.R. Irving, Metal matrix composites can claim advantages over monolithic metals in terms of desirable properties, *Metalwork. News,* Feb. 22, 1988, pp 17-18.

H. Jiang, Study on Al_2O_3-particle reinforced composite materials of Al-based metal, Jilin Univ., China, *J. SAMPE,* **25** (5) (1989).

J.M. Papazian, P.N. Adler, Tensile properties of short-fiber-reinforced SiC/Al composites: Part 1. effects of matrix precipitates, *Met. Transactions A,* **21A,** 401 (1990).

D.J. Lloyd, Metal matrix composites- an overview, Adv. Struct. Matls. Internat. Symposium, Montreal, Canada, Aug 28-31, 1988, ed. D.S. Wilkinson, Pergamon Press, pp 1-22.

S. Miller, A.D. Tarrant, Metallurgical design of novel metal matrix composites for aerospace applications, TMS mtg., Session on Lightweight Alloys for Aerospace Applications, 2/27/89, Las Vegas, NV, 39 p.

K. Upadhya, Developments in ceramic and metal-matrix composites, Structural Matls. Div. of TMS, 1992 TMS Annual mtg., San Diego, CA, March 1-5, 1992, 371p.

Y. Ishiguro, An overview of the fine ceramic industry in Japan, Japan 21st Century, 21 Apr 1992, pp 63-79.

D.W. Petrasek, High temperature strength of refractory metal wires and consideration for composite applications, NASA Rep. TN-D-6881.

J.G. Morley, Fibre reinforcement of metals and alloys, *Int. Met. Rev.*, **21**, 153 (1976).

N. Mykura, Casting MMCs by liquid pressure foaming, *Advanced Matls.*, pp 52-56, Technology Inter., 1992, Sterling Publ., London, Gt. Brit.

A.A. Zabolotsky, Structure and properties formation of metal matrix composites, *Composites Sci. and Tech.*, **45**, 233 (1992).

L.K. English, Fabricating the future with composite materials, Part 1: the basics, *Mater. Eng.*, **4** (9) 15 (1987).

J.E. Schoutens, Introduction to metal matrix composite materials, MMC No. 272, DOD MMC Information Analysis Center, Santa Barbara, CA, 1982, 602 p.

J.W. Weeton, D.M. Peters, K.L. Thomas, Engineers guide to composite materials, ASM, Metals Park, OH, 1987, 397 p.

T.W. Chou, R.L. McCullough, R.B. Pipes, Composites, *Sci. Am.*, **255** (4), 193 (1986)

U.S. Department of Commerce, International Trade Administration. A comparative assessment of selected reinforced composite fibers, U.S. GPO 491-097/36677, 1985, 48 p.

W.R. Mohn, Your MMC product will stay put, *Res. and Dev.*, **29** (7) 54 (1987).

N. Saka, N.K. Szeto, Friction and wear of fiber-reinforced metal-matrix composites, *Wear,* **157** (2) 339 (1992).

T. Vasilos, E.G. Wolff, Strength properties of fiber-reinforced composites, *J. Met.*, **18** (5) 583 (1966).

M.H. Stacey, *Mat. Sci. and Techn.*, **4**, 227 (1988).

Advanced composite design guide, 3rd ed, Air Force Matls. Lab., WPAFB, OH, 1973.

S.J. Harris, *Mat. Sci. and Techn.*, **4**, 231 (1988).

C.T. Lynch, J.P. Kershaw, in Metal matrix composites, *CRC Press,* 1972.

A.R. Champion, W.H. Kruger, H.S. Hartmann, et.al., in ICCM-II, The Metallurgical Soc., Inc., 1978, pp 882-904.

E. Joseph, V. Krukonis, A.W. Hauze, Exploratory development and evaluation of low-cost boron aluminum composites, 3rd Qtr. Rept., Contr. F33615-74-C-5082, ASD, Nov 1974.

W.H. Hunt, Jr., O. Richmond, R.O. Young, in ICCM-VI and ECCM-2, *Elsevier Applied Sci.,* **2,** 209 (1987).

Y. Fugita, H. Fukumoto, Y. Kurita, in ICCM-VI and ECCM-2, *Elsevier Applied Sci.,* **2,** 340 (1987).

T.W. Clyne, in ICCM-VI and ECCM-2, Elsevier Applied Sci., **2,** 275 (1987).

J. Dinwoodie, E. Moore, C A.J. Langman, et.al., in ICCM-V, The Metallurgical Society, Inc., 1986, pp 671-85.

F.K. Chi, R.D. Maier, T.W. Krucek, et.al., in ICCM-VI and ECCM-2, Elsevier Applied Sci., **2,** 449 (1987).

Hybrid fibers improve composite strength, *AM &P,* **133** (1) 87 (1988).

A. Mortenson, J.A. Cornie, M.C. Flemings, Solidification processing of metal matrix composites, *J. Met.,* **40** (2) 12 (1988).

A.E.R. Singer, S. Ozbek, Metal matrix composites produced by spray codeposition, *Powder Metall.,* **28** (2) 72 (1985).

New way makes better near-net-shape parts, *Metalworking News,* Oct 26, 1987, p 37.

S.B. Lasday, Unique approach to manufacture of ceramic composite components, *Ind. Heating,* **55** (4) 14 (1988).

Ceramic-matrix composites, *AM&P,* **132** (1) 28 (1987).

D. Richter, Commercial alternatives in metal matrix composites, *Advanced Matls.,* pp 57-60, Technology Inter., 1992, Sterling Publ., London, Gt. Brit.

M.C. Flemings, R. Mehrabian, *Met. Trans.,* **5,** 1899 (1974).

M.K. Surappa, P.K. Rohatgi, *J. Mat. Sci.,* **16,** 983 (1981).

S. Naka, M. Thomas, T. Khan, Potential and prospects of some intermetallic compounds for structural applications, *Mater. Sci. and Technol.,* **8** (4) 291 (1992).

J.W. Pickens, et.al., Fabrication of intermetallic matrix composites by the powder cloth process, NASA TM-102060, Jan 1989.

E. Aigeltinger, et.al., Microstructure property relations in ductile phase toughened gamma-TiAl alloys, *Proc. ASM Symp. on Intermetallic Materials,* Los Angeles, CA, Mar 21-23, 1989.

D.M. Dimiduk, D.B. Miracle, C.H. Ward, Development of intermetallic materials for aerospace systems, *Mater. Sci and Technol.,* **8** (4) 367 (1992).

C.T. Liu, A.I. Taub, N.S. Stoloff, et.al, editors, High-Temperature Ordered Intermetallic Alloys III, *MRS Symp. Proc.,* MRS, Pittsburgh, PA, **133,** 1989.

M.V. Nathal, R.D. Noebe, I.E. Locci, et.al., Hi Temp Review 1988, NASA CP 10025, p 235.

CHAPTER II – JOINING

2.0 INTRODUCTION

When design and material engineers select an appropriate composite material from the wide range of materials at their disposal – plastic, metal, and ceramic composites (each has strengths and weaknesses that affect its applicability for a given application), price and property performance are not the only factors that determine final selection.

Complex market issues outside of the designer's control can often influence this decision, however, in materials selection, there is only one driver: acceptable performance for the least cost.

Machining, joining and finishing, etc., often are the largest contributors to the total cost of a composite.

In dealing with the various composite there will be joining and machining problems. Ceramic composites cannot be joined with conventional fasteners or machined as easily as some MMCs. To this end, engineers have developed methods and techniques to overcome the aforementioned problems whether it be for RMC, MMC, or CMC. They have even turned to techniques in which one-piece shapes can be made to near-net or net shape. Producing single pieces in one step eliminates the need for joining different parts of a component, and making parts to near-net or net shape eliminates the need for much post-machining. The challenges here will be developing processes that can handle small or large parts, such as a fuselage component of a wing section, and improving processes to create the optimum composite structure.

2.1 MECHANICAL FASTENERS

When two or more components are joined together, the advantages which composite materials have (high strength-to-weight and high stiffness-to-weight) tend to be diminished.

It is all but impossible to transfer the full strength or stiffness of a composite through a joint. Even when something approaching full properties can be achieved, the price is usually a gain in weight. This gain results from the use of metal bolts or other fasteners, thickening or reinforcing structures in the area of the joint, etc.

For all practical purposes, the joint designer is limited to three joining techniques:

Mechanical – Mechanical joining involves the use of bolts, screws, rivets, or other fasteners. Bolting is common enough that we will generally refer to mechanical joining simply as "bolting."

Adhesive – In most cases, components are bonded together adhesively after they have been cured. However, the adhesive layer is sometimes cured along with the components in the process known as co-curing. Most persons will generally refer to adhesive joining, simply as "bonding."

Combined – The components are joined both mechanically and adhesively. Most will call this approach "bolting-bonding."

45

Table II-1 Comparison of Three Joining Methods

Criterion	Co-Cure	Bonding	Mechanical Fasteners
Production Cost	1	2	3
Ability to Accommodate Manufacturing Tolerances and Component Complexity	3	2	1
Facility and Tooling Requirements	3	2	1
Reliability	1	2	3
Repairability	3	2	1

Welding/Brazing/Diffusion Bonding – The components (RMC, MMC and/or CMC) are joined by one of these three methods.

The choice of one technique over the other depends entirely on the design requirements of the case at hand. Each approach has its strong and weak points. For example, Table II-1 compares mechanical fasteners with both normal bonding and co-curing for several design criteria.

The material on which Table II-1 is based is a Kevlar®/Ep composite. However, the results would apply to most other fiber composites as well.

2.1.1 Bolted vs. Bonded Joints

Mechanical joints offer several advantages over the adhesive variety. The most important of these advantages are the following:

- Bolted joints generally are less affected by temperature or humidity.
- Bolted components are readily disassembled.
- Inspection of bolted joints is relatively easy.
- Little or no surface treatment of components is needed before they are bolted together. Extreme cleanliness is rarely important.

As for disadvantages, the following are characteristic of mechanical joints:

- Bolts or other metal fasteners add weight and bulk to a joint.
- Holes for bolts or screws cut through fibers and weaken the components at the joint.
- Mechanical fasteners themselves create stress concentrations which can lead to joint failure.
- Corrosion can be a problem with metal fasteners.
- In general, metal fasteners are relatively expensive.

By comparison, bonded joints are superior in the following areas:

- They save weight over mechanical fasteners. For secondary structures, overall weight savings can be as high as 25%. For primary structures, weight savings of from 5 to 10% are more typical.
- With bolts, loads are transferred through the relatively small areas around the bolts themselves. Bonding typically involves a relatively large area over which loads can be transferred.
- Bonding requires no holes, so the components remain intact and unweakened.

46

- For large area joints, bonding is generally less costly than mechanical fastening.

The main disadvantages of bonded joints are as follows:

- Adhesive bonds, in general, have low shear strengths.

- The bond may be degraded by temperature and humidity conditions.

- Inspection for internal bond integrity is difficult.

- The bonded components cannot be disassembled without gross damage.

- Bonding generally demands careful surface treatments and extremely clean conditions. In some cases, this can mean white-glove handling, positive-pressure clean-rooms, and similar precautions.

What about the third approach, bolting and bonding? Shouldn't this approach combine the best features of both joints? To a certain extent, it does. Bolted-bonded joints out perform either type of joint alone. The bolt and the bond complement one another. The bolt reduces the bond's tendency to fail in interface shear. The bond decreases the bolted joint's tendency to shear out.

Put another way, bolting-bonding simplifies the failure-mode picture of the joint. For bolting or bonding alone, several different failure modes exist. For bolted-bonded joints, failure generally occurs in one of two modes (or in the two combined). These modes are:

- Tensile failure through a section including a bolt (or other fastener).

- Interlaminar shear failure in the composite itself.

Bolting-bonding offers several other advantages. The relatively large bonded area improves load distribution. The bond also provides a significant safety margin against fatigue failure of the metal fastener. At the same time, such a joint usually is designed so that the bolts can carry the load if the bond fails.

Unfortunately, bolted-bonded joints also share some of the disadvantages of each joint type. For example, the bolt adds weight and bulk. The bond makes inspection difficult and disassembly impossible. Except for problems like these, bolting-bonding might well be the universal design choice for structural composite parts.

2.1.2 Bolts and Other Mechanical Fasteners

Metallic fasteners have succeeded very well with fiberglass, boron, and Kevlar® fiber composites. However, graphite reinforcements present problems. Some metallic fasteners in contact with graphite composites are prone to corrosion. Aluminum is the worst in this respect. Stainless steel is somewhat better. Nickel and titanium alloys show excellent compatibility with graphite. but are expensive.

This kind of corrosion especially concerns aircraft designers because of the growing use of graphite. Other than relying on the more expensive metal fasteners, two main solutions are possible.

The first is to insulate the graphite from the metal. This can be done by surrounding an aluminum or steel fastener with fiberglass or adhesive with scrim. (Scrim is a low-cost reinforcing fabric made from continuous filament yarn in an open-mesh construction.) While this approach prevents corrosion, it complicates fastener installation.

The second solution is to use composite fasteners rather than metallic ones. Gl-Ep and Gr-Ep do not corrode galvanically. They also save weight. A Gr-Ep fastener is five times lighter than a stainless steel fastener of similar size. Gl-Ep is four times lighter. Tests by one helicopter manufacturer indicate that fiberglass-composite fasteners are sound for light-to-medium loads, although they are somewhat degraded by humidity.

In the end, however, design loads and stresses dictate the choice of fasteners. A given high-performance aircraft structure may require titanium fasteners despite the cost.

2.1.2.1 Failure Modes and Prevention

As discussed earlier, a number of failure modes are possible with bolted joints. Before going into these modes. one needs to define some terms widely used in describing such joints. Among these terms are bearing area, bearing stress, and bearing strength.

All of these terms are based in one way or another on the size of a hole used for a bolt or other fastener. The bearing area is the diameter of the hole D times the thickness t of the material. The bearing stress is the applied load divided by the bearing area.

Bearing strength is generally defined as the stress required to elongate the hole by a certain percentage of its original diameter. For composites, a 4% elongation is usually considered the cut-off point for bearing strength. Bearing strength generally is inverse to the ratio D/t. In other words, bearing strength decreases as the width of the hole relative to the thickness of the material increases.

For fiberglass composites, a conservative ratio D/t would be 1. A hole 12.7 mm in diameter in a laminate of the same thickness would provide this ratio. At ratios of four and above, bearing strength is greatly reduced. Thus, very thin laminates (less than 0.75 mm) should be reinforced at the point of attachment to increase thickness relative to hole diameter.

Edge and side distance are also prime considerations in mechanical joint designs. Edge distance is defined as the distance from the end of the joint to the center of the closest hole. Side distance is the distance from the side of a joint to the center of the nearest hole. Both distances can be expressed as ratios to the hole diameter. Table II-2 gives some generally accepted edge and side-distance ratios for various composite thicknesses.

Figure II-1 illustrates the most common failure modes for bolted joints. Two or more of the failure modes shown in Figure II-1 may occur in combination. Figure II-1 does not illustrate another relatively common failure mode, bolt failure, which involves simple shear failure of the bolt itself.

Most composites are quite brittle, that is their strains to failure are low. Thus, stress concentrations (and strains) caused by the bolts themselves can lead relatively easily to failure of the surrounding composite.

Table II-2 Recommended Fastener Distances for Various Composite Thicknesses (Based on Fiberglass Epoxy)

Thickness of Laminate mm (in)	Edge Distance Ratio (Edge Distance Hole Diameter)	Side Distance Ratio (Side Distance Hole Diameter)
3.18 or less	3.0	2.00
3.18 - 4.76	2.5	1.50
4.76 or greater	2.0	1.25

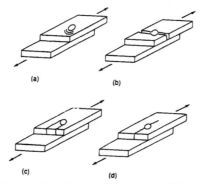

Figure II-1. Bolted-joint failures. (a) bearing, (b) net tension, (c) shear out, and cleavage.

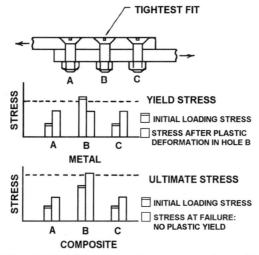

Figure II-2. Effect of bolt tightness stresses in metals and composites.

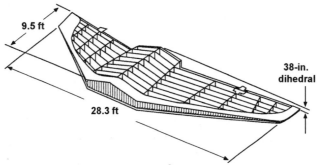

Figure II-3. AV-8B wing assembly.[8] 25.7 mm = 1 in

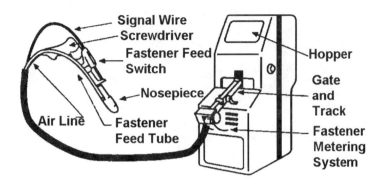

Figure II-4. Fastener installation subsystem.[8]

One way to understand this is to compare composites with metals. When a bolt is inserted into a bolt hole in aluminum or steel which is too small, stresses rise in the surrounding metal. If these stresses are great enough, the metal will yield and some of the load will shift to the adjacent bolts. At the same time, stress will be reduced around the original hole.

Brittle composites, on the other hand, do not have yield points. If one bolt is tighter in its hole than others, the bearing stress at that hole simply remains higher. If loading increases this stress beyond the ultimate, the composite will not yield but fail. At that point all of the load will fall on adjacent bolts, possibly causing failure there as well. Figure II-2 illustrates the differences between metal and composite bolt-stress behavior.

One way to alleviate this problem is to "soften" the composite in the bolt area. This is done by using staggered fabric plies to reduce brittleness. Another approach is to coat the bolt with uncured resin before inserting. The resin flows into the open area between the bolt and the composite. Upon cure, it helps to distribute loads more evenly. The cured resin is less brittle than surrounding fibers such as graphite. Finally, bonded metal inserts can be used, so that the bolt is actually in the metal, not in the composite. However, this approach is both costly and adds weight.

A poor approach to the problem is to reinforce the hole with more high-modulus fibers, for example, graphite. This only increases brittleness around the hole, making the joint more likely to fail under bolt-induced stresses.

2.1.2.2 Special Consideration for Fasteners

As in metallic joints, the prime considerations for composite joint design are bearing, tensile, and compressive strengths in the laminate, and shear and bending stresses in the fasteners.

Advanced composites can be delicate materials to fasten. Ordinary fasteners are not designed to protect the fibrous panels and work within the strength limitations. These materials are sensitive to radial expansion of the holes, which would create undue stresses, and to pressure on the back or blind side, which might result in fastener pull-through. The pressure required during installation can cause buckling or delamination. Problems that may occur during use include delamination, crushing, or pull-through.[1-7]

Composites require special treatment for several reasons. Since they have lower thickness compression strength than metals, blind fastener heads should expand more for proper fastening. Too small a clamp area can cause excessive load on the bearing area resulting in crushing or delamination around the hole. The pull-up length of the fastener should precisely match the thickness of the panels being joined to avoid local crushing. The fastener should not expand within the hole, which might cause delamination. A larger clamp area may prevent pull-through. In addition, holes must be drilled cleanly minimizing splintering and fraying, and subsequent fiber separation.[7]

To control compressive stresses, engineers have designed various fasteners with large bearing areas to distribute the load over a large area. Installation problems have also been reduced or eliminated through customized fastener designs and the use of special installation tools.

2.1.2.3 Materials

Titanium alloy 6Al-4V is the favorite fastener material for joining carbon-fiber-reinforced materials. It has ultimate tensile and shear strengths of 1,104 MPa and 650 MPa, respectively.

In answer to the exacting weight and strength requirements for composite fastening on the European Fighter Aircraft (EFA), a new Ti-Matic Blind-Bolt, an all-titanium blind fastener, has been developed, Figure II-5. In this bolt are combined the desirable strength to weight properties of titanium with other composite-friendly features, to produce a light, versatile product, capable of replacing many existing fastener types and offering considerable weight savings.

By their nature, blind fasteners offer significant productivity gains over two-piece threaded fasteners and also involve a lower skill factor for installation. Thus, resulting in a lower installed cost. The grip overlap capability of the Ti-Matic allows the operator to use one fastener throughout, where with other types, different grip sizes may have been needed. This is an important factor in parts-planning and offers the benefit of reduced parts holding. Unlike finished metal, the surface of composites is inevitably irregular, leading to variation in hole thickness. Without a grip overlap capacity, an operator cannot be expected to accurately forecast which fastener (grip) size is required. The Ti-Matic's minimum blind side clearance also enables the designer to reduce the thickness or depth of a structure and thereby reduce weight and airframe drag if the situation permits, Figure II-6.

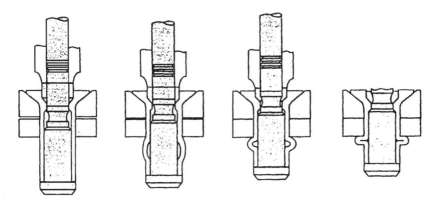

Figure II-5. Installation sequence.

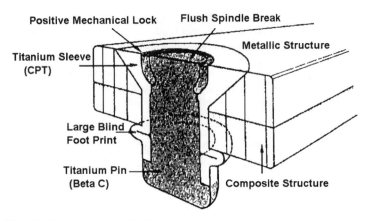

Positive Mechanical Lock   **Flush Spindle Break**

Metallic Structure

Titanium Sleeve (CPT)

Large Blind Foot Print

Titanium Pin (Beta C)

Composite Structure

Figure II-6. Section through installed fastener.

A drilled hole in composites is a potential trouble spot. During the blind side formation of the fastener, considerable forces may be exerted on the area around the hole. With some fasteners, there is always a risk of delamination of the composite lay-up. Delamination, or fraying of the strand ends, can weaken the joint and/or lead to a site for water absorption. Ti-Matic eliminates this problem area with the off-sheet bulbing feature.

The Ti-Matic fastener is specifically designed to tolerate a 7° sloping (blind) surface and requires only a minimum of blind side clearance. Both these factors give the fastener the ability to accommodate the variations and frangibility experienced in composite material structures. The bolt's mechanical values meet or exceed the requirements of the industry benchmark specifications for blind fasteners and a 656 MPa shear strength that is equivalent to threaded fasteners.

For enhanced mechanical properties, Inconel 718 or A-286 corrosion-resistant steel (CRES) materials may be cold-worked to develop tensile strength up to 1,517 MPa, and shear strength up to 758 MPa. Multiphase alloys MP35N and MP159 can be used to produce fasteners which have minimum tensile strengths of 1,800 MPa and are also resistant to stress-corrosion cracking and hydrogen embrittlement. Fasteners of these materials are available in diameters up to 38 mm. Among the applications for MP35N are as wheel bolts for use on NASA's Space Shuttle Orbiter. The alloy has also been used to produce fasteners for critical joints on the Space Shuttle, Table II-4.

Programs such as National Aerospace Plane (NASP) also present new and unique challenges to fastener engineers. While the severe environmental conditions that will be encountered by NASP are not completely understood, engineers agree that atmospheric re-entry profiles will require fasteners to withstand very high temperatures (possibly in excess of 1,370°C). The fasteners must also be able to operate at extremely low temperatures for extended periods of time.

Researchers are currently testing fasteners made from materials such as Haynes Alloy 230 (good to approximately 1,000°C), and Ti1100 high temperature titanium (good to about 600°C) for certain NASP applications. Other considerations include using coated refractory alloy fasteners, however, compatibility of the coating with the C-C structures of the NASP is a concern. Engineers are also examining experimental materials such as titanium aluminides, and SiC-re-

Table II-3 Fastener Characteristics

Fastener Type	Contact Area	Installed Profile	Clamp Force Range	Clamp Force Control	Shark Clearance	Diameter Range	Crip Range	Available Reuses	Cost Per Fastener	Number of Components	Access Requirements	Removal Efficiency	Installation Efficiency	Proven Technology
Low-Profile, Single- Sided	U													
Low-Profile, Single-Side & Washer	E	A	E	A	A	A	A	A	A	A	A	E	A	E
Screw, Washer & Nut	E	E	E	E	E	E	E	E	E	A	A	A	A	E
Standard-Profile, Single-Side & Washer	U													
Pull Rivet												U		
Blind Bolt												U		
McAir Design, Single-Side	A	A	E	A	A	A	A	A	P	E	E	E	E	P
Spring-Loaded, Single-Side		U												

E = excellent A = accepatable P = poor U = unacceptable

inforced titanium and aluminum as fasteners. Fasteners made of C-C materials may resolve material compatibility problems; however, much research and development still remains to be done.

A carbon-reinforced composite requires that the fastener material be near carbon on the electromotive scale. An aluminum- or steel-alloy fastener would corrode rapidly in the presence of carbon and an electrolyte. Salt spray tests show titanium alloys, austenitic stainless, and certain multiphase and Inconel alloys (discussed earlier) to be the most compatible fastener materials.

A fastener must be compatible with all materials in the joint. For example, an uncoated titanium fastener can be used to join a carbon-fiber-reinforced composite with titanium. When pairing a carbon-fiber-reinforced material with aluminum, coat the titanium fastener with aluminum for galvanic protection. A faying surface sealant or a protective-barrier coating also must be used between the aluminum and carbon-reinforced materials.[6-7]

Carbon-, aramid-, and boron-fiber-reinforced composites require precision hole-making techniques. The fastener hole should be straight and round, typically within 0.01 mm. Two common

Table II-4 Chemical Composition of Selected High-Strength Alloys.[10]

Alloy	UNS Number	Typical Chemical Composition (%)
A-286	K66286	15 Cr, 25.5 Ni, 1.3 Mo, 2.1 Ti, 0.06 b, 0.30 V, Balance Fe
Alloy 718	N07718	52.5 Ni, 19 Cr, 3.1 Mo, 5.1 Cb, 0.90 Ti, 0.05 Al, 18 Fe
MP35N	R30035	20 Cr, 35 Ni, 35 Co, 10 Mo
MP159	R30159	19 Cr, 36 Co, 25 Ni, 7.0 Mo, 0.50 b, 2.9 Ti, 0.20 Al, 9.0 Fe

problems involve delamination of the exit side (usually caused by drilling without a back-up), and overheating caused by excessive feed or tool wear. Delamination can reduce joint strength. Overheating, which may not be visible or may become visible only after installing the fastener, is a worse problem in that it can result in a lower strength and increased elongation of the joint.[6-7]

There are a few companies which make fasteners out of advanced composites for use with advanced composites. Several of these are illustrated in detail in Refs. 1, 2, and 7. A Huck Comp GP[2] lock-bolt titanium fastener is reported to save about 1.75 kg of weight for every 1,000 fasteners. Another lock bolt is expanded into an interference fit during installation to provide intimate contact to form a water barrier. This eliminates the need for sealant, in addition to providing fuel tightness, fatigue resistance, and a conductive path for electrical-charge dissipation.

A Composi-Lok blind bolt[1-2] has been designed for composite-to-composite and composite-to-metal applications where the blind bearing side is against the composite sheet. The fastener insures structural integrity under severe vibration.

A Live Lock fastener[2] is recommended for access panels, engine cowlings, and any removable composite sections, while Comp-Tite[2] fasteners are primarily intended for composite-to-composite fastening applications, such as in empennage and wing sections.

Fastener firms have now developed a group of fasteners, Comp-Fast,[2] that can be customized by design and manufacture to handle various applications. They can be used to fasten composites to metals in skin fastening, structural-panel fastening, and in applications along with rivetless plate nuts.

Preliminary efforts have been made toward automatic installation in composites, and automation is expected to become even more prevalent. An automated system for drilling and installing Huck Comp fasteners was reported[2] whereby 70,000 lock bolts have been used in the V-22 Osprey (VTOL) aircraft. The system is expected to have a machine cycle time of only about 9 seconds.

Another firm, Monogram,[2] has developed end effectors for robotic installation of its blind bolts. The fairly simple pneumatic tools for the rotary-actuated blind bolts weigh about 15.5 kg, making it easier for the robotic system to maintain placement accuracy. According to Monogram, installation of blind bolts is easier to automate because they require access to only one side.

Cherry-Textron and several other firms[2] have introduced an all-composite fastener. The ACP pin-and-nut fastening system of Cherry-Textron is produced in eight composite materials: 1) PEEK-long carbon fiber, 2) PEEK-short carbon fiber, 3) VECTRA-glass fiber, 4) PEI-glass fiber, 5) PI-glass fiber, 6) Epoxy-carbon fiber, 7) PI-carbon fiber, and 8) Epoxy-glass.

In external applications, composite pins reduce radar reflection and lightning-strike risk. They also are sonic vibration resistant and eliminate potential corrosion problems.

In aircraft, the pins and fasteners can be used in structural and nonstructural applications to reduce weight by up to 80%. In addition, they prove useful in electronically sensitive applications such as radar and instrument housings.

The performance of specific composite light-weight fasteners depends on the resin and fiber combination. A fastener of AS-4 fiber with epoxy resin results in a 275 MPa shear strength.

Manufacturing fasteners out of composite materials is still a relatively recent venture which presents an entirely new set of challenges to fastener engineers. Not only is the method of manufacturing completely different, but the ways in which the fasteners are inspected and tested are unique. Developments in fasteners made from composite or composite-compatible materials are continuing as more aircraft are built from these materials.

2.1.3 Mechanical Fasteners and RMCs

In order to meet the manufacturing requirements for the intermediate sections of the B-2 stealth bomber, a computer-controlled robotic system for automated drilling and fastening of the composite/metallic aerostructure was installed by the LTV Aerospace and Defense System Group. The automated five-axis robotic system performs the drilling and fastening operations of large, highly-contoured aircraft parts which combine composite materials with titanium or aluminum substructures.

The aircraft sections are fixed in the Robotics for Major Assembly (RMA) cell while the computer-controlled robot moves over the assembly drilling holes and installing fasteners in one pass in conjunction with specially-designed coolant-feeding drill/countersink tools. To date, the robotic cell has installed more than 50,000 fasteners in B-2 parts, improving quality of fastener operations, decreasing production time, and reducing manual labor.

In another advancement in materials development for commercial airliners, the Boeing Co. is using a polyetheretherketone (PEEK) rivet to secure aircraft bay liners. The main reason for the selection of PEEK was that it exhibits extraordinary flame retardance and withstands exposure to high temperatures. Although PEEK is not a new material, its use in fasteners is new. While it is expensive, Boeing's specifications could only be satisfied by this material.

The AV-8B Harrier wing assembly is shown in Figure II-3. Most spars and ribs composing the structure are made of a C/Ep composite material. The remaining members are aluminum or titanium machined members. Most of the C/Ep spars and ribs incorporate a sinusoidal-shaped web. The upper and lower highly contoured, one-piece, C/Ep skins are fastened to the substructure with approximately 6,700 flush head screws and Hi-Lok fasteners.[1]

To handle this fastening problem, an Automated Drilling System (ADS) composed of two numerically controlled, six-axis, gantry-type machines was developed to produce close-tolerance, outer mold-line (OML) fastener holes in highly contoured parts. Numerous studies concluded 1) that drilling holes with the ADS and installing fasteners to provide the clamping force was the most practical solution, 2) installing a fastener in every 10th hole would provide sufficient clamping action to install 600 holes immediately after each one is drilled without causing excessive reduction in rate capability and increase in cost, 3) the best fastener characteristics and their effect on the application, Table II-3,[8] and 4) best combination of installation method and fastener types, Figure II-4.[8]

Fasteners come in all forms (even hybrids). The Aerospace Div. of SPS Technologies[9] had a unique fastening problem with an advanced composite component for an aircraft fuselage. The problem was to form a blind head using a blind fastener in a 25.4 mm or 50.8 mm honeycomb joint between two composite parts. The fastener consistently pulled through the hole on the blind side and damaged the honeycomb material. The solution was a specially designed fastener made of titanium, two types of stainless steel, and a proprietary high-strength, corrosion-resistant alloy. The part didn't use the surface of the joint to form the blind side since it had a coiled split washer that pulled up an angled ramp on the body of the fastener, simultaneously forming a sleeve in the fastener. This was accomplished without contacting the joint.

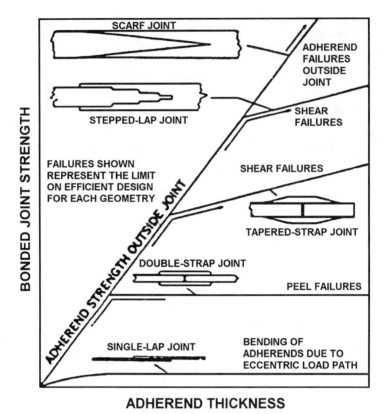

ADHEREND THICKNESS

Figure II-7. Relative uses of different bonded joint types.[15]

Virtually all fasteners require holes of some kind. But with some high-performance composites, drilling a hole may delaminate the structure. This is especially true with carbon and graphite composites. If the hole isn't molded in when the composite is formed, grommets should be used for hole support. A recently introduced threaded metal stud can be ultrasonically welded into thermoplastics. The stud saves the cost and cycle time of making preformed holes.

2.1.4 MMCs

Large vertical tail boxes have been fabricated from the largest sheets of MMCs. Large SiC/Al sheets, each 2.3 mm thick and measuring in excess of 1,780 x 590 mm, were fabricated in an Air Force Flight Dynamics Lab. demonstration program. Assembly tooling for the project was similar to tooling used for metal-bodied airplanes. The only difference in handling the composite materials was in machine cutting where feeds and speeds were adjusted somewhat. The sheets were handled via the normal production process. Fixtures and fittings were those typically used on metal-skinned airplanes. Skins for the tail boxes were cut with water jets. Once the skins were cut to size, they were fastened to MMC spars with standard steel fasteners as a means of reducing costs. Four vertical tails were fabricated. Two had skins of SiC whiskers mixed with aluminum, with spars made of SiC fiber mixed with aluminum. The other two had tails with skins and spars made with SiC fiber mixed with aluminum. The basic aluminum was

56

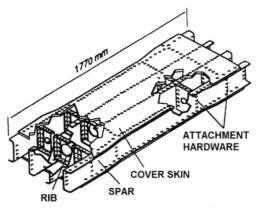

ATTACHMENT
HARDWARE

COVER SKIN

SPAR

RIB

Figure II-8. C-SiC box for thermal structure.[11]

2124, with whisker content of 15 vol%. The SiC fiber-aluminum composite had a SiC fiber content in the 6061 aluminum of 42 vol%.

2.1.5 CMCs

Joining may be required when the desirable properties of a CMC need to be combined with those of a metal or an MMC, as in the electronics industry, where both are essential to the functioning of a component, or device. Joining is also required for components where design, or economic constraints, restrict the use of CMCs to the most arduous service conditions and allow the use of cheaper materials elsewhere.

2.1.5.1 Joining Processes

There are many possible techniques for joining ceramic materials and CMCs. The CMCs technologies range from mechanical attachment which introduces no bonding medium and creates no interface, to direct bonding processes such as brazing and diffusion bonding which involve the formation of an intimate and continuous interface across which chemical and, to some extent, mechanical bonding occurs. One reason why so many methods have been proposed is that CMC technology has the potential to be applied across many industries including electronic, automotive, aerospace, and biomedical.

2.1.5.1.1 Mechanical Fastening

A range of mechanical methods can be used to form a joint between similar and dissimilar materials. These are often quite simple in design and, in many cases, employ bolts, screw threads, or an interference shrink fit. Unfortunately, a ceramic and/or CMC screw thread usually involves expensive machining which can result in cracking and excessive tightening of bolts can also lead to cracking and premature failure of the component.

Shrink and taper fitting which utilize the difference in thermal expansion between ceramics/CMCs and metals are widely used. In general, metals expand more than ceramics/CMCs on heating and contract more on cooling. The joint is formed at temperature, with the metal sufficiently expanded to surround the ceramic/CMC. On cooling, it contracts and clamps. The joint is strong and can be made gas tight by using compliant interlayers of metal such as aluminum. The interlayer serves two major purposes, first, it distributes the contact loads to minimize stresses and, secondly, it allows for some movement. The major drawback is that the joint operating temperature must be considerably lower than that at which it was prepared. Shrink fitting has been applied in the production of spark plugs, where a compliant layer is used to seal the metal base to the CMC insulator. In this case, the metal base is heated and forged around the CMC and interface material forming a tight joint as it cools.

2.1.5.2 Application

The C-SiC composite structure, Figure II-8, representing the Hermes winglet is a first of its kind. This test article was used to reflect mature technologies developed in materials and processes, assembly techniques, and high temperature fasteners. In order to produce a strong thin-skinned optimum weight part, the Societe Europeene de Propulsion (SEP) developed Skinex®, a 3D carbon fiber texture preform with large surface area. Outer skins and spar bars were manufactured and the C-SiC parts were secured by very high temperature resistance metallic fasteners (< 1,300°C). However, the wing box assembly will not withstand very high heat fluxes and, thus, a limitation is placed on the C-SiC part thermal capabilities. Currently underway is the development of advanced ceramic composite fasteners which are designed to optimize the high temperature characteristics of C-SiC structures when mechanically assembled. Prototype C-SiC screws, hinges, and attachment systems have been fabricated by SEP and are in evaluation.[12]

2.2 ADHESIVE BONDING

Currently, the two most successful methods for joining composite structures are mechanical fastening and adhesive bonding. Eliminating as many joints as possible is one of the most important goals in designing composite structures. Joining of RMCs has been largely limited to adhesive bonding and mechanical fastening.[13]

There are overwhelming reasons to adhesive bond composites to themselves and to metals, and, conversely, there are substantial reasons for using mechanical methods such as bolts or rivets, Table II-5.

Whenever bonding is planned, adhesive selection is of primary concern. Two-part, medium-viscosity, non-slumping, room-temperature-curing adhesives are preferred because of their user-friendly properties. The most desirable properties are 1) a 1:1 mix ratio of two different color components that combine to give a third distinct color, thus signifying a complete mix, and 2) a quick cure for rapid handling strength. However, the adhesive must possess a pot life long enough to allow time to complete the application. There is typically a trade off between rapid cure and long pot life.

Table II-5 Reasons For and Against Adhesive Bonding

For	Against
Higher strength-to-weight ratio	Sometimes difficult surface preparation techniques cannot be verified 100% effective
Lower manufacturing cost	Changes in formulation
Better distribution of stresses	May require heat and pressure
Electrically isolate components	Must track shelf life and out time
Minimize strength reduction of composite	Adhesives change values with temperature
Reduce maintenance costs	May be attached by solvents or cleaners
Reduced corrosion of metal adherend (no drilled holes)	Common statement, "I won't ride in glued-together airplane."
Better sonic fatigue resistance	

Epoxies, urethanes, polysulfides, and acrylics are adhesives that have been used successfully in structural applications. Ultimately, the choice of adhesive depends on the performance characteristics required to make an effective attachment. Characteristics such as installation environment, end use environment, cure time, pot life, application equipment, desired strength level, optimal failure mode, and cost are all factors to consider when identifying an appropriate adhesive.

When adhesive bonding, surface preparation is the major concern for producing a satisfactory joint. The extent of surface preparation required for a reliable adhesive bond depends on the material being bonded and the adhesive used. From a manufacturing stand point, it is desirable to perform the least amount of surface preparation possible that will result in a reliable bond. There are standard procedures that will always enhance bonding regardless of adhesive selection. A good solvent wipe of the substrate and the fastener base is considered to be a minimum in surface preparation. Abrasion of the surfaces by means of a scuff pad, sandpaper, or grit blasting, prior to solvent wiping, provides a surface condition that is excellent for successful adhesive bonding.[5]

The studies on adhesives, surface preparation, test specimen preparation, and design of bonded joints reported for the PABST program,[14] give much more credibility to the concept of a bonded aircraft and provide reliable methods of transferring loads between composites and metals or other composites.

"Successful bonding " implies consistent strength and long-term reliability. To achieve consistent strength, the bondline thickness must be controlled so that reproducible joints with similar performance characteristics are achieved.[15]

A number of decisions must be made before bonding to a composite is attempted:

1. Surface preparation technique

   Composite: Peel ply, use manual abrasion, or, if possible, co-cure.

   Metal: For aluminum, phosphoric anodize is the accepted, preferred airframe method. For titanium, several types of etches are available. For steel, the time between surface preparation and bonding is of major concern.

2. Type of adhesive

   Film adhesive: Reproducible chemistry and frequent quality assurance testing (because resin is premixed by the manufacturer) require special storage and generally cannot accommodate varying bond-line thicknesses.

   Paste adhesive: Longer shelf life and fewer or no changes in storage accommodate varying bond-line thicknesses, but quality assurance is performed after the structure is bonded.

3. Cure temperature of adhesive – Unless they are co-cured, the maximum cure temperature of the adhesive should be below the composite cure temperature. The cure temperature or the upper use temperature of the adhesive may dictate the maximum environmental exposure temperature of the component. There are several cure temperature ranges:

   Room temperature to 107°C: Generally for paste adhesives for noncritical structures.

   107 to 140°C: For nonaircraft critical structures. Cannot be used for aircraft in general because moisture absorption may lower the heat-deflection temperature (HDT) below the environmental operating temperature .

177°C and above: For aircraft structural bonding. A higher cure temperature may mean more strain discontinuity between adhesive and both mating surfaces (residual stresses in bond).

4. Joint design – The primary desired method of load transfer through an adhesive-bonded joint is by shear. This means that the design must avoid peel, cleavage, and normal tensile stresses. One practical application of a method avoiding other than shear stresses is the use of a rivet-bonded construction. The direction of the fibers in the outer ply of the composite (against the adhesive) should not be 90° to the expected load path.

2.2.1 Joint Configuration

There are three basic types of bonds- primary, secondary, and tertiary. A primary bond is a bond between an uncured laminate and a fresh laminate. A secondary bond is between an uncured laminate and a cured laminate. A tertiary bond is between two cured laminates.

Whatever the bond type, the designer can choose from a great variety of joint configurations. Among the most widely used are the lap and strap joints shown in Figure II-7. Joints featuring straps and doublers impose a slight weight penalty, (although not as much as a bolt). They are unacceptable only when a smooth uninterrupted surface is necessary.

Also widely used are the scarf joints and other configurations shown in Figure II-7. Scarf joints, by definition, involve tapered joining surfaces. In this way, they represent a design middle ground between lap joints and the end-to-end attachment of butt joints.

There are also two kinds of joints: mechanical and bonded. Mechanical joints are created by fastening the substrates with bolts or rivets. Bonded joints use an adhesive interlayer between the adherends. Adhesively bonded joints can distribute the load over a larger area than the mechanical joints, require no holes, add very little weight to the structure, and have superior fatigue resistance.

Kim and associates,[16] after the above evaluation, selected the adhesive-bonded lap joint for work on a torque transmission joint. Included in their investigation were the effects of adhesive thickness and adherend surface roughness on the fatigue strength of a tubular adhesive-bonded single lap joint. From the fatigue experiments, it was found that the optimal arithmetic surface roughness of the adherends was about 2 μm and the optimal adhesive thickness was about 0.15 mm. Using the optimal adhesive thickness and the optimal adherend roughness, the prototype torsional adhesive joints for the power transmission shafts (0.66 mm) of an automobile or a small helicopter were manufactured and statistically tested under torque. The tests were performed on the single lap joint, the single lap joint with scarf, the double lap joint, and the double lap joint with scarf. One adherend was high strength carbon fiber epoxy composite and the other was S45C carbon steel. From the experiments, it was found that the double lap joint was the best among the joints in terms of torque capacity as well as cost of manufacture.

2.2.2 Surface Preparation

A major cause of bond failure is contamination of the surface of the adherend. A number of conventional treatments are in use to improve the adhesion of surfaces. These treatments include, but are not limited to, mechanical abrasion, solvents, acid or caustic etching, and corona treatment. All have a number of limitations, therefore, the need for alternative treatments is great.

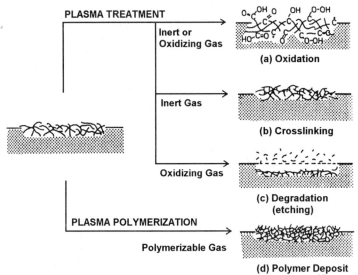

Figure II-9. Schematic of plasma modifications of polymer surface.[20]

Researchers at Lockheed Aeronautical Systems Co. (Burbank, CA) completed studies indicating that treatment of PEEK /Gr composites with chromic/sulfuric acid etch, gas plasma, or a Kevlar® peel ply provided good surface preparation for epoxy-adhesive bonding.[17]

Scientists at the Ministry of Defense in Israel used argon fluoride (ArF) excimer lasers and 10.6 μm CO_2 lasers to improve bonding of engineering plastics such as polycarbonate and PEI. The radiation dissociates some surface molecules into reactive fragments. Researchers believe the ArF laser massively alters surface chemistry; the CO_2 laser mainly softens the thermoplastic surface.[17]

Researchers at Air Products in the United Kingdom have been investigating the pretreatment of polyethylene, polypropylene, and acetal with fluorine to improve adhesive joint strengths. Using fluorine/nitrogen and fluorine/oxygen/nitrogen mixtures for pretreating low-density and high density polyethylene films, they found that large increases in adhesion were possible with all mixtures containing fluorine.[17]

2.2.2.1 Plasma Etching

In recent years, researchers[18-19] have found that surface activation by plasma treatment prepares plastic and other substrates for adhesive bonding. The article to be treated is placed in a chamber which is evacuated by vacuum pump. A specific gas, such as oxygen, argon, or air, depending on the treatment desired, is flowed through the chamber and drawn out the exhaust. The gas is then excited with radio-frequency (RF) energy produced by applying power to the chamber electrodes. This strips electrons from the gas molecules, forming free radicals which react with the surface and creates sites for bonding, Figure II-9. Water, oils, and other organic materials on the surface are also attacked and broken down by free radicals. Their volatile remains are swept away by the vacuum system, leaving the surface ultra-clean. The glow associated with plasma treatment is a result of stripped electrons falling back to a lower energy state, emitting light as in neon, fluorescent bulbs, and lightning.

61

The effect of plasma treatment is to reduce the contact angle (a function of surface energy) of the substrate being treated. This reduction in contact angle results in improved wetting of adhesives on the treated surface and higher bond strengths. When the treated substrate is exposed to air contact angles begin to increase again, so adhesive should be applied to treated surfaces as soon as possible after plasma treatment.

Gas plasma is a good surface preparation for the adhesive bonding of plastics because it involves no solvents, chemical fumes, disposal problems, or heat, which could harm the plastic surface. The plasma reacts to a depth of 0.01 to 0.1 μm (100 to 1,000 Å).[21] Joining carbon/PEEK composites is an important technological issue[22] and surface preparation is particularly critical.[23] A comparison between different surface preparation methods showed that plasma treatment, albeit with particularly "hard" parameters (long treatment time and use of a CF_4/O_2 mixture), scored remarkably well.[22] Plasma treatment is particularly interesting since it is reasonably uniform over the whole part, compared with flame or corona treatments, and can be made environment-friendly by an appropriate choice of gases.[24-27]

Occhiello et.al.,[23] assessed the efficiency of plasma treatments in improving the adhesion of carbon/PEEK composites with epoxy adhesives. Carbon/PEEK composites were treated by plasma to improve adhesion with conventional epoxy adhesives. Oxidizing (oxygen, air) and inert (nitrogen, argon) mixtures were used. An interesting finding was that, for plasma treatments lasting more than 5-10s, (much lower than the 5 min quoted in Reference 22 or the 2 min of Reference 25) there is no real increase in pull strength with treatment time. Also, RF power seems a rather mild requirement, since 20 W treatments provide excellent adhesion. Oxygen provides consistent improvements in adhesion even at very low treatment times. Argon is most efficient for treatments lasting at least 30 seconds, nitrogen is somewhat in between. Moreover, the ultimate pull strength is close for all gases, a behavior reminiscent of that observed with other oxidation-resistant polymers, such as polyarylsulphones, which was attributed to surface cross-linking induced by plasma treatment.[27] Another important factor could be mechanical interlocking, favored by the easier penetration of the adhesives into surface imperfections made hydrophilic by the treatment.

Plasma treatments introduced oxygen-, and, in some cases, nitrogen-containing functionalities at the surface of the composite, and increased wettability. While untreated samples showed no pull strength and adhesive failure, plasma-treated samples reached high pull strengths and cohesive failure even after very short treatment times (< 30s). Pull strength values were relatively insensitive to the nature of the treatment gas and plasma parameters; furthermore, aging did not affect adhesion but, in some cases, improved it. These affects suggest the occurrence of plasma-induced cross-linking of the surface layer.[23]

2.2.2.2 Pretreatment of Metal Surfaces (Aluminum).

A great deal of research has been conducted on the adhesive bonding of aluminum alloys and the main consensus is that some form of pretreatment of the aluminum prior to bonding is essential for durability performance.[28-29] However, the exact reasons as to why different pretreatments of aluminum substrates result in various degrees of bonding performance still elude the adhesion scientist.

Many theories have been advanced to explain why a particular pretreatment of aluminum results in superior or inferior performance of an adhesive joint. These vary from macrosurface roughness factors, surface oxide chemistry, surface oxide hydration resistance, and weak layers within the oxide.[29] Other factors resulting from the use of adhesives with surface oxides have also been investigated covering wettability, homogeneity of the adhesive properties in the joint and adhesive chemistry.

Davies and Ritchie[30] have recently tried to identify the reasons for the environmental failure of adhesive joints using aluminum alloy substrates and the potential for improvement in performance by the development of modified anodizing treatments.

They demonstrated that adhesive penetration occurs in the porous surface oxides created by phosphoric and chromic acid anodizing (PAA and CAA) of aluminum alloys. The extent of this penetration is determined by the oxide morphology produced. In the case of PAA oxide, extensive penetration results and only partial penetration is evident within the CAA oxide. However, the CAA oxide morphology can be modified to allow extensive adhesive penetration to occur. From the durability tests, there would seem to be some correlation between the extent of adhesion penetration into the surface oxide and the resulting durability performance.[30]

2.2.3 Adhesives and Processing

Research on high-temperature resin systems has intensified during the past few years. Structural applications include: 1) engine nacelles involving long-time exposure (thousands of hours) to temperatures in the 150 to 300°C range. 2) supersonic military aircraft involving moderately long exposure (hundreds of hours) to temperatures of 150 to 200°C, and 3) missile applications involving only brief exposure (seconds or minutes) to temperatures up to 500°C and above, Tables II-6, 7, and 8.[6,15,31]

Adhesive bonding can be carried out at the composite lay-up stage with the adhesive curing during the laminate process schedule. This is described as "co-curing." Alternatively, adhesives may be used solely at the assembly stage, where finished components are permanently joined together or to other parts of the overall structure. This is known as secondary bonding. The choice between co-curing or secondary bonding as a method requires consideration of a number of parameters at the design stage as well as during manufacture. These are shown in Table II-9.

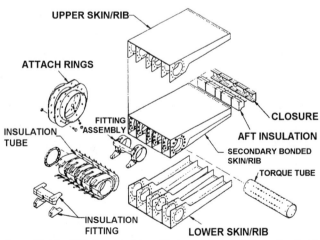

Figure II-10. C/C elevon with oxidation-resistant SiC coating holds tight tolerances in webs on top/bottom skins, torque tube, and other fittings.

Table II-6 Maximum Service Temperature of Adhesives.[15]

Adhesive Type	Typical Maximum Service Temperature (°C)*
Epoxy Epoxy-Polyamide Epoxy-Nitrile Epoxy-Novolac Epoxy-Phenolic	93 149 121 176 204
Polyamide	300
Silicone	300
*Short Term/Dry	

Table II-7 Examples of Materials Joined by Adhesive Bonding for Aerospace Applications.[15]

Material Type	Properties	Materials	Typical Lay-Ups	Uses
Carbon Fiber/Epoxy	Ultra High Modulus	GY70	0°/90° 0°/60°/120° Thickness < 1 mm Normal	Antenna Dish Skins and Satellite Constructions, Where Stiffness and Dimensional Stability Properties Dominate
	High Modulus	HMS	0°/ + 60°/90°/-60°/0° and 0°/±60°/90°/±60°/0°	Stiffness/Strength Property Compromise
	High Strength	T300/5208 XAS/914C T300/976	0°/±45°/0° Sandwich Skins ±45°/0₄°/±45°/0₄° ± 45° Wing Sections of 0°, ± 45°, and 90° up to 38 mm in Thickness	Space Shuttle Payload Doors Ariane 4 Interstage 2/3 Aircraft Assemblies. Strength Dominated Properties
Aramid Fiber/Epoxy	High Specific Tensile Strength. Poor Compressive Properties. Toughness. Impact Resistance	Kevlar® 49/ Phenolic or Epoxy	0°/90° Antenna Dish Skins Either UD of Fabric. 0°, 90°, and 45° Lay-Up for Aircraft Fairings. Mixture of UD and Fabric. Multiple Angles for Filament Winding.	Pressure Vessels, RF Transparency, Aircraft Fairings and Radomes Honeycomb Core Electrically and Thermally Insulative
Glass Fiber/Epoxy	Low Stiffness	E-Glass S2-Glass/Epoxy, Phenolic	Thin Sections <2 (mm) Combining 0°, ± 45°, and 90°	Casings on Launch Vehicles Localized Areas Near Joint Honeycomb Core
	Thermally and Electrically Insulative. Low Dielectric Constant	D-Glass		Gemini and Apolo Sacraficial Ablative Shields, RF Transparancy Radomes
Nomex Fiber/Matrix Combinations	Nomex Fiber has Modest Mechanical Properties. Thermally and Electrially Insulative	Nomex/Phenolic Nomex/Epoxy		Honeycomb Cores Low Expansion Characteristics Electrically and Thermally Insulative Shuttle Payload Doors Antenna Dishes
Aluminum Alloy	Specific Strength	5052-H39 5056-H39 2024-T3 2024-T81		Honeycomb Cores Construction/Antenna Dishes
		6061-T6		Bonded Coupling (Allowing Mechanical Fastening)
Titanium Alloy	Specific Strength	Ti-5Al-4V (Corona 5) (Beta III)		Bonded Couplings (Allowing Mechanical Fastening)

64

Table II-8 Emerging Materials with Increased Temperature Performance with Potential Future Use and Their Applicability for Adhesive Bonding.[15]

General Material Category	Detail of Material	Temperature (°C)		Joined by Adhesive Bonding[‡]
		Maximum Upper Limit	Perceived Service	
Warm Advanced Composites	Carbon Fiber Reinforced Bismaleimide Matrix Polyimide Matrix Thermoplastic	250 300 250	‡ ‡ ‡	Yes. Some Existing Adhesive Systems Plus New Adhesives
Advanced Metals	Aluminum Lithium Alloys	‡	110 NOM*	Possible with Existing Adhesives
Metal Matrix Composites	Silicon Carbide, Carbon, and Boron Fiber Reinforced Aluminum Matrix Titanium Matrix	450 650	110 NOM* ‡	Possible with Existing Adhesives at >110°C <250°C with Improved Adhesives. Not at Upper Maximum Limit
High Temperature Composites	Carbon, Silicon Carbide, and Aluminua Reinforced Carbon Matrix Glass Matrix Glass/Ceramic Matrix	>1000 1000 >1000	‡ ‡ ‡	Possible with Refractory Cements Not Organic Based Adhesives
High Temperature Metals	Superalloy Fiber Reinforced Superalloy Refractory Metals	1100 NOM[†] 1100 NOM 1200 NOM	1000 NOM 1000 NOM 1000 NOM	Not Possible

One significant advantage of using advanced composites is that single components can be produced with complex geometries in one operation. This reduces the number of parts in comparison with a metal assembly of equivalent performance.

With careful design, composite lay-ups are created so that the part can be manufactured as a single entity. This can be achieved by

- Combining the various "lay-ups" without an adhesive.

- Combining the various "lay-ups" with an adhesive layer.

Both of these techniques are known as co-curing. It should be remembered that differences exist between these methods and are shown in Table II-10.

2.2.4 Applications – RMC to RMC

Although joining composites presents many of the same challenges as joining metal structures, successful bonding of composite substrates differs in that composites vary in fiber, surface finish, and resin, whereas metals, with essentially identical properties, are obtainable from several manufacturers.

Although a lot of it is done, composites present an awkward, though not impossible, fastening problem. However, bonding is a much better way to create joints between composites .

An example in Figure II-11 shows the Navy F-18 aircraft wing which is bonded to the fuselage. This is the ultimate in primary structure dependence on a bonded attachment. The wing joint

Table II-9 Comparison of Co-Curing and Secondary Bonding as an Assembly Technique[15]

Co-Cure	Secondary Bonding
Parts to be joined are both composites	Parts to be joined are composite, metal, or a combination of both
Required mechanical performance of joint region can be met by either: composite resin matrix or adhesive	Adhesive is compatible with all adherends
Adhesive is compatible (chemically + cure) with composite	Required mechanical properties of joint region can be met by an adhesive
Composite lay-up has been designed forco-curing	Composite lay-up has been designed for adhesive bonding
Adjacent plies of preferred orientation without compromising strength/stiffness of composite requirments	Surface preparation methods and equipment are available
Tooling is available for complex geometry required and is cost effective	Jigs and tools are available
Process machinery (layp-up and consolidation) is available to: total component size	Process machinery is available
Dimensional tolerances on component can be met	Adhesive cure schedule is known not to degrade adhesives
The thickness ofthe "adherends" in the joint overlap is accpetable	Inspection andtext requirements can be met
Wide variations in thicknesses within a single component may result in a change in the composite cure schedule	
The total component size can be handled without subsequent damage	
Inspection and test requirments can be performed on complex total component size	
Finished component is a single-source operation	

consists of carbon-fiber-reinforced composite top and bottom wing cover skins bonded in a stepped lap-shear configuration to titanium fittings at the sides of the fuselage. The configuration removes the load gradually from the carbon composite in a series of shear areas.

As manufacturers increase their use of composite materials and improve their understanding of how to design adhesively bonded structures, continued growth in the use of adhesives will occur. This will be spurred by the increasing demand for lighter-weight, more complex aerodynamic structures, which can best be constructed with adhesive bonding.[32] Two other aircraft utilizing composite structures are the F-15 and C-17. The F-15 has B/Ep composites

Table II-10 Comparison Between Two Co-Curing Methods of Producing Joints Between Composite Materials.[15]

Without Adhesive	With Adhesive
Composite resin matrix is responsible for load transfer between two or more composite parts	Adhesive is responsible for load transfer between two or more composite parts
Mechanical performance of joint region can be met by resin matrix, Strength Stiffness Stain to failure	Mechanical performance of joint region can be met by adhesive, Strength Stiffness Stain to failure
No or low incidence of shock loadings	The adhesive is chemially compatible with the composite matrix resin and fibers
	The adhesive cure schedule, to give required mechanical performance, is compatible (temperature and pressure) with composte resin matrix cure schedule

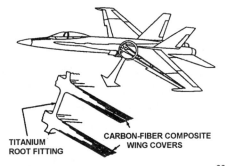

Figure II-11. F-18 wing-to-fuselage attachment.[22]

and horizontal stabilizers with composite materials which transmit flight loads to the fuselage through primary structural adhesive bonded joints. In addition, the aircraft's composite speed brake is an all-bonded structure. The C-17 aircraft now entering production contains extensive applications of composites and adhesive-bonded structure. Figure II-12 shows some of the areas proposed for bonding applications according to T. Reinhart.[33]

2.2.4.1 RMC and CCC

In the mid 1970s, interest developed in a one-piece composite driveshaft to replace a two-piece steel shaft used in light trucks. The steel configuration with the two short sections had a sufficiently high natural rotational frequency to detune the driver train vibrations. Analysis showed that a composite of a single-piece construction, if it were stiff and lightweight, could perform as well. The challenge was to develop a material combination and fabrication

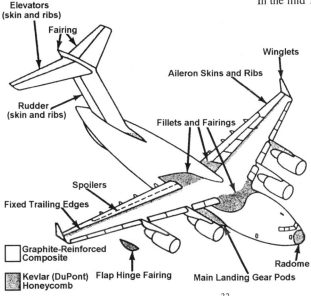

Figure II-12. Bonding applications for the C-17.[33]

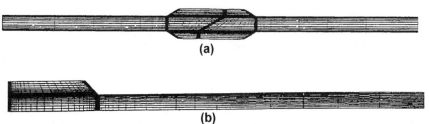

Figure II-13. a) Finite element model of the scarf joint.[34] b) Finite element model of the butt joint.[34]

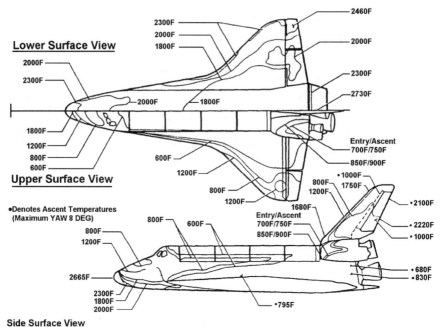

Figure II-14. Orbiter isotherms-trajectory.[40]

method that could be cost competitive with the steel assembly. Limited production runs of a filament-wound composite driveshaft occurred during the mid-1980s, and the shafts were used on Ford Econoline and Astrostar models. Two distinctly different designs developed. The first, made by Hercules, combined a near-axial wrap of carbon fiber with a high angle wrap of glass in vinyl ester resin. Steel and sleeves were wound and adhesively bonded and mechanically pinned into the composite shafts, and steel yokes were welded to the sleeves. The second shaft, made by Ciba-Geigy, combined two fibers into a single wind. The resin was epoxy. Steel end sleeves were welded to the yokes, attached to the composite shaft with an outer ring. In both cases, filament winding was the manufacturing mode. A third design, being evaluated for a General Motors van, utilized filament winding and a pultrusion and an aluminum core. Aluminum yokes were electron beam welded to the aluminum core.

Figure II-10 shows a CCC elevon demonstrated in a structure fabricated by LTV. The structure uses stiffer, higher modulus T-300 carbon fiber in the 71.12 cm long x 49.53 cm wide elevon with temperature resistance to 1,649°C. The oxidation-resistant SiC coating is based on a system developed for the Space Shuttle, and offers improved match-up with the CCC substrate. A secondary bonding technique for joining the top and bottom halves of the elevon was developed in layup, along with C-C fasteners, Figure II-10.

Thick laminated composite hulls for armored vehicles are being developed to provide lighter, quieter vehicles that will meet all service requirements. According to Young et.al,[34] test programs were used to determine data for use in the finite element models for both the adhesives and composites.

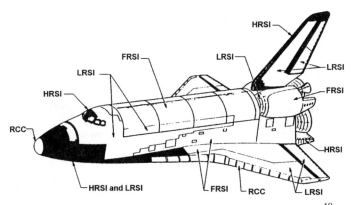

Figure II-15. Columbia (OV-102) TPS (thermal protection system) locations.[40]

A high-performance adhesive was selected to withstand loads that caused failure in the composites before the adhesives failed in nearly all tests. The joint design was expected to meet all requirements.

Testing showed that two joint designs for thick composite laminates could meet design goals. The scarf joint was the better of the two designs based on the analytical results using the proposed criterion. Ultimate failure occurred at a load about 1.5 times higher than the butt joint. The analysis method predicted failure initiation at 1.45 times the butt joint load capability. Predicted laminate failure was the actual failure mode in these joints. The adhesive scarf joint was an effective alternative to mechanically fastened joints in thick laminate vehicle structures, Figures II-13a and 13b.

Thermoplastics vs thermosets offer engineers several advantages however, to the manufacturing engineer it's rapid processing.

Engineers at McDonnell Douglas examined and evaluated a glass/PPS fairing and Gr/PEEK structural skins for the vertical fin of the AH-64 Apache helicopter and found thermoplastic composites offered slightly better strength-to-weight ratios than thermosets with the same reinforcing fibers.

A design and demonstration (build) program was developed whereby a thermoplastic horizontal stabilator could be compared directly with the metal tail on production Apaches. The demonstration main spar and three assembled stabilators were completed and weighed 21.39 kg, 18% less than the 26.22 kg metal stabilator.

Despite its apparently conventional internal structure, the thermoplastic stabilator was not a copy of the original. It followed the same loft lines, but the only metal parts were the pivot fitting and attachment. Adhesive bonds replaced individual fasteners- the 3,300 fasteners in the original structure were reduced to just 148. The metal stabilator had 36 pieces, while the thermoplastic version had just 25.[35]

Other thermoplastic composites, adhesives that are suitably appropriate for the specific thermoplastic composite and the process being used have been developed.[36-42]

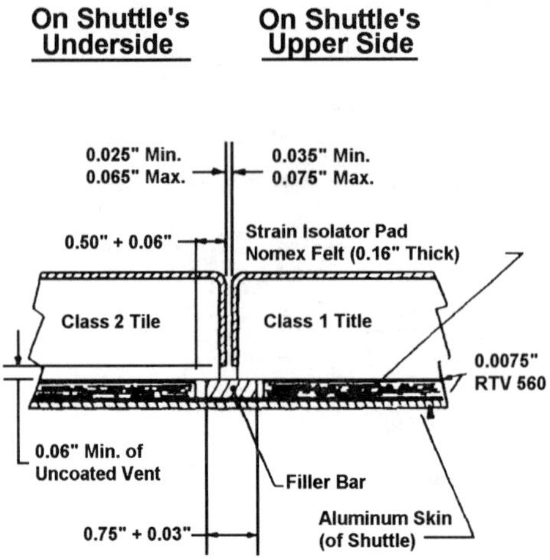
On Shuttle's Underside
On Shuttle's Upper Side
0.025" Min.
0.065" Max.
0.035" Min.
0.075" Max.
0.50" + 0.06"
Strain Isolator Pad
Nomex Felt (0.16" Thick)
Class 2 Tile
Class 1 Title
0.0075"
RTV 560
0.06" Min. of
Uncoated Vent
Filler Bar
Aluminum Skin
(of Shuttle)
0.75" + 0.03"

Figure II-16. Installation placement.[40]

2.2.4.2 MMC

The ceramic or graphitic reinforcements that give MMCs their great strength and modulus at elevated temperatures make these materials uncommonly difficult to machine and form. Even within the airframe community, which is accustomed to paying a premium for exceptional materials, MMCs have made few in-roads into production. An effort to reduce the cost of transforming MMCs from mill form to final products was undertaken by Lockheed Aero. Systems under contract to AFFDL to fabricate, demonstrate and test a full-scale MMC vertical tail for a modern fighter aircraft.

Two types of MMC skin materials were used:

- 15v/o SiC_w /2124-T6 Al, and SiC/6061 Al.

The purpose of the two different side skin materials was to test the effects of a directional (as opposed to isotropic) reinforcement laminated within a more ductile matrix. Thus, whisker-reinforced in two vertical tails and continuous fibers in the other two. Each of the skin panels was a two-piece, adhesive bonded assembly-a constant thickness outer panel bonded to a tapered inner-panel and precut epoxy adhesive film (FM 300K) was used.[38]

2.2.4.3 CMC

Yttrium SiAlONs are ceramic materials that have the form Y-Si-Al-O-N. They have been shown to be one of the most promising advanced engineering ceramics in that they exhibit high strength and toughness and show good wear and chemical resistance. At present, slip casting and injection molding are the main techniques for shaping of complex products; however, these techniques are limited in the size and shape of the component that can be formed. The obvious alternative to these fabrication techniques is to use a joining method. However, techniques such as diffusion bonding, welding, soldering, or brazing cannot be directly applied to SiAlON ceramics because of their tendency to decompose rather than melt at high temperatures. However, Y-SiAlON ceramics contain a small percent of an intergranular glassy phase that melts at approximately 1,350°C (eutectic temperature in the Y-Si-Al-O-N system[39]). Thus, joining should be possible if this liquid-forming component of the ceramic is utilized.

Walls and Ueki [39] developed an adhesive that differed from the pure-glass and composite adhesives developed in the past [39] in that they contain more solid phase than liquid phase at the joining temperature. The main constituent of the present adhesive is α-Si_3N_4 (45 wt%); the remainder is made up of a mixture of oxides (Y_2O_3, Al_2O_3, and SiO_2). The advantage of this type of adhesive is that it is closer in composition to the ceramic being joined, and, therefore, the properties of the joint are not very different from those of the parent material. Also, at the joining temperature, i.e., 1,600°C, the α -Si_3N_4 reacts to form β-SiAlON, which has an acicular nature that reinforces the joint. Material joined with pure-glass forming adhesives is more likely to break at the joint, because the layer of glass between the adherends has a low resistance to crack propagation.

Adhesive bonding with an RTV 60 adhesive has been used extensively to hold Class I and II tiles to a strain isolator pad on the Space Shuttle orbiter. Figure II -14 shows the orbiter and the temperatures reached during reentry in a typical trajectory. During reentry of the Shuttle into the earth's atmosphere, the ceramic tile surface reaches 1,260°C. Barring mechanical damage, the tiles can survive 100 flights without replacement due to degradation of the ceramic fibers or binders.

The basic Shuttle insulation material is a very fine glassy fiber of SiO_2. Figure II-15 shows the approximate layout of the tiles on the orbiter surface. HRSI is the high-temperature tile designed for the 1,260°C temperature regime. LRSI is the low-temperature tile designed for temperatures in the range 399 to 649°C. The insulation protects the aluminum structure of the vehicle so that the aluminum never exceeds 177°C. The HRSI and LRSI differ only in the surface coating. The high-temperature material is coated with 15 mils of a high-emissivity black reaction-cured borosilicate glass. The high emissivity serves to reradiate the heat efficiently during reentry. The low-temperature material is coated with a white silica-aluminum oxide coating designed to reflect the sun's radiation during the orbiting phase of the mission.

Figure II-16 shows the assembly of the tiles on the skin and the RTV 560 silicone adhesive. The strain isolator, as its name implies, isolates the strains of the aluminum skin from the relatively rigid tile to avoid cracking the tile during deflections of the orbiter structure. The tiles are small, on the order of 6 in $^{2.}$, because larger tiles would be cracked during normal deflections of the airframe and the adhesive joins them to the isolator pad.[40]

2.3 REPAIR

In order to make composites fully competitive with other engineering materials, field mainte-nance and repair materials, equipment, and procedures have been developed and are being improved.

In-flight, in-field, and in-service use of advanced composites must be accompanied by the ability to repair them. Incorporating repairability at the design stage is the ultimate goal which is to be achieved by relatively easy-to-use repair systems and procedures which do not require substantial equipment. Making these systems more reliable, more shelf-stable, easier to use, and able to deliver a higher percentage of the original component's properties are today's targets.[45-60]

As composites gain wider acceptance on aircraft, there is a need to understand the requirements of the commercial airliners, general aviation, and the military regarding repair. The economic benefit realized by the airplane manufacturers from the improvements obtained with composites can present the users with difficult repair options.

In current practice, each composite part must be repaired using the original material specified by the manufacturer.[43] For thermoset based composites, each component has a limited shelf life (6-12 months) and must be stored in a freezer. Waste is high due to the difficulty in obtaining small amounts of material from the manufacturers in a timely manner. Repair kits are available, but are only effective if a part to be repaired is in a depot or the aircraft is on a maintenance cycle. To help eliminate waste of these materials, a need to standardize repair materials is widely recognized by everyone.[61-62]

Primary load-bearing structures are frequently built up from combinations of prepregs from various suppliers, honey-comb cores with different configurations, assorted film or foam adhesives, surface veils, skins, etc. To repair damaged components with the same type of materials as originally used (attempting to replicate the original structure as closely as possible) can lead to onerous materials handling and inventory problems.

Allowable material substitutions must be determined by each engineering staff in coordination with the aircraft manufacturers.[43] To avoid some of the problems of the past, an effort involving the material manufacturers, airline, military, and aircraft manufacturers is underway with the objective of reducing the amount of duplication and increasing the reliability of composite repairs.[61]

References regarding the general use and selection of adhesives for specific uses such as ship repair and in the automotive industry are available.[63-65] Also, previous studies have related the material properties of adhesives to joint performance.[66-69] Armstrong [42-43] did an extensive study relating adhesive properties to lap shear performance of joints. His report also expressed the need for standardized materials and for manufacturers to provide more complete material information. Papers on repair give a summary of materials that were used for specific applications or specific aircraft, but generally do not address the issue of standardization.[70-72]

In the aircraft and helicopter industries, most repair relates to two types of damage: environmental factors (such as hail, lightning and bird strikes, and debris kicked up on takeoff or landing) and "hangar rash" (mishandling of aircraft or components on the ground). Puncture-type damage and microcracking of composites are common; they can damage face skins and allow moisture ingress to the underlying honeycomb. Large-part or large-area repair can require total replacement of the part, or removal of it for remanufacturing. For damage over a smaller area, operators use prepregs and wet-layup techniques in bonded, bolted, or flush aerodynamic patching.[41]

Still another problem is that of personnel training and turnover. Evaluation of the type and extent of damage to a structure, and feasible approaches to repair within local material and equipment availability constraints demands an expertise that comes as much from hands-on experience as from formal training. This is compounded by the wide variety of composite structures already in use, each with its own layup schedule, making virtually every structure a special case.[42]

2.3.1 Design Decisions

There may be damage to composite structures which may not be visible, but is of concern. The damage to a composite structure may be more than that incurred by a comparable metal structure under an identical impact due to the lack of strain capability of the composite. While metal may yield to show the impact site, a composite may delaminate internally and not reveal the damaged area. In-plane tensile strength may not be always compromised by impact damage to the matrix, but a damaged matrix cannot stabilize fibers under compressive bending or shear stresses, and compression is usually the critical loading mode in aircraft structures. A modest impact to a composite can result in severe undetected internal damage in the form of delamination.

The objectives of most repair and maintenance programs are as follows:

1. Investigate and quantify the extent of damage

2. Determine the extent of the repair effort and where the repairs will be performed

Table II-11 Summary of Available Composite Repair Techniques

Method	Advantages	Disadvantages	Ease of Repair	Structural Integrity
Bolted Patch	No surface treatement; no regriferation, heating blankets, or vacuum bags required	Bolt holes weaken structure; bolts can pull out	Fast	Low
Bonded Patch	Flat or curved surface; field repair	Not suitable for high temperatures or critical parts	Fast, but depends on cure cycle of adhesive	Low to Medium
Flush Aerodynamic	Restores full design strength; high-temperature capability	Time-consuming; usually limited to depot; requires refrigeration	Time-Consuming	High
Resin Injection	Quick; may be combined with external patch	May cause plies to separate further	Fast	Low
Honeycomb; fill in with body filler	Fast; restores aerodynamic shape	Limited to minimal damage	Fast	Low
Honeycomb; remove damage; replace with synthetic foam	Restores aerodyanmic shape and full compressive strength	Some loss of impact strength, gain in weight	Relatively Quick	High
Honeycomb;' remove damage; replace with another piece of honeycomb	Restores full strength with nominal weight gain	Time-consuming; requires spare honeycomb	More Difficult	High

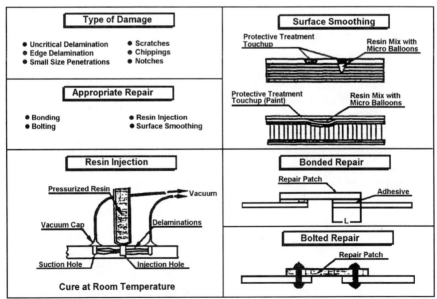

Figure II-17. Strength wise uncritical repairs.[73]

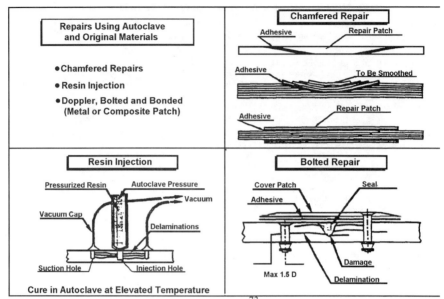

Figure II-18. In house repair of secondary structure.[73]

a. Repair to full properties (with all flaws detected and corrected), or

b. Repair to acceptable properties, which will allow the structure to be usable at some pre-determined percentage of full-scale operation, or

c. Repair to some emergency level, which will allow use at a low percentage of operational effectiveness (such as fly damaged aircraft to depot maintenance base)

3. Select the repair configuration

4. Define the materials and processes to be used, namely, adhesives and cure cycle limitations

Most aircraft repairs to the composite involve damaged and possibly wet honeycomb. The first priority is preparing the composite for repair, which may involve drilling holes for a bolt-on patch or reducing the moisture to a level which will allow heating the composite and the adhesive for cure.

Repair can involve the simple injection of a wet resin through drilled holes into the delaminations, then adding blind fasteners to delay the onset of local buckling. Or it can involve the use of wet or dry prepregs in conjunction with adhesives to form a plug, or flush bonded-on patch. A summary of the available techniques for the repair of solid and honeycomb laminates is given in Table II-11.

Bolted repairs are the most common for battle-damage repairs because of their reliability. Alternative bonding methods require more rigorous control of surface treatments, storage of heat-sensitive bonding materials, and special equipment and/or expertise.[42]

Though prone to many idiosyncrasies, field repair environments generally restrict users to vacuum-bag cure techniques (which yield low bonding pressure), localized heat sources such as heat blankets or pads, induction curing (IC), infrared lamps, or hot air guns, and generally

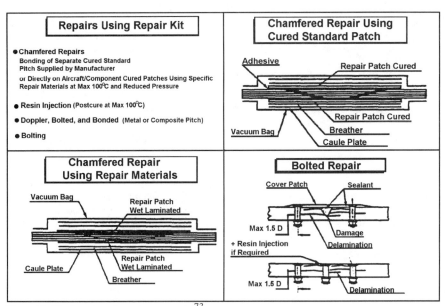

Figure II-19. On aircraft or depot repair.[73]

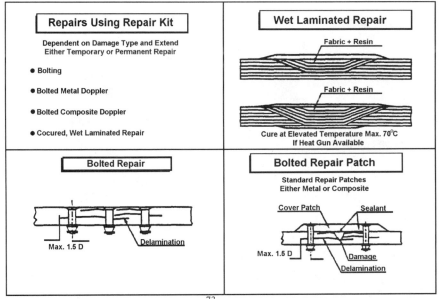

Figure II-20. Field or battle damage repair.[73]

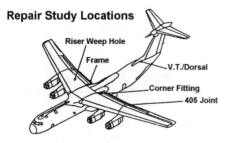

Repair Study Locations

Riser Weep Hole

Frame

V.T./Dorsal

Corner Fitting

405 Joint

Figure II-21. C-141 repair using composite materials.[80]

poor storage conditions. The materials to be used must be room-temperature storable for 1 year, cure at 177°C or less within 2 hours, and, ideally, be a one-component system. Typical repair information is shown in References 6 and 45-62.

2.3.2 Process and Material Selection

Thermoplastics have been shown to be tougher and more damage resistant, but have not been fully specified for repair from an adhesive-bonding stand-point and for reprocessing parameters. Economic tradeoffs will again come into play with use of thermoplastics for repair-with the rapid, low-cost processability balanced against the significant capitalization already invested (but not yet amortized) for thermoset processing equipment such as autoclaves.[41]

2.3.3 Repair Solutions

The key element for repairs is the selection of a suitable material combination which meets the demands of the composite material system used for production. For 125°C and 175°C curing epoxy thermoset matrices Hysol EA 956 material is a very promising candidate. EA 956 features a room temperature cure, the capability for low, medium and high temperature post cures, and a low viscosity to allow for low porosity laminates and resin injection. For most structural repairs, graphite, glass, and aramid fibers are used as the reinforcement, Figures II-17-20.

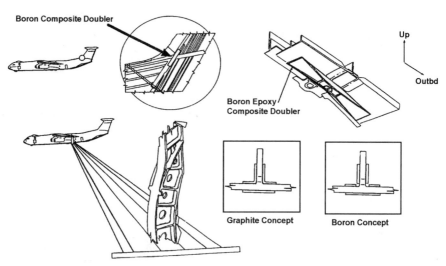

Boron Composite Doubler

Up

Outbd

Boron Epoxy Composite Doubler

Graphite Concept

Boron Concept

Figure II-22. Upper left, C-141 dorsal/vertical stabilizer composite repair; lower left, C-141 FS 998 mainframe; upper right, C-141 WS 405 rear beam splice; lower right, C-141 weep hole composite repair concept.[80]

Despite differences of opinion about repair priorities in materials, techniques, and economics, solutions are at hand. Various associations, technical societies, and military agencies have formed committees and are compiling preliminary repair station advisory circulars that will set forth definitive requirements for repair facilities, including equipment and tooling, training, documentation, and parts and materials inspection.

2.3.4 Applications

Heating (fusion bonding) methods offer significant advantages for repair of TPCs over conventional adhesive bonding and mechanical fastening. Induction heating is fast, efficient, and precise, and has the potential for a healing type of repair as well as a patching repair. Resistance heating is equally fast and relatively simple. By controlling current density, heating time, and consolidation pressure, resistance heating can produce lap shear strengths above 20 MPa in APC-2 thermoplastic material (PEEK-based). Resistance heating is most adaptable for patching repair. Induction and resistance heating also allow control of the cooling rate which may affect the crystalline structure of a semi-crystalline thermoplastic and, in turn, the mechanical properties of the composite.

The use of composite patch repairs on cracked metallic aircraft structures is an accepted technique for improving fatigue life and maintaining high structural efficiency. The use of both B/Ep and Gr/Ep composite patches, bonded with either room-temperature-cured or high-temperature-cured adhesives, results in sufficiently high fatigue lives to meet the damage tolerance requirement of many aircraft.[74] In fact, the B/Ep patch with high-temperature adhesives gave better fatigue lives than the Gr/Ep patch with room temperature adhesives. There have been several successful applications reported for this technique, e.g., Mirage III fighters, C-130 transporters and F-111 fighters.[75-79]

A recent first in-service composite material repair was applied to the metal structure on USAF C-141B operational aircraft.[80] Five locations were evaluated representing both fatigue and corrosion cracking susceptible areas on the C-141B. Repairs using either B/Ep or Gr/Ep were beneficial for crack retardation, end fastener effect relief, and field stress reduction. The most obvious time saving feature of these repairs was that no fastener removal or installation was involved. Extensive structural analysis utilizing finite element modeling was done to ensure that the repair design not only would have the requisite strength and durability but, also, would reduce the inspection burden after the aircraft was returned to service.

A test program was conducted on the following components of the aircraft seen in Figures II-21 and 22. As seen in Figures II-21 and 22, B/Ep materials have been fully qualified for not only primary structural applications on the F-14, F-15, and B-1 aircraft, but, also, as a patch and doubler on metallic structures to stop crack propagation and increase fatigue lives.

The use of bonded B/Ep doublers for the repair of metallic aircraft structures has considerable experience on military aircraft. In fact, the first application of the technique was on the F-111 fighter/bomber wing pivot fitting. This application dates back to 1971. Since then, there have been many more applications of this technique. In addition, the U.S. Air Force Materials Laboratory validated the concept and determined a minimum of 20-25% increase in overall fatigue life through the addition of bonded B/Ep doublers to cracked and damaged aluminum structures.[81] The use of bonded B/Ep composites, proven in the military, will offer the commercial airframe manufacturer and airline maintenance facility a cost-effective technique to extend the economic lives of their aircraft.

B/Ep composite has been used to repair and structurally reinforce the forward titanium longeron of the B-1B bomber. The original remedy – drilling a hole at the end of each crack to inhibit propagation and bolting on an aluminum doubler – failed on 17 of 37 aircraft. Drilling and bolting increased stress concentration, exacerbating the problem. The composite, applied in the form of prepreg tape and heat-cured on the aircraft was attached by adhesives, precluding drilling and reducing stress concentrations 15-20%. To reinforce the B-1B longeron, two doublers about 64 mm wide x 508 mm long, were used. Repair time was less than two days, markedly less than other methods and substantially reducing downtime.

Many helicopters are going to composite blades so repair can be performed similar to metal blades. Metal blades are repaired using carefully developed adhesive bonding techniques and, thus, the procedures for repairing composite and metal blades have much in common. The primary stages are removal of damage, surface preparation, application of adhesive and repair material, and bonding under controlled cycle conditions. With both metal and composite blade repairs, conventional heating and vacuum or pressure bagging techniques are used.

Just as with aircraft, eventually, metal blades may be repaired using composite patching techniques. The best hope here appears to be boron since the material is transparent to eddy current NDT, and has high strength and modulus. Application in helicopters is currently in the demonstration and evaluation phase.[82]

2.3.5 Future Repair Concepts

Currently, no depot or field level repair techniques exist for the repair of high temperature resin matrix composites. High temperature resins, such as PMR-15 polyimide, are material systems used on structures with operation temperature beyond the capability of epoxies and bismaleimides. Current repair materials cannot be applied due to the higher service temperature requirements and current repair procedures and methods cannot be applied to PMR-15 due to the unique processing requirements of this material. The absence of repair materials will necessitate the removal, replacement, and scrappage of damaged parts. In a series of tests performed by the NADC[83], materials for high temperature repair have been evaluated. The effort was directed at the development of patch materials which could be used for depot and field level repair of 288°C composite structures. The initial conclusions that could be drawn from this study were that processing concepts could be developed for the fabrication of preconsolidated formable patch materials for high temperature composites. PMR-15 staged laminates could be produced, however, the material quality would be strongly dependent on the staging conditions used. The powder prepregged material form provided the best laminate properties from the processes studied. The acetylenic PMR material showed promise as a vacuum cure polyimide, but additional work was required for the development of suitable staging conditions. The PT resin composites were found to be easily processable, but the laminate properties were poor even after fiber surface treatments. Finally, tests of the CPI-2310 prepregged composites showed that processability, storage, and handling concerns were reduced because this material doesn't contain MDA.[83]

BMI based composite materials are being applied extensively on naval aircraft structures requiring elevated temperature service. Repair concepts have not been developed for these materials. An approach taken and findings from a program conducted to develop and verify BMI repair concepts has been reported.[84] Two repair concepts, one bolted and one bonded, were successfully developed and demonstrated in this program. The bolted repair built on field level maintenance experience with bolted repairs to metal structures. The bolted concept utilized a rapid curing elevated service temperature polythioether base sealant. A full scale bolted repair specimen was successfully tested. The bonded repair concept employed patch materials which are storable at room temperature and which could be processed with the support equipment currently in the field. The approach taken in development of bonded repair concepts was to fabricate fully consolidated, partially reacted, vitrified patch materials through a combination of advanced staging and non-autoclave processing procedures. Several conclusions drawn from the study are:

1. Nonautoclave processing was successfully applied to a number of BMI composite materials and the laminates produced by this technique were comparable in microstructure and mechanical properties to those produced with autoclave processing techniques.

2. The non-autoclave/staging procedure provided a technique for the fabrication of ambient temperature storable, high quality patches for repair of BMI composite components.

3. Evaluation of the two repaired specimens showed that this patch concept restored full structural integrity to BMI laminate materials. The bonded repair test specimens were successful in restoring greater than design ultimate strain to the damaged specimen (170% of design, ultimate strain).

4. The bolted repair was also successful in restoring strains up to 0.005 in/in which exceeds the design ultimate strain requirement of 0.004 in/in.[84]

2.4 BONDING

Thermosets such as epoxies or pheonlics cannot be welded. Once formed, thermosetting plastics cannot be reformed by heat and pressure. They can only be joined by mechanical fastening and adhesive bonding.

Of the two types of thermoplastics, amorphous or semicrystalline, polymers are generally considered to be more weldable than the latter. Amorphous materials have no defined melting point. In other words, when heated, they gradually pass from a rigid state, through a glass transition, into a rubbery state, followed by a liquid flow in a true molten state. Crystalline materials, on the other hand, have a sharp melting point; they remain solid until they reach their melting temperatures, then, immediately, become liquid.

Table II-12 Comparison of Rapid Adhesive Induction Bonding with Conventional Bonding of Pultruded Aerospace Composites.[86]

Composite Material	Overlap Shear Strength at Room Temperature, MPa	
	Conventional Bonding	Toroid Induction Bonding
Cr-Vinyl Ester	20.7	26.2
Gr-Ep	20.7	31.1
Gr-PEEK	31.1	38.0
Gr-PEI	27.6	35.2
Gr-LaRC = TPI	33.0	33.5

2.4.1 Rapid Adhesive Bonding (RAB)

Most adhesive bonding is performed in heated platen presses or autoclaves which have considerable thermal mass, limiting the heating and cooling rates of the work considerably. A few minutes to achieve a viscosity low enough to obtain flow and wetting of the adherend surfaces is sufficient for thermoplastic adhesives. Because of the cost savings possible, the adhesive bonding processes could be accomplished in minutes, using advanced induction heating technology at kilohertz frequencies. These concepts are designated as RAB for components and field repair.[85] The technology utilizes a combination of aerospace adhesives and induction heating to produce lightweight, compact, energy efficient prototype equipment.

RAB equipment is based on a self-tuning solid-state power oscillator, which may be powered from a variety of sources, feeding kilohertz power to a ferrite toroid induction heater. The toroid geometry produces a uniform, concentrated magnetic flux into the specimen or component to be bonded, causing eddy currents to flow in a ferro-magnetic susceptor and/or paramagnetic adherends. These currents heat only the bond line or its vicinity and the equipment operates at 30-80 kHz. The power required to heat 6.5×10^2 mm bond area to temperatures above 427°C within ~ 1 min is 300 W. Maintaining the bondline temperature at 177 to 427°C typically requires less than 200 W. Since no large fixtures are being heated, the cooling rate of the work is rapid, typically less than 2 min from the bonding temperature to below the glass transition temperature of the adhesive, at which time the bonded component maybe removed from the RAB equipment.

Researchers [85-86] have also designed a machine for seam welding composites by the toroid induction method. Bonding tests were conducted with the toroid induction welder and the results of these tests were compared to the strength of bonds made using a conventional bonding technique, Table II-12.

2.4.2 Dual-Resin Bonding (D-RB)

A process which uses a low-melting-point analog of the high-temperature thermoplastic used in the composite matrix, D-RB involves melting and bonding the analog without melting the composite.[18] The Thermabond process utilizes PEEK composites which are coated with PEI film during part consolidation, prior to joining. Because the amorphous PEI has a Tg of 210°C, parts can now be fused together, without sophisticated tooling, at temperatures well below the PEEK melt temperature of 340°C. For this process to be successful, adhesion of the amorphous layer to the composite is essential. This is accomplished by selecting an amorphous polymer which is compatible with the composite matrix resin, and/or by migration of some of the fibers from the surface of the laminate into the amorphous layer.

Several aircraft companies have evaluated this system for several thermoplastic materials. Lockheed reports[18] lapshear strengths of 28 to 35 MPa with this dual-resin system with an additional layer of neat PEEK film at the interface. Joining was accomplished at 260 to 282°C at pressures of 0.33 to 0.60 MPa. After wet conditioning, lap shear strengths, when tested at 149°C, exceeded 21 MPa. These joints survived three lifetimes of fatigue cycling before the fixtures broke.

One weakness of D-RB is the reliance on an amorphous interlayer that can be attacked by hostile solvents. Lap shear strengths with Gr/PEEK laminates, for instance, decrease by ~ 1/2 after 24-h immersion in dichloromethane. However, development work is currently evaluating a thermoplastic interlayer resin that can be crystallized in an annealing process after D-RB has

Table II-13 The Thermal, Friction and Electromagnetic Methods for Welding Thermoplastics

Welding of Thermoplastics		
Thermal	**Friction (Mechanical)**	**Electromagnetic**
Hot Gas Welding	Spin Welding	Resistance (Implant) Welding
Extrusion Welding	Virbration Welding (100 - 250 Hz)	Induction Welding (5 - 25 MHz)
Hot Tool Welding	Ultrasonic Welding (20 - 40 Hz)	Dielectric Heating (1 - 100 MHz)
Infrared Welding		Microwave Heating (1 - 100 GHz)

been done.

In another program,[87] the performance of carbon fiber reinforced PEEK subassemblies joined using a D-RB approach was conducted. The program focused on the properties of joints between square tubes, formed from AS4 carbon fiber reinforced polyetheretherketone (APC-2/PEEK) prepreg tape, and half cruciform joints woven from commingled AS4 carbon fiber/PEEK yarn. The joints were assembled using a bonding polymer, PEI, which is miscible in the PEEK matrix. The conclusion from the tensile lap tests was that the bond strength was greater by ~ 50% than the failure stress recorded. The average lap shear strength value was 16.29 MPa, a conservative estimate of the actual bond strength. This work indicates that the use of PEI/PEEK as a miscible combination for bonding of structural PEEK matrix composites results in excellent joint performance with minimal susceptibility to surface contamination during bonding. D-RB offers an efficient method for rapidly assembling space structures from standardized components. Further, the characteristics of thermoplastic matrices permit joining techniques with the potential for reconfiguring the structure to accommodate design changes without scrapping existing components. This would also allow reconfiguration of structures in space without the need to lift additional materials to orbit. Finally, the results indicate that the joints show good reproducibility in bond strength over the entire test temperature range. However, the use of the lower melting point PEI bonding material does decrease the upper service temperature of the resulting structure.

2.5 WELDING

There are three groups of fusion-welding techniques which deal with the basic problem of localizing heating to the bondline so that part configuration and structural integrity are maintained in the process, Table II-13.

The first technique has heat directly applied to the individual surfaces to be joined. The surfaces are quickly brought into contact, held under pressure, and allowed to cool, forming a bond. Methods used to heat the surfaces include heated tools and platens, hot gas, focused infrared, and laser beams.

In the second method, parts are first put in contact with each other, and then heat is applied at the bondline. In the case of spin welding, for instance, two concentric parts are assembled, and one is held stationary while the other is rotated. Heat is generated by friction welding the parts. Other examples include vibration, ultrasonic, induction, and resistance welding, all of which have been evaluated with advanced TPCs.[18]

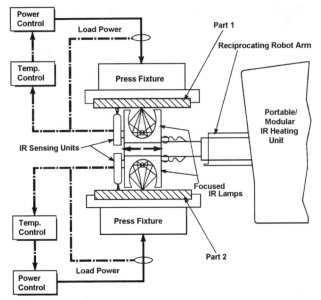

A third series of welding processes which are both quick and efficient includes induction, microwave, dielectric, and implant resistance welding. The technique used a sacrificial heating element placed between two APC-2 laminates. By applying pressure and passing an electric current through the element, heat is generated and a bond produced.

Figure II-23. The FIRE process consists of moving a pair of linear, focused infrared lamps back and forth along the bond lines of the material surfaces that are to be joined. Reciprocation rate of the lamps is selected by the melt-flow temperature of the material. Temperatures of the surfaces are monitored to control power applied to the lamps. When the required bonding temperature is reached, the lamps are retracted, and the two surfaces are pressed together to make the bond.

Induction welding is based on the principle that a magnetic field passing through a conductor generates eddy currents which dissipate in the form of heat. If the conductor is also magnetic (graphite fibers, for instance, are conductive but not magnetic), hysteresis losses also dissipate as heat. In practice, a thin layer of metal-filled adhesive or a thermoplastic-film-laminated metal screen is placed at the joint line. This metal susceptor localizes heat in the joint area once the magnetic field is applied. Lap shear strengths of 42 to 49 MPa have been reported for Gr/PEEK laminates welded in this fashion.

Some PEEK, PPS, and PEI composites of graphite without metal susceptors have been successfully induction-welded[18] with lap shear strengths of 38.5 to 48.3 MPa by using APC-2 laminates and PEEK film at the interface. Some Gr/PPS composites have been welded with lap shear strengths of 23.8 to 30.8 MPa, while a Gr/PEI composite was joined with lap shear strengths of 31.5 to 41.3 MPa.

Resistance-welded Gr/PEEK composites have generated lap shear strengths up to 52.5 MPa.

2.5.1 Focused Infrared Energy (FIRE)

The FIRE welding system can be mounted on to existing commercial hot-plate welding equipment. It has the dual advantages of surface melting without contamination and joining rates double that of hot plate welding. Low distortion in welded joints and low cost make it an attractive alternative to hot plate or linear vibration joining processes.

The focused infrared heaters replace the hot plate, Figure II-23, and heat the joint area. Infrared sensors are used to check that the surfaces have reached the required temperature. The components are then forged together in an operation similar to hot plate welding. This technique, however, avoids the problems that arise from having the heater plate in direct contact with the work-pieces, namely contamination and disturbance of the fiber layup.[88-90]

2.5.2 Heated Tool and Platen Welding

Heated tool welding, more commonly called hot plate welding, is the simplest welding technique used with plastics. It is popular both in mass production and for large structures.

A heated plate is clamped between the surfaces to be joined until they soften. The plate is then withdrawn and the surfaces are brought together under a controlled pressure for a specific period. The fused surfaces are allowed to cool, forming a joint which normally has at least 90% of the strength of the parent material.[91] Work conducted on hot plate welding of APC-2 indicated that welds can be achieved with a lap shear strength up to 50 N/mm^2.

Hot plate welding large components with wall thicknesses above 25 mm currently presents difficulties.[91] TWI has designed and built a test-bed machine which has an electrically-heated hot plate and hydraulically applied axial force in common with commercial equipment, but these parameters can be varied over wide ranges. The maximum hot-plate temperature is 650°C and force is adjustable from 1 to 150 kN, so that all current thermoplastic high-temperature composites, can be welded in sizes up to the dimensions of the hot plate (650 x 650 mm). Typical areas of current research include joining large diameter (up to 630 mm) advanced thermoplastic composite parts for aerospace use.

Although considerable progress has been made in understanding the hot-plate process, the factors that influence weld quality and the weldability of different materials are poorly documented according to Girardi and TWI. They have attempted to gain a better understanding of material behavior during the welding process and have conducted or been involved in the following research programs:

- Influence of melt viscosity on bead formation and weld strength;
- Effect of crystallinity in butt fusion weld joints;
- Improved productivity;
- Effect of defects in butt fusion weld joints;
- Finite element analysis of the welding process.[91]

2.5.3 Hot Gas Welding

In hot gas or hot air welding, filler material is fed into a prepared joint and a stream of hot gas (generally air) is used to heat filler and parent material. The equipment can be manual or semi-automatic. Normally, the filler material is a solid rod which is forced under light pressure into the joint. The filler rod is not melted in the gas stream, but is sufficiently softened to allow it to fuse to the parent material.[92-95]

The filler rod can be circular in section, but, recently, triangular section rods have been used as they allow multi-run welds and fillet welds to be made more easily. The filler rod, for best results has the same composition as the substrate. Usually, only the surface is fused. Gas temperatures will depend on the polymer but, typically, fall in the range of 200 to 300°C and

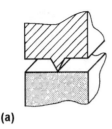

(a)

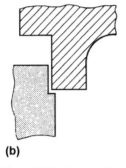

(b)

Figure II-24. Forms of energy director: a) projection; b) shear.[9]

gas flow rates range from 15 to 60 L/min.

Gases used include nitrogen, air, carbon dioxide, hydrogen, and oxygen. Compressed air is popular because it is satisfactory for many purposes and economical. For smaller, manual-welding operations, results depend on the skill of the operator.[91-96]

Hot gas welding is used mostly for butt and fillet welding, but, also, for lap welding of thin sheet. In lap and fillet welding, no joint preparation is needed. In fillet welding, the filler is fed directly into the 90° angle of the fillet, and, for lap welding, no filler is used.[96]

2.5.4 Friction Welding

The various types of friction welding include spin, rotational, and vibration welding.

Heat is produced in friction welding through mechanical rubbing of two surfaces in contact under an applied axial load. The technique is ideally suited for joining thermoplastics because the frictional heat developed at the joint line is sufficient to cause rapid surface melting without significantly raising the temperature in regions away from the rubbing surfaces. Because plastics have a relatively low thermal conductivity and melting point, (compared with most common metals) short welding times (generally 0.5-3.0 sec) can be used to provide joints with properties that approach the parent material strength.

Spin Welding (also known as spin bonding) – This process uses the heat from friction to melt substrates together. Circular thermoplastic parts are bonded by this method. Part joints or filler rods may be spun to create the weld joint. Once the melt cools, the bond is completed.

Major welding parameters include rotational speed, friction pressure, forge pressure, weld time, and melting length. Rotational speeds range from 1 to 20 m/s, friction pressures from 80 to 150 MPa, forge pressures from 100 to 300 kPa, and weld times from 1 to 20 s. Because the heating effect depends on relative surface velocity, maximum heat is generated at the outer edge in solid components. The differential generation of heat can result in weld zone stresses. Hollow sections with thin walls are more satisfactory for this joining process.[6]

Vibration (linear friction) Welding – Vibration welding involves the rubbing of two thermoplastics together under pressure at a suitable frequency and amplitude until enough energy is expended to melt the polymer. The vibration is stopped at that point, the parts are aligned, and the molten polymer allowed to solidify, creating the weld.[96] The process is similar to spin welding except that the motion is linear rather than rotational.[96-97] The process is rapid, (weld times of 1-5 sec are generally used) the vibration is typically 100-240 Hz at 1-5 mm amplitude, while pressures are similar to those used in rotary friction and ultrasonic welding (1-4 N/mm^2). The main advantage of vibration welding is its ability to weld large complex linear joints at high production rates. Other advantages include: the capability of welding a number of components simultaneously; the simplicity of tooling; and the ability to weld most thermoplastic materials.

Vibration welding is more difficult to apply to continuous fiber materials, because of the possibility of fiber displacement. In some tests on APC-2 material, the effect of weld times between 1-6 sec was investigated with weld pressures in the range 1-2.5 N/mm^2 and vibration amplitude between 2-3 mm. The direction of vibration amplitude was parallel to the length of the specimen. Using low pressure and vibration amplitude resulted in little or no bonding of the surfaces. Increasing either parameter resulted in higher strength welds with a maximum of 17 N/mm^2 being attained (35% of parent material). At all welding conditions, the weld strength tended to increase with weld time. However, the degree of flash and fiber displacement also increased. This distortion/displacement is a major disadvantage of the process when applied to APC-2.[91, 98-99]

Vibrational welding has joined PEEK laminates yielding lap shear strengths of 39.9 MPa. The linear friction welding is well-suited and developed for nonreinforced thermoplastic materials and works well for unidirectional composites. Results indicate it may not be suitable for large parts or for cross- or angled-ply laminates because fibers in adjacent plies may cut each other, decreasing strength at the joint.[6]

Ultrasonic Welding – Ultrasonic welding is a bonding process which uses high-frequency mechanical vibrations, i.e., ultrasonic vibrations, as a source of energy. Heat is generated by a combination of surface and intermolecular friction.

Joint design influences the quality of ultrasonic welds. The most important factors are a loose fit and the provision of an energy director. A slip fit is essential since the welding process depends on movement between the two parts as well as friction and high pressure. The energy director is a small triangular ridge, typically 0.25-0.8 mm high, depending on the material. This concentrates the applied power to provide rapid melting of the material contained in the director. Molten material from the energy director flows across the joint interface and fuses with the two components to form a weld. Figure II-24a illustrates a simple butt joint modified with an energy director. A shear joint, Figure II-24b, provides a small initial contact area and, then, a controlled interference along the joint as the parts collapse together.[91] Pressure is applied to the parts being joined, and a welding horn is applied to the area to be bonded. The horn delivers high-frequency (20- to 40-kHz), low-amplitude (20- to 60-μm) vibrations which are concentrated by the energy directors, localizing heating and joining the thermoplastic parts. Welding time is normally 3 to

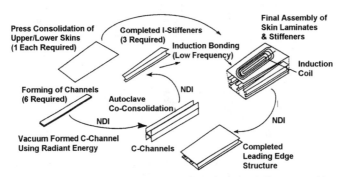

Figure II-25. Despite requiring 14 separate manufacturing steps, Grumman Aircraft Systems demonstrated that an aircraft stabilizer could be produced more economically from thermoplastic prepreg with electromagnetic induction bonding than with one-step autoclave co-curing of a thermoset.

4 s, with parts held in place to cool for a total cycle time of 10 to 15 s. This was successfully done with Gr-PEEK composites and lap shear strengths of the APC-2 material were virtually equal to the strength exhibited in compression-molded laminates (78.4 MPa).

APC-2 is a trade name of ICI Americas Inc. It is a graphite reinforced TPC usually expressed as APC-2/PEEK. The thermoplastic, ICI APC-2 graphite/PEEK, has been made as a prepreg tape and subsequently into laminates. The graphite in APC-2 has usually been (AS4). APC-2/PEI has also been produced where PEI was a film. Fiberite has also produced APC-2/PEEK.

Size and power of the welder limit the area that can be bonded in one operation. Large parts can be joined by sequential or scan welding. In sequential welding, a bond is made, the part is indexed, and adjoining sections welded until the part is fully joined. In scan welding, parts travel under the welder at a constant speed until bonding is complete. In both cases, lap shear strengths equal to 80% of the compression-molded control laminates have been realized. For continuous fiber composite materials, the main barrier to the use of ultrasonic welding is the difficulty of providing energy directors on sheet components and the consequent risk of fiber disruption at the interface under the high deformation necessary to obtain a satisfactory bond.[98-104]

2.5.5 Induction Welding

Induction welding of carbon-fiber-reinforced composites involves generation of induced eddy currents in the material. A work coil provides the energy input to the material and this coil converts energy from a high frequency power supply into a dynamic magnetic field. The shape of the work coil and the geometry of the composite component being welded determine the pattern of eddy currents generated by the magnetic field in the material.[105-106] This principle for induction welding is the same as for resistance implant welding; an entrapped metal implant at the joint line is heated by a high frequency electromagnetic field (2-30 MHz) to produce local fusion and the joining of the work pieces. Using empirical and computer modeling techniques, it is possible to optimize the work coil design to achieve maximum eddy current generation in the vicinity of the weld.

Induction welding can use the amorphous bonding technique [106] which involves making a joint by melting a layer of the amorphous polymer PEI at the weld line. PEI has a melting temperature of 270°C, whereas PEEK has a melting temperature of 345°C and it is this difference which allows a joint to be made by melting the PEI while leaving the PEEK composite still fully consolidated. Induction welds using the amorphous bonding technique have achieved lap shear strengths of 40 N/mm^2.

An interesting case study demonstrated the advantages of electromagnetic induction bonding over autoclaving for joining TPC sections. Grumman fabricated an all-thermoplastic horizontal stabilizer out of prepregs of carbon fiber and PAS-2 polyarylene sulfide resin (known commercially as Ryton S). The thermoplastic was compared in two variations, one with conventional autoclave bonding of the outer skins to the I-beam stiffeners, and one with induction bonding of the two using 3M's AF-191 amorphous thermoplastic film adhesive, and a vacuum bag to apply pressure. Induction heating acted directly on the carbon-fiber reinforcements, eliminating the need for any metal susceptor material.

Grumman was quite encouraged that the induction-bonded thermoplastic component showed slightly greater bond strength and 10% greater stiffness than its autoclave-bonded counterpart, Figure II-25.

2.5.6 Resistance Welding (Resistive Inplant Welding)

Resistive inplant welding is based on the principle of trapping a conducting, usually metallic, implant between the two parts to be joined and then heating the insert by resistive heating. The heat causes the surrounding plastic material to melt and the weld is effected by subsequent cooling under pressure. The conductive implant remains within the joint and as such affects the final strength of the weld. To avoid use of a metallic implant and introduction of foreign material into the thermoplastic joint, prepreg carbon fiber tape, specifically unidirectional fibers set in PEEK, can be used as a compatible implant material. Welds of up to 127 x 25 mm can be achieved with a lap shear strength of up to 50 N/mm^2.

Because welding times are short, up to 20 seconds, for the largest components, resistance welding offers other advantages as a bonding technique for TPCs. It requires little surface preparation (sanding of surfaces), costs are low, and the process is relatively simple and fast, since joints can be consolidated in minutes at room temperature, compared to hours at 177°C for many epoxy adhesives.[6, 97, 107-109] Fusion bonding of APC-2 by resistance welding and compression molding has been investigated.[104-105, 108-109] Lap shear strength from resistance-welded samples rapidly increases and asymptotically approaches the compression-molded baseline in much shorter process times. Resistance-welded APC-2 with PEEK film yielded a maximum lap shear strength of 44.8 MPa, with fiber motion occurring at the joint. The APC-2-PEI resistance-welded specimens achieved a maximum strength of 35.9 MPa, with no fiber motion detected. The baseline for the APC-2-PEI was determined from isothermal processing (1,200 s at 293°C) to be 43.6 MPa.

Recently completed studies and investigations have shown that:

a) Within a fairly wide processing window, resistance welding can reliably and consistently provide strengths approaching the compression-molded baseline; bond strength is observed to increase rapidly with time in the melt for a given power level; and the time to melt (t_m) and time in the melt (t_o - t_m) to achieve a desired strength decrease with increasing power levels.[110]

b) Joint strength for welded Gr/PEEK TPCs is extremely sensitive to the non-isothermal process history.

c) Design methodology that combines traditional stress analysis with a non-isothermal process model for joint strength allowables can predict the performance of a skin/core structure.[111]

d) The process for resistance welding of APC-2 composite has been developed with the aid of thermal analysis using two-dimensional finite element modeling. The numerical prediction was used as a guide for selecting the processing parameters. Single-lap shear testing and microstructure viewing were employed for bonding evaluation. Resistance welded joints achieved a shear strength 33.9 ± 2.3 MPa and retained this strength after hot-wet conditioning. The fiber orientation adjacent to the bondline layer was found to have an influence on the failure mode, and, thus, the shear strength.[112]

2.5.7 Dielectric Heating (Radio Frequency Welding)

Dielectric (radio-frequency) welding involves the use of radio-frequency radiation (27 MHz) to activate polymers that have a dipole and causes rapid oscillation of the polymer molecules. Frictional heat from this movement causes the polymer to become molten. The process differs from induction welding in that no conductive materials are involved and it operates at a higher frequency. Dielectric welding has been used mainly for sealing operations.[91]

2.5.8 Other Welding Techniques

Electromagnetic Welding – Electromagnetic welding uses micrometer-sized particles of iron oxide, stainless steel, ceramic, ferrite, and graphite which respond to the radio-frequency (3- to 40- MHz) magnetic field. These (opaque) powders or inserts must be molded in the polymer matrix and remain in the final weld. As the high frequency field passes through the magnetic materials, they are induced to become hot. The surrounding polymer melts, forming the bond. The induction coil must be located as close to the joint as possible if rapid bonds are to be made. Nonmetallic tooling must be used for alignment. Reinforced, coextruded layers, sandwiches, and non-thermoplastic-coated composites may be bonded. Some are bonded by incorporating magnetic particles in hot melts, liquid adhesives, or films at the joint area.

The electromagnetic bonding process uses induction heating to reach fusion temperature by a heat-activated electromagnetic adhesive layer between two abutting thermoplastic surfaces. Lap shear values of 26 MPa have been obtained. The electromagnetic material at the bonding interface consists of a dispersion of finely divided metal particles in a thermoplastic matrix. When the interface is subjected to a high-frequency alternating current, fusion temperature is instantly achieved. Under slight pressure, a bond is formed.[6]

Heated Press Joining – The heated press technique was used to thermally join carbon fiber reinforced PEEK. By employing the heated press technique, lap shear strengths in excess of 50 MPa were obtained and, by investigating the variation of joint strength with welding temperature, it was shown that, with a suitable choice of contact time, it is possible to join carbon/PEEK successfully at 350°C. Below this temperature, the degree of molecular interdiffusion appears limited and the resulting average lap shear strengths are very low.[113]

Polymer Coated Material (PCM) – Induction heating is used to join dissimilar materials such as thermoplastics to aluminum with a maximum single lap shear strength of 30 N/mm^2 to be achieved for a joint between APC-2 and L113. The PCM welding technique has been extended to include joining of TPC to alumina future work will involve Radel (reinforced polyarylsulfone) and Cetex (reinforced polyetherimide).

Work coils positioned a few millimeters above the materials to be joined generate eddy currents in the metal and consequently produce a heating effect in the materials. Welding pressure is provided by a vacuum pump mounted within the machine and throughout the weld cycle the two materials are pulled into intimate contact by this means. The weld cycle lasts about a minute during which the thermoplastic material melts and forms a perfect seal which in this case is on the aluminum alloy with the help of a film interlayer.

PCM offers a number of advantages over existing joining techniques:

- Speed – joints can be made in less than one minute and require no curing;

- Simplicity – the process is particularly easy to use and has significant potential for automation;

- Improved QC – the process is simple to monitor and control.[91]

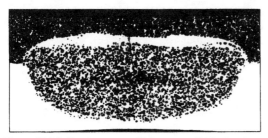

Figure II-26. Resistance weld between 6082 aluminum alloy and 6082 alloy reinforced with SiC (x 25).

2.6 JOINING MMCs

In assessing the use of MMCs for specific applications, joining has been a major concern. The underlying issue is that the heat typically used to fabricate metal-to-metal joints has the effect of promoting reactions between the matrix and the reinforcement of MMCs. This can cause both structural and metallurgical problems in the product. Structural problems include cracking at the matrix-reinforcement interfaces, and for continuous-fiber reinforced materials, cracking of the fibers themselves. Metallurgical problems include loss of protective coatings on the fibers (for continuous-fiber reinforced materials), partial or complete dissolution of the reinforcement, and potential detrimental phase formation caused by the matrix alloying with the reinforcement material.[114-115]

2.6.1 Joining Issues

Joining issues involve assessing microstructural changes in welds, examining alternate welding processes, and developing new processes tailored to these materials. Additionally, the application of rapid thermal cycle processes to minimize particle-matrix interactions with regard to particulate-reinforced aluminum composites is being investigated. The concern has been that, during fusion welding (particularly with Al-SiC composites), the SiC particles break down in the superheated molten aluminum, allowing, on cooling, the subsequent precipitation of detrimental phases. Large aluminum-carbide plates dominate the structure. Scientists and welding engineers have postulated which mechanisms create these changes and have attempted to predict which welding processes would be most applicable.

The application of solid-state processes has been considered advantageous since there is no melting to allow dissolution of the reinforcing particles. The resulting structure shows essentially a base metal distribution of MMC particles all the way to the bond line. As for continuous-rein-forced titanium materials, considerable work has been done developing solid-state variations of existing rapid thermal cycle processes, including resistance spot diffusion bonding and resistance brazing. These approaches have the advantage of achieving a bond at the contacting surface but avoiding the high heat that would degrade the reinforcing fibers. Application of other processes, including ultrasonic welding and resistance implant welding, are also under consideration.

The particles, whiskers, or continuous fibers that are used for reinforcement, have a volume fraction typically between 10 and 60%. Because these materials usually have high strength and high moduli, and retain their strength to high temperatures, operations such as forging and machining are difficult. As a result, joining simple parts provides an attractive method for making more complex products.

There are a number of problems peculiar to MMCs. Particulate reinforcements may have different densities from the matrix, and this can lead to severe segregation effects. Second, undesirable chemical reactions can take place between the particle and the matrix, especially when the matrix is molten. This has been shown to be a particular problem in aluminum alloys

89

Table II-14 Mechanical Properties of Extruded Composite Base Metal, Typed (B), and Friction Welded Composite Metal, Typed (a to h) for Two Different Temper Conditions (i.e., T3 and T6).[118]

Temper/(w.nr.)	R_{pQ2} (MPa)	R (MPa)	C (%)	Vickers HV_{strain}
T3/(B')	109.9	212.2	11.4	60
T3/(B')	111.4	211.9	10.6	60
T3/(a)	100.8	177.2	15.9	60
T3/(b)	100.8	179.0	14.5	60
T3/(c)	313.8	331.8	1.7	135
T3/(d)	308.1	347.0	2.6	135
T6/(B')	315.1	351.9	3.7	135
T6/(B')	315.1	351.2	3.5	135
T3/(e)	207.7	265.6	2.7	90
T3/(f)	206.6	270.2	3.2	90
T3/(g)	311.1	346.7	3.1	135
T3/(h)	314.6	348.4	—	135

reinforced with SiC, and can lead to formation of undesirable aluminum carbides. Third, the ductility of the materials is very low compared with that of the matrix material, and thus the risk of cracking in the weldment, particularly in fusion processes, is higher.

These problems can be overcome by careful optimization of the process and procedure, and by applying a detailed understanding of the metallurgical phenomena peculiar to these materials to development of the joining procedure. One aspect of the problem is illustrated in the micrograph in Figure II-26, which shows the joint line of a resistance weld between a 6082 aluminum alloy and a 6082 alloy reinforced with SiC particles. Segregation of the particulates can be clearly seen, but no evidence was found of reactions between the matrix and SiC particles. Correct manipulation of the welding parameters can minimize segregation.

2.6.2 Capacitor-Discharge (C-D) Welding

Capacitor-discharge (C-D) resistance-spot welding is a joining process which offers potential for the joining of aluminum, magnesium and titanium MMCs. This process utilizes the electrical energy discharged from a bank of capacitors to rapidly heat the weld zone via I^2R heating.* Compared with conventional resistance welding processes, C-D resistance welding is characterized by an extremely short weld thermal cycle, typically less than 10-20 ms in duration. This short cycle promotes the concentration of the weld in a small zone near the weld interface, rather than allowing it to be conducted away from this region into the surrounding base material. The process is also characterized by an extremely rapid cooling rate, which further minimizes heat-effect to the surrounding material.

The C-D welding process has been successfully applied to the following aluminum and magnesium MMCs.[115]

---

*It should be noted that during the resistance-spot welding of titanium, I^2R heating occurs principally within the titanium sheets (ie., bulk heating) due to the high resistivity of titanium and not at the faying surfaces due to a high interface resistance (in contrast to steel and aluminum for which heating at the faying surfaces is much more important.

90

6061 Al- up to 40% SiCp

6061 Al- 48% SiCf

6061 Al- 30% B4Cp

2024 Al- 30% B4Cp

AZ61 Mg- 40% B4Cp

The above weld microstructures exhibited no porosity and no observable alteration of the SiC and B4C reinforcement.

A recently completed study and program by Baeslack III, Cox, Zorko, et.al., showed that solid-state and fusion welds have been produced between sheets of monolithic and SiC fiber-reinforced Ti-6Al-4V using C-D resistance-spot welding.[116]

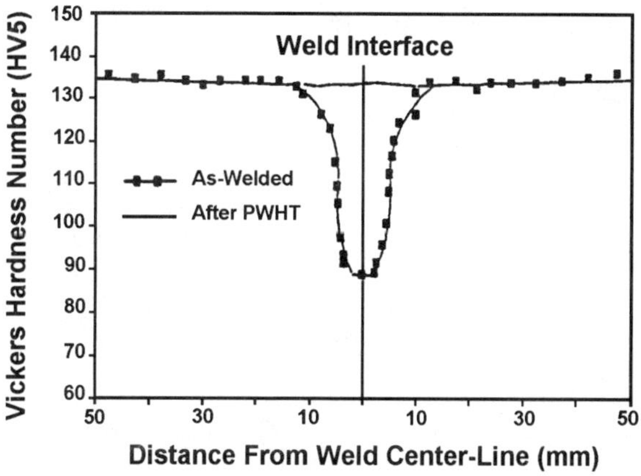

Several conclusions were:

- Capacitor voltage (and corresponding current) and electrode force can be controlled to generate reproducibly solid-state and fusion spot welds in both monolithic and fiber-reinforced Ti-6A1-4V sheet.

- Solid-state and fusion welds in Ti-6A1-4V sheet exhibited tensile-shear strengths above minimum requirements set forth in Mil. Spec. MIL-W-6858D. Shear fracture in the solid-state welds occurred through the weld interface region while fracture in the fusion welds occurred by nugget pullout.

- Solid-state welds were produced in the fiber-reinforced sheet which exhibited complete welding across the weld interface with no evidence of fiber displacement or degradation. Increased voltage promoted the initiation of melting at the fiber/matrix

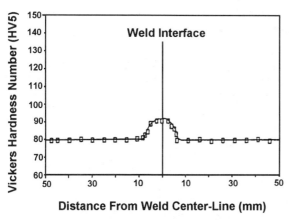

Figure II-27. Measured HAZ hardness profiles after friction welding of T6 heat treated composites.[118]

Figure II-28. Measured HAZ hardness profiles after friction welding of T3 as-extruded composites.[118]

interface and, at sufficiently high voltage, across the entire weld interface. Excessive interface melting promoted appreciable fiber dissolution and displacement.

- Optimized solid-state welds in the fiber-reinforced material exhibited tensile strength and shear stress levels 60 and 80%, respectively, of similar welds produced in the monolithic material. This reduction in fracture strength was attributed to fracture along a layer of fibers in the weld HAZ adjacent and parallel to but remote from the weld interface.

- Dissimilar solid-state and fusion welds were produced between monolithic and fiber-reinforced sheets. The average tensile-shear strength and stress of the welds produced at 160 volts were 70 and 80%, respectively, of similar welds produced in the monolithic material. Fracture of all welds occurred remote from the weld interface along an adjacent layer of fibers.

2.6.3 Inertia Welding

Ongoing studies have been investigating inertia welding (IW) of SiC-6061 aluminum and Al_2O_3-6061 aluminum [117] and friction welding (FRW) an Al7SiMg matrix alloy (A357) with 13 vol% 20 μm SiC particles.[118-119]

The latter investigation[118] found the following:

- The continuous drive FRW process has proved to be a useful method for joining of SiCp/Al MMCs, Table II-14 shows the heat affected zone (HAZ) yield and tensile strength, ductility and Vickers hardness in the friction welded composites. Comparable data for the base materials are included.[118]

- Compared to conventional welding methods, the FRW of aluminum based composites resulted in no pores or clusters of SiC particles in the HAZ.

- In general, the plastic deformation occurring during the welding operation had no significant effect on the overall distribution of SiC particles in the material. This, in turn, implies that the stiffness is maintained across the joint.

- It follows that the resulting HAZ strength level is mainly controlled by reactions taking place within the aluminum matrix during the weld thermal cycle. As a result, a full HAZ strength recovery can be achieved by the use of an appropriate post weld heat treatment, which involves solution heat treatment at 535°C followed by artificially aging at 160°C for 10 hours, Figures II-27 and 28.

In Reference 117, the 6061 aluminum alloy was reinforced with a discontinuous Al_2O_3 particulate and formed into various products that were well-suited for joining applications. In one investigation,[117] IW, because of its solid-state nature, was explored as an alternative to fusion welding particulate-reinforced aluminum MMCs. The purpose of this investigation was

Table II-15 Summary of Mechanical Properties After Welding[121]

Property	Comral-85	6061
Tensile Yield (MPa)	150	115
Bending Yield (MPa)	260	250
Charpy Impact (J)	6	17
Fatigue (Cycles) at 25°C with Maximum Stress 75 MPa	1.6×10^5	4×10^4

to identify the procedural variables that influence the weldability of aluminum alloy 6061 reinforced with Al_2O_3 particulates during IW. As a result of the study, IW was selected over fusion welding particulate-reinforced Al MMCs. Al MMC extruded rods, composed of a 6061 aluminum alloy matrix reinforced with 15 vol% Al_2O_{3p}, were IW. The Al MMC material was joined in the extruded and annealed condition as well as the fully heat-treated condition (T6). Tensile tests revealed joint properties of approximately 40 and 50% of the parent materials, respectively. Metallographic analysis of the weld region showed good interfacial mixing at the bondline and Al_2O_3 particulates were continuous along the bond interface. The bond region appeared to be sound and free of any defects.

Work continues examining the process variables in order to optimize parameters, thereby approaching joint efficiencies closer to that exhibited by the parent material.[117]

2.6.4 Fusion Welding

Gas Metal Arc Welding (GMAW). G.H. Reynolds[120] examined welding procedures for 6061/Al_2O_{3p} composite base plates using primarily the GMAW process. Continuous composite welding electrodes (1.14 mm diameter) were successfully produced by extrusion/rolling/drawing. Welds were prepared using short circuit transfer and spray transfer GMAW procedures employing conventional 4043 and composite 4043/10% Al_2O_{3p} welding electrodes. The welding performance of both the conventional and composite 4043 welding electrodes on the composite base plate was unsatisfactory. There was little retention of Al_2O_3 in both the conventional and composite electrode welds and extensive weld metal porosity and cracking in the composite electrode welds made by either the short-arc or spray transfer welding process. This was largely due to poor wettability of the base plate and the dispersed Al_2O_{3p} by the 4043 filler metal composition. A Mg-containing filler metal composition provided significant improvements in both weld metal soundness and Al_2O_{3p} retention. The as-welded mechanical properties of welds produced with a conventional Mg-containing 5356 electrode were found to be superior to those achieved with the conventional 4043 electrode. Mg additions to welds produced with the 4043/Al_2O_{3p} composite electrode improved both soundness and Al_2O_3 retention but not to a sufficient degree to make these welds suitable for mechanical testing. Mg-containing filler metal compositions such as 5356 appear to be promising starting points for the construction of composite electrodes for GMAW welding of Al/Al_2O_{3p} composite base plates and future work. Other compositions which might be suitable on the basis of their wetting characteristics toward Al_2O_3 particulates are Cu, Si-containing compositions such as A390.

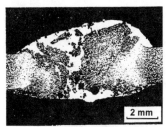

Figure II-29. Macro section through a GTA weld in 8090/20% SiC, using a 4047A filler metal alloy.[122]

Gas Tungsten Arc Welding (GTAW). D. Bhattacharyya et.al.,[121] initiated a study to determine the influence of Al_2O_3 microsphere reinforcement on the mechanical behavior and wettability of a 6061 aluminum MMC.

Although the composite has superior properties compared to those of the matrix, the fracture toughness appears to be much lower. It is evident that the structure of the reinforcement particles greatly influences the material behavior. Tests showed that the composite material had good weldability using both GTAW and GMAW processes. However, larger heat affected zones and increased porosity in the weldments were common in the composite and the Al_2O_3 microspheres appeared to contribute largely to these phenomena. Comral-85, a standard heat treatable 6061 aluminum matrix reinforced with Al_2O_3 microspheres was selected. The reinforcing particles had an average diameter of 20 μm, and a nominal 25% volume fraction. Tensile strength and stiffness of Comral-85 weldments were found to be greater than those of 6061 weldments. The fatigue strength was also superior for Comral-85, as shown in Table II-15. Compared to 6061 aluminum alloy, Comral-85 MMC is about 10% stronger in yield at 25°C and more than 100% stronger at 200°C. It is also 25% stiffer but its toughness is significantly lower than that of 6061. It can be readily welded by both GTAW and GMAW processes. However, it displays a large heat affected zone due to its lower thermal conductivity. Additionally, welded specimens, compared to 6061 welded specimens, show higher levels of tensile, bending and fatigue strength but very low toughness. Finally, weldments display extensive porosity probably due to gas bubble attachment via surface tension forces onto reinforcement particles.

In studies conducted by Gittos and Threadgill of TWI,[122] it was found that molten aluminum MMC materials are very viscous, even at temperatures considerably in excess of the melting point. This means that it is difficult to use conventional filler alloys. The two liquids do not mix satisfactorily in the few seconds that the pool remains molten, and care must be taken to minimize the melting of the MMC material. This effect may not be a problem if low strength welds are required, but it has significant implications for design of welded structures which incorporate MMCs. A section through a GTAW weld in an 8090/20% SiC alloy, using a 4047A filler metal, Figure II-29, shows an extreme case of the lack of mixing. However, a prime aluminum producer, Duralcan, has demonstrated that aluminum MMCs can be GTAW or GMAW with procedures chosen to minimize dilution, thus overcoming the poor miscibility of the molten MMC. A second feature is rejection of the particulate by the solidifying interface. Most MMCs are reasonably homogeneous, but following fusion welding, ceramic particles agglomerate in the interdentritic regions, reducing their efficiency as a reinforcement. The magnitude of this effect and its significance are clearly influenced by the particle size and the cooling rate, which determines the dendritic cell size. In large weld pools, gravitational effects may also lead to enhanced macrosegregation, although there are no known investigations of this.

Duralcan has a major research program under way involving the welding of 101.6 mm diameter composite drive shafts to aluminum yokes for light trucks. Individual welds are completed in 11 to 12 s at speeds of 152.4 cm/min. The seamless tubes that make up part of the drive shafts consist of 6061 aluminum reinforced by particulate of Al_2O_3. The choice of Al_2O_3 over SiC, the material normally used to reinforce aluminum, is based on the former's superior weldability. To keep the Al_2O_3 particles from adhering to each other while the matrix material is in the molten state, at least 3.5% Mg should be present. A filler metal that is often used in such situations is 5356, which contains 5% Mg. The Al_2O_3 in the weld bead is introduced by dilution from the base metal due to its fusion during welding. SiC-reinforced aluminum can be welded. However, rapid weld passes generated by low heat input processes should be used. The SiC in the weld pool must be kept cold in order to prevent the material from decomposing and forming aluminum carbide. Resultant mechanical properties are just as good with welded 6061 aluminum-matrix composites reinforced with Al_2O_3 particulate, as they are with welded 6061 aluminum. Welded structures have achieved 96.133 x 10^3 MPa modulus from MMCs.

Laser Beam Welding (LBW) – To make practical complex engineering components from MMCs, a technique for joining MMCs to each other and to monolithic materials is a key technology. In the welding of MMCs, the fusion zone and heat-affected zone (HAZ) should be minimized to avoid degradation of mechanical properties due to fiber-matrix reaction at elevated temperature. From this point of view, LBW seems to be the most feasible fusion

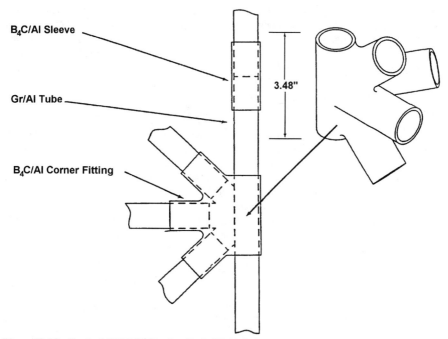

B₄C/Al Sleeve

Gr/Al Tube

B₄C/Al Corner Fitting

3.48"

Figure II-30. Typical P100/6061 tube, B4Cp/6061 fitting truss structure design.

welding process, because the laser beam can provide a controllable heat source with high energy density.

Hirose and his associates at Osaka University[123] joined three- and ten-ply SiC fiber-reinforced Ti-6Al-4V composites using a laser beam. With a 300 m m thick Ti-6Al-4V filler metal, fully penetrated welds without apparent fiber damage, were obtained in welding directions both parallel and transverse to the fiber direction by controlling the welding heat input. Excess heat input resulted in the decomposition of SiC and subsequent TiC formation, and also caused a

Table II-16 Listing of Material Combinations Investigated During Program[126]

Aluminum Matrix Combinations
6061 Al to 6061 Al
P100 Gr Fiber Reinforced 6061 Al to B₄C particulate reinforced 6061 Aluminum
P100 Gr Fiber Reinforced 6061 Al to 6061 Al
B₄C Particulate Reinforced 6061 Al to 6061 Al
Magnexium Matrix Combinations
AZ31B Mg to AZ31B Mg
B₄C Particulate Reinforced AZ91 Mg to 6061 Al
P100 Gr Fiber Reinforced AZ91 MG with AZ61 Cover Sheets to Itself
B₄C Particulate Reinforced AZ91 MG to Itself
P100 Gr Fiber Reinforced AZ91 MG with AZ61 Cover Sheets to B₄C Particualte Reinforced AZ91 Mg
P100 Fiber Reinforced AZ91 Mg with AZ61 Cover Sheets to SiC Particulate Reinforced ZK60A Mg

decrease in joint strength. The welding of the three ply composite in which full penetration was achieved at lower laser power, exhibited higher flexibility in heat input than that of the 10 ply composite. Heat treatment at 900°C after welding improved the joint strength because of the homogenization of the weld metal and decomposition of TiC. The strengths of the transverse weld joints, after the heat treatment, were approximately 650 and 550 MPa for the three and 10 ply composites, respectively. With the welding direction parallel to the fiber direction, the strengths both parallel and transverse to the weld joint were equivalent to those of the base plate.

Cast Welding: A new method, cast welding, has been developed for metallurgically bonding inserts into a cast component during a normal casting operation. Recently, there has been interest in making pistons or other cast components which are strengthened in critical areas by ceramic reinforcement (e.g. SiC particulates or Al_2O_3 short fibers). This method consists of making a preform of the ceramic reinforcement, placing this in the die, and squeeze casting or applying high pressures in order to infiltrate the ceramic preforms and form a composite component.[124-125] This is often unsatisfactory owing to preform deformation, positioning, component size limitations, and expense. The only other method currently used for producing a component made from two different materials is to fabricate the cast and composite portions separately and to perform a subsequent welding or brazing operation. However, it is very difficult to weld MMCs and the long times involved in brazing can cause severe degradation of the properties of both the composite and the cast structure. Also, it may be impossible to access the joint location in order to perform the weld. Successful bonds have been obtained between cast Al-12 wt% Si and inserts of Al-12 wt% Si, (Al-12 wt% Si) -55 vol% SiC, (Al-12 wt% Si) -15 vol% SiC, aluminum alloy 2014-15 vol% Al_2O_3, aluminum alloy 6061-15 vol% Al_2O_3, A356-15 vol% SiC, ZA-12, ZA-12-55 vol% SiC, and ZA-12-15 vol% SiC, between cast ZA-12 and inserts of Al-12 wt% Si, (Al-12 wt% Si)-55 vol% SiC, ZA-12, ZA-12-55 vol% SiC and ZA-12-15 vol% SiC and between cast Al-12 wt% Si-1.5 wt% Mg-1 wt% Ni-1 wt% Cu and inserts of (Al-12 wt% Si)-55 vol% SiC and ZA-12-55 vol% SiC.

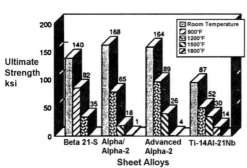

Figure II-31. A comparison of some candidate matrix alloys.[131]

Keys to control of the cast-welding process for bonding MMC inserts into a cast component

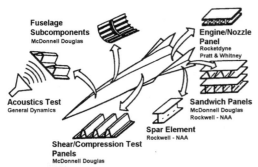

Figure II-32. Subcomponent tests will demonstrate fabrication methods and structural performance.

96

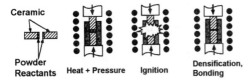

Ceramic

Powder
Reactants   Heat + Pressure   Ignition   Densification, Bonding

Figure II-33. Lightly loaded splice subcomponent will demonstrate a complex joint detail.

during a casting operation are:

- The flow of cast material over the insert surface must be controlled (by gating, insert geometry, and insert preparation) so as to melt a surface layer of the insert and to peel away the oxide skin which prevents bonding.

- Any large temperature fluctuations within the die from one insert location to another must be eliminated.

- The inserts must be free of entrapped gases which will expand on heating and cause excess porosity.

Diffusion Bonding (DB) – A recently completed development program conducted by Rosenwasser, Auvil, and Stevenson[126] documented the results of low-temperature, solid-state bonding (LTSSB) technique to produce high quality joints between MMC parts. These materials, Table II-16, are shown in Figure II-30. Composite materials could be bonded at temperatures and pressures low enough to prevent dimensional change (deformation) and/or damage to the composite materials. Thus, net shape bonding could be accomplished without the need for subsequent machining. The bonding could be performed at locations (such as in space) where subsequent machining would be impractical or impossible. The room temperature shear strengths of the resultant bonds ranged from about 13.8 to over 1,380 MPa. Except for aluminum or magnesium alloys bonded to themselves, the LTSSB joints were always stronger than the shear strengths of the composites themselves. The measured shear strengths were much higher than those reported for adhesively bonded or brazed joints, permitting greater design freedom and a larger margin of safety for space structure designers. The shear failures of composite specimens bonded using the LTSSB approach occurred in the composite materials even when tested at -157°C or 121°C (nominal space conditions). The LTSSB joints exhibited good low cycle fatigue properties. The thermal conductivity of the LTSSB joints was very high, as much as a factor of 10 greater than adhesively bonded joints. This higher thermal conductivity is important in that it reduces the thermal gradient across the joint and increases heat transfer. This is important for thermal management applications such as space radiator components and for threat survivability requirements.[126]

Dunford and associates[127] examined the aluminum-lithium alloy MMC sheet material containing up to 20 wt% SiC and found adequate formability of 8090 MMC is clearly possible at 530/540°C as well as subsequent diffusion bonding. Their tests indicate that the 20 wt% SiC_p/Al-Li 8090 alloy MMC can be diffusion bonded with the possibility of combining SPF with DB in the fabrication of multisheet structures. This MMC has the advantage of a lower density matrix compared with MMCs based upon the 2XXX or 7XXX alloys.

In another series of tests by Dunford and Partridge[128-130] a particulate reinforced MMC containing 17 vol% SiC_p in an Al-Li 8090 (wt% Al-2.4 Li-1.2 Cu-0.6 Mg) alloy matrix in sheet form was manufactured using a proprietary thermomechanical processing route. They found solid state DB of Al-Li 8090 alloy without additional interlayers produced joints with shear strengths and microstructures similar to the base metal. For DB/SPF structures, it was predicted that sheet thicknesses up to 2 mm will produce strain without peel fracture.[128]

The NASP materials and structures program has a series of titanium matrix composite (TMC) development activities which entail the development of fiber/matrix materials, SCS-6, SCS-9,

Sigma and Figure II-31. In order to demonstrate the material and structural performance of TMC, several subcomponents have been fabricated and tested. These subcomponents are representative of X-30 structural types including fuselage, wing, and nozzles, Figure II-32. Probably the best example of one of the subcomponents is the Lightly Loaded Splice Subcomponent, Figure II-33. It is representative of lightly loaded hat-stiffened structures such as a fuselage. It is made of both SCS-6, and SCS-9 fibers and Beta 21-S matrix and is assembled by DB and mechanical fastening. It includes a complex splice joint which contains manufacturing details such as joggles, build-ups, and single sided assembly.[131]

Many different approaches have been used to incorporate reinforcing fibers in metal matrices, some more successful than others. One of the most challenging scenarios is the synthesis of MMCs for high temperature structural applications. These MMCs are generally composed of brittle ceramic fibers in high strength, creep resistant alloy matrices. Bampton, Cunningham, and Everett[132] have developed a method for synthesizing this type of MMC. It utilizes a thin, transient liquid layer at a critical stage of composite consolidation. The so-called "foil/fiber/foil" (F/F/F) MMC consolidation approach has proven very effective in several MMC systems, most recently in the titanium alloy/SiC fiber system β 21-S/SCS-6.[132] These MMC systems all have the prerequisites for successful F/F/F processing, namely, ductile matrix alloys and large diameter, monofilament fibers. This Transient Liquid Consolidation (TLC) is a fabrication method which could be employed in joining MMCs as well as CMCs, aluminides (γ -TiAl (Al_2O_3) and superalloy/Al_2O_3).

Larker, Nissen, and associates[133] joined, by solid state pressing (HIP) a CMC (SiC/SiC) to two superalloys, HastelloyX and Incoloy 909. The HIP pressure was 200 MPa in all joining cycles, and the temperature (dwell time) were either 800°C/15 min, 900°C/1 h or 1,000°C/1h. The reaction zones formed consisted of a thin layer of carbides surrounded by several layers containing silicides and free carbon. The thicknesses of the reaction layers increased with increasing temperature, but were more affected by the composition of the alloy. With more carbide formers in the alloy, the thickness of the reaction layer decreased. The SiC composite was found to be considerably more prone to reactions with these superalloys during HIP as compared to Si_3N_4 under similar conditions.[134] Finally, the rough surface of the composite can, combined with a proper design, be useful through mechanical interlocking by plastic deformation of the superalloy during DB by HIP.

2.7 CMC JOINING

Brazing – Interfacial materials behavior during brazing of CMCs is an area of critical need. This class of materials serves a wide range of technologically important applications, including heat engine components, wear-resistant parts, metal extrusion dies, heat exchangers, machine tools, and medical products. Ceramic composites are often most easily introduced into an application by joining them to a metal support structure. Brazing is a widely employed method for producing such joints. Silicate brazes offer advantages such as ease of wetting the ceramic and the potential for corrosion-resistant refractory joints. In contrast, metallic brazes offer ease of metal wetting, ductility, and lower processing temperatures.

Silver-copper eutectics containing several percent titanium and other minor alloying additions, are among the most successful active filler metal brazes.

Work by Lannutti, Park, and Cawley[135] has focused on joining SiC_w/Al_2O_3. The low thermal mass of the materials allows for close control of the temperature during the brazing anneal. As

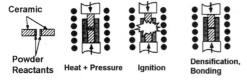

Figure II-34. Schematic diagram of the tape casting process used to fabricate thin layers of the powder joining reactants on the surface of the ceramic workpieces.[149]

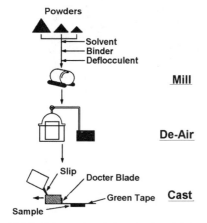

Figure II-35. Schematic of the reaction joining technique in which the joint material is synthesized and bonding is accomplished using in situ ignition of an exothermic combustion reaction.[149]

an active metal braze material Incusil ABA was chosen. This metal consists of 59.2 wt% Ag, 27.1 wt% Cu, 12.1 wt% In, and 1.2 wt% Ti. This braze material was chosen for its low melting point and the inclusion of titanium which, even in the low percentage present, has a strong effect on the brazing process. Indium is needed as a melting point depressant. Its low surface energy also enhances the extent of wetting produced by beneficial interfacial reactions. Analysis by TEM/EDS and SEM/EDS are underway to examine and determine layer composition and use of etchants for delineating microstructural details.

Braze Filler Material – The addition of 20 vol% nickel-coated carbon fibers (400 mm long) to Ag-Cu alloy powder provided a mixture that was applied as a filler material for the brazing of ceramics. Ceramic joints were made by hot pressing at 830 to 870°C and 10 MPa for 40 to 60 min in a nitrogen atmosphere. A nickel coating was needed to enhance the wetting of the fibers by the alloy. This new concept was developed by Cao and Chung[136] for filler metal improvement, i.e., a concept of using a composite filler metal of low thermal expansion. The brazing material was applied in the form of a mixture of short carbon fibers and Ag-Cu alloy powder and when 20 vol% bare carbon fibers was used, the strength for Al$_2$O$_3$-to-Kovar brazed joints tested under shear increased by 65% relative to the joint strength when no fiber was used. The joint strength was increased by ~ 300% when 20 vol% Ni-coated carbon fibers were used. The optimum fiber content was 20 vol% for Al$_2$O$_3$-to-Kovar brazed joints.

Diffusion Bonding (DB) – In the fabrication of large, complex ceramic components for structural applications, bonding is an important practical technique. DB is advantageous because the strength of the bonding parts almost equals that of the base material. In DB, the bonding surfaces must contact closely for the atomic attraction between the two materials to occur.

Recently, the superplasticity of Y$_2$O$_3$-stabilized tetragonal ZrO$_2$ polycrystals (Y-TZP)[137] and of ZrO$_2$/Al$_2$O$_3$ composites was discovered,[138-139] and the superplastic bonding of Y-TZP was demonstrated.[140] Using the superplasticity of ceramics, as well as the DB of superplastic metals,[141-143] ceramics can be diffusion bonded at lower bonding pressure and temperature,

Figure II-36. Inert gas shielded resistance brazed joint in which the molten interlayer infiltrated into the pores of the composite material.

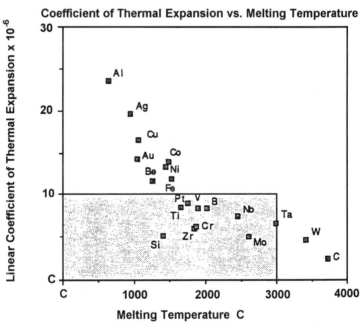

Figure II-37. Guideline for the selection of interlayers. Linear CTE vs. melting temperature plot of carbon compared to other metallic elements. Shaded region indicates some of the interlayers used during brazing experiments.

to achieve superior bonding strength after short periods of bonding. In addition, SPF/DB accomplishes pressing and bonding at the same time, reducing manufacturing costs.

In a recently completed investigation, Nagano and Kato[144] diffusion bonded ZrO_2/Al_2O_3 composites with 0% to 100% Al_2O_3 content at 12.5 MPa for 30 min in the temperature range 1,450 to 1,500°C. Under appropriate bonding conditions, a bonding strength greater than 1,000 MPa was attained between dissimilar materials with different thermal expansion coefficients.

Other findings included:

- Bonding strength increased with deformation in bonding.

- Bonding strength was influenced by bonding temperature, bonding pressure, and the flow stress of the materials in the superplastic deformation.

- The bonding interface adhered closely by superplastic deformation; for example, in the bonding of 79.9 wt% ZrO_2/20 wt% Al_2O_3 to itself, a bonding strength of 1,340 MPa was attained.

- The bonding of dissimilar materials was strong if the thermal expansion difference between the materials was less than $8.0 \times 10^{-7}/°C$.

- It was difficult to bond between 99.7 wt% Al_2O_3 and 99.7 wt% Al_2O_3, but it was possible to bond coarse-grained Al_2O_3 with materials that had low deformation resistance (19.9 wt% ZrO_2/80.0 wt% Al_2O_3, 99.9 wt% Al_2O_3).

- The mechanical strength of HIP ZrO_2/Al_2O_3 composites decreased by annealing in air.[144]

In Situ Reaction Joining – SiC fiber-reinforced SiC matrix composites (SiC-SiC) produced by CVI are being developed for use in structural applications at temperatures approaching 1,000°C.[145-147] These composites contain about 40 vol% SiC fibers (Nicalon) and are infiltrated to about 85% of the theoretical density with SiC. In order to fully realize the advantages of these materials, practical joining techniques are being developed. Successful joining methods will permit the design and fabrication of components with complex shapes and the integration of components parts into larger structures. These joints must possess acceptable mechanical properties and exhibit thermal and environmental stability comparable with the composite which is being joined.

The method developed for joining SiC/SiC composites by Rabin[148-149] utilizes an exothermic combustion reaction, Figure II-34, to simultaneously synthesize the joint interlayer material and to bond together the ceramic workpieces. The method uses Ti-C-Ni powder mixtures that ignite below 1,200°C to form a TiC-Ni joining material. Thin layers of the powder reactants were prepared by tape casting, Figure II-35, and joining was accomplished by heating in a hot-press to ignite the combustion reaction. During this process, localized exothermic heating of the joint region resulted in chemical interaction at the interface between the TiC-Ni and the SiC ceramic that contributed to bonding. Room-temperature four-point bending strengths of joints produced by this method have exceeded 100 MPa.

2.8 ALUMINIDES
AND CCCs JOINING

Welding FeAl-Type Alloys – One of the major considerations in developing ductile intermetallic alloys for structural applications is weldability, i.e., the ease with which the alloys can be welded or joined. Since these alloys achieve their unique properties from their ordered crystal structure, the behavior during a weld thermal cycle is very critical. Weldability has been found to be a strong function of composition and welding parameters.[150-154] Solidification behavior, atomic mobility, ordering kinetics, and the resulting mechanical behavior can influence the weldability and properties to a great extent. An alloy of interest is the iron-aluminide (Fe_3Al type). David and Zacharia[155] found that they could produce successful welds without cracking in the alloy using the electron beam process. This was attributed to the highly concentrated heat source and possible refinement of the microstructure. On the other hand, GTA welds revealed a cracking tendency in all of the Fe_3Al alloys tested except one with chromium, niobium and carbon alloy additives. This alloy also measured 172 MPa for a minimum threshold stress which is within the material range of some commercial stainless steels.

Resistance Brazing CCC – CCC materials have various fiber orientations and may contain inhibitors for oxidation protection. Because of these factors, they will have some effect on the wetting behavior of the interlayer. Wetting behavior during resistance brazing is important because if the molten interlayer does not wet the substrate, a bond will not form. Figure II-36 reflects a strong joint produced by resistance brazing. A semi-continuous interlayer region as well as infiltration of the interlayer into the pores of the composite material is illustrated. If this infiltration can be localized, the molten interlayer, Figure II-37, will interact with several layers within the composite, and a stronger joint may result.

Diffusion Bonding CCC – Solid-state DB of CCC using boride and carbide interlayers was investigated by Dadras and Mehrotra.[156] The primary objectives were to obtain joint shear strengths at 2,000°C that exceeded the interlaminar shear strength of the composite material, and to perform the bonding operations at T, < 2,000°C so that degradation of the CCC could be avoided. Single-phase borides (TiB_2 and ZrB_2) and boride-carbide eutectic systems (22 mole percent TiB_2 - 78 mole percent B_4C, and 57 mole percent ZrB_2 - 43 mole percent ZrC) were used as interlayers under the bonding conditions T, < 2,000°C, t, < 30 min, and P, < 69 MPa. Only weak points could be made with these interlayers, due to excessive cracking of the ZrB_2 bonds, and poor sinterability in the other systems. Bonding by solid-state sintering of B_4C + Ti + Si interlayers produced relatively strong joints that consisted of the reaction products TiB_2 + SiC + TiC. The highest joint shear strength was obtained for joints with no excess B_4C. The variation of joint shear strength with bonding temperature was determined and the maximum strength was achieved for bonding at 2,000°C. The joint shear strength increased with the test temperature, from 81 MPa at room temperature to an average value of 144 MPa at 2,000°C.

REFERENCES – CHAPTER II

1. W.H. Schwartz, Fasteners hold their own, *Assembly Engr.,* pp 24-5, April 1989.

2. N. Albee, Aspects of mechanically fastening advanced composites, *Advanced Composites,* pp 54-60, Nov-Dec 1988.

3. Fasteners for Composites, *Aircraft Eng.,* pp 2-3, July 1988.

4. L.H. McCarty, Lightweight fasteners reduce unit stresses on composites, *Des. News,* pp 186-87, Sept. 1989.

5. Manufacturers are developing improved structural systems, *Av. Wk. & Space Tech.,* pp 67-72, Dec. 1988.

6. M.M. Schwartz, *Composite Materials Handbook, 2nd edi.,* McGraw Hill Inc., New York, NY, 500 p, 1992.

7. J. Rausch, Fastening fundamentals for fabricated composites, *Assembly Eng.,* p 34-6, Oct 1988.

8. Automation in Aerospace Fastening, *Aerospace Engineering,* pp 13-8, Aug. 1992.

9. P.E. Teague, Fasteners that mate with new materials, *Des.News,* pp 98-102, Sept. 1992.

10. L. Trego, Fasteners for aerospace structures, *Aerospace Engineering,* pp 19-23, Nov. 1991.

11. A. Lacombe, C. Bonnet, Ceramic matrix composites, key materials for future space plane technologies, AIAA-90-5208, 14 p, Oct. 1990.

12. A.V. Pocius, R.P. Wenz, *SAMPE Journal,* **21** (5) 50 (1985).

13. A.B. Carter, III, Preserving the structural integrity of advanced composite materials through the use of surface mounted fasteners, *SAMPE Journal,* **25** (4) 21 (1989).

14. A.L. Seidl, Repair of composite structures on commercial aircraft, presented at the 15th Ann. Advanced Composites Workshop, N. California Chapter of SAMPE, Jan. 1989.

15. Adhesive Bonding Handbook for Advanced Structural Materials, European Research & Technology Centre, Netherlands, ESA PSS 03 210, Issue 1, Feb 1990, Rev O.

16. K.S. Kim, W.T. Kim, D.G. Lee, et.al., Optimal tubular adhesive-bonded lap joint of the carbon fiber epoxy composite shaft, *Composite Structures,* **21** (3) 163 (1992).

17. Materials Technologies 1991: The Year in Review, SRI International, D92-1627, March 1992.

18. T. Stevens, Joining advanced thermoplastic composites, ME, pp 41-5, March 1990.

19. J. Ostendorf, R. Rosty, M.J. Bodnar, Adhesive bond strength and durability studies using three different engineering plastics and various surface preparations, *SAMPE Journal,* **25** (4) 15 (1989).

20. Surface modification of matrix polymer for adhesion, Chap. 3, Adhesion and Bonding in Composites, ed. by R. Yosomiya, K. Morimoto, A. Nakajima, et.al., Marcel Dekker, Inc. New York and Basel, pp 49-79.

21. P.W. Rose, E.M. Lison, Treating plastic surfaces with cold gas plasmas, *Plast. Eng.,* pp 41-5, Oct. 1985.

22. E.M. Silverman, P.A. Griese, Joining methods for graphite/PEEK thermoplastic composites, *SAMPE Journal,* **25** (5) 34 (1989).

23. E. Occhiello, M. Morra, G.L. Guerrini, et.al., Adhesion properties of plasma-treated carbon/PEEK composites, *Composites,* **23** (3) 193 (1992).

24. D.C. Goeders, J.L. Perry, Adhesive bonding PEEK/IM-6 composite for cryogenic applications, in *Proc. 36th Int. SAMPE Sym. and Exhibition,* ed. by J. Stinson, R. Adsit, F. Gordaninejad, p 348, 1991.

25. T.-H. Yoon, J.E. McGrath, Adhesion study of PEEK/graphite composites, ibid, p 428.

26. E. Occhiello, F. Garbassi, Surface modifications of polymers using high energy density treatments, *Polymer News,* **12,** 365 (1988).

27. H.V. Boenig, Plasma Science and Technology, Cornell University Press, Ithaca, NY, 1982.

28. A.J. Kinloch, Adhesion and Adhesives, Science and Technology, Chapman and Hall, London, 1987.

29. G.D. Davies, J.D. Venables, Durability of Structural Adhesives, A.J. Kinloch, ed. (Applied Sci. Publishers), London, p 43, 1983.

30. R.J. Davies, M.D. Ritchie, Future design concepts for the development of new pretreatments of aluminum alloy/metal matrix composite for adhesive bonding, *J. Adhesion,* Gordon and Breach Sci. Pub, S.A., United Kingdom, **38,** 243 (1992).

31. R.E. Politi, Recent developments in polyimide and bismaleimide adhesives, High Temperature Polymer Matrix Composites, T.T. Serafini, ed., Noyes Data Corp., Park Ridge,NJ, p 123-27, 1987.

32. N. Albee, Adhesives for structural applications, *Adv.Composites,* pp 42-50, N/D 1989.

33. T.J. Reinhart, Structural bonding needs of aerospace vehicles, *Adhesives Age,* pp 26-8, Jan. 1989.

34. N.K. Young, C.C. Chen, W.L. Cheng, Joining sections of thick fiberglass laminate, *Fourth Ann Conf on Adv. Composites,* ASM Int , SPE, and ESD, Dearborn MI, pp 209-16, Sept. 1988.

35. F. Colucci, Trying thermoplastics, *Aerospace Composites & Materials,* **3** (5) 11 (1991).

36. B.R. Bonazza, Adhesive bonding of improved polyphenylene sulfide thermoplastic composites, 35th Int SAMPE Sym, pp 859-65, April 1990.

37. K. Spoo, Properties and performance of improved polyphenylene sulfide pultruded thermoplastic composites, ibid, pp 866-75.

38. D.M. Ginburg, Metal matrix composite vertical tail fabrication, EM90-438, SME Metal Matrix Clinic, Anaheim, CA, 18 p, Nov. 1990.

39. P.A. Walls, M. Ueki, Joining SiALON ceramics using composite β-SiALON-Glass adhesives, *J. Am. Ceram. Soc.,* **75** (9) 2,491, (1992).

40. S. Musicant, Space Shuttle Insulation Tiles, TransCon Technologies, Inc., Paoli, PA, Marcel Dekker, Inc., NY, Basel, Hong Kong, pp 147-51.

41. V.P. McConnell, In need of repair, *Adv. Composites,* pp 60-70, 1989.

42. L.K. English, Field repair of composite structures, ME, pp 37-9, Sept. 1988.

43. K.A. Armstrong, Conference Bonding and Repair of Composites, Jul 1989.

44. J. Thorbeck, Advanced Materials: Cost Effectiveness, Quality, Control, Health and Environment, SAMPE/Elsevier Sci. Publish., 1991.

45. L.C. Cook, The repair of aircraft structures using adhesives bonding-A survey of current UK practices, The Welding Institute, 392/1989, Feb 1989.

46. S.R. Hall, M.D. Raizenne, D.L. Simpson, Proposed composite repair methodology for primary structure, *Composites,* **20** (5) 479 (1989).

47. C. Ramsey, Battle damage repair of composite structures, *34th SAMPE Int Sym and Exhib.,* Reno, NV, May 1989.

48. T. Reinhart, Composites supportability in the U.S. Air Force, ibid.

49. M. Cichon, Repair adhesives: Development criteria for field level conditions, ibid.

50. T. Steelman, Repair technology for thermoplastic aircraft structures (REPTAS), ibid.

51. B. Purcell, Battlefield repair of bonded honeycomb panels, ibid.

52. C. Hartup, Battle damage repair of thermoplastic structure, ibid.

53. G. Sivy, P. Briggs, Rapid low-temperature repair system for field repair, ibid.

54. J. Connolly, D Vannice, Development of PMR-15 repair concepts, ibid.

55. M. Livesay, N. Smith, E. Castenada, Fast simple system for field repair of composite armor and structures, ibid.

56. M. Diberardino, J. Dominquez, R. Cochran, Bonded field repair concepts using ambient temperature storable materials, ibid.

57. R. Meade, Repair of thermoplastic composites, ibid.

58. D. Bittaker, Manufacturing technology for bonded repair procedures, ibid.

59. J. Mahon, Induction bonded repair of aircraft and structure, ibid.

60. M.S. Sennett, Field repair of composite materials in Army service; Planning for future, MTL, TR-89-45, FR5/89, May 1989.

61. Guidance Material for the Design, Maintenance, Inspection and Repair of Thermosetting Epoxy Matrix Composite Aircraft Structures, Int'l Air Transport Association, 1991.

62. J.A. Fenbert, J.C. Seferis, Fundamentals of composite repair, EM92-100, SME Composites in Manufacturing '92, Anaheim, CA, 9 p, Jan. 1992,.

63. R. Mohan, *Plastics Engineering,* Feb. 1990.

64. R.C. Allen, J. Bird, J.D. Clarke, *Materials Sci. and Techn.,* **4,** Oct. 1988.

65. W.A. Lees, *Materials Sci. and Techn.,* **4,** Dec. 1988.

66. H.N. Cheng, *Adhesives Age,* Dec. 1988.

67. D. Satas, *Tappi Journal,* Sept. 1989.

68. D. Bamborough and P.M. Dunckley, *Adhesives Age,* Nov. 1990.

69. C.O. Arah, D.K. McNamara, H.M. Hand, et.al., *SAMPE Journal,* **25** (4) (1990).

70. F. Lee, S. Brinkerhoff, S. McKinney, *20th Int. SAMPE Techn. Conf.,* Sept. 1988.

71. R. Raghava, *J. of Polymer Science,* Part B Polymer Physics, **26,** 1988.

72. Assessment of Fatigue Damage in MMC and their Joints, CRP Project at EWI, Columbus, OH, **7** (1) 8 (1993).

73. H. Eschbaumer, Repair procedures for advanced composites for helicopters, ERF91-88, MBB-UD-0606-91-PUB 17, European Rotorcraft Forum, Berlin, Germany, 23 p, Sept. 1991.

74. C.-L. Ong, S.B. Shen, The reinforcing effect of composite patch repairs on metallic aircraft structures, *Int. Adhesion and Adhesives,* **12** (1) 19 (1992).

75. A.A. Baker, Repair of cracked or defective metallic aircraft components with advanced fiber composites - An overview of Australian work, *Composite Struct.,* **6,** 153 (1984).

76. A.A. Baker, R.J. Callinan, M.J. Davis, et.al, Repair of Mirage III aircraft using the BFRP crack-patching technique, *Theo. Appl. Frac. Mech,* **2,** 1 (1984).

77. F.A. Sandow, R.K. Cannon, Composite repair of cracked aluminum alloy aircraft structure, Report AFWAL-TR-87-3072, Air Force Wright Aeronautical Laboratories, Dayton, OH, 1987.

78. A.A. Baker, R. Jones, Bonded Repair of Aircraft Structures, Martinus Nijhoff Publishers, Dordrecht, The Netherlands, 1988.

79. L. Molent, R.J. Callinan, R. Jones, Design of an all boron/epoxy doubler reinforcement for the F-111C wing pivot fitting structural aspects, *Composite Struct.,* **11,** 57 (1989).

80. T.F. Christian, Jr., D.O. Hammond, Composite material repairs to metallic airframe components, *J. of Aircraft,* **29** (3) 470 (1992).

81. R.P. Caruso, Boron/epoxy composites for metallic aircraft repair, World Aerospace Technology '92, Sterling Publications Ltd., UK, 1992, pp 48-50.

82. G. Marsh, Repairing rotor blades, *Aerospace Composites & Materials,* **4** (2) 26 (1992).

83. M.F. DiBerardino, P.A. Mehrkam, T.M. Donnellan, Repair concepts for polyimide composite structures, NADC, Structural Composites, *Proc. of the 6th Ann. ASM/ESD, Adv. Comp. Conf.,* Detroit, MI, pp 659-66, Oct. 1990.

84. M.F. DiBerardino, E.L. Rosenzweig, T.M. Donnellan, Repair concepts for bismaleimide composite structures, ibid, pp 641-50.

85. J.D. Buckley, R.L. Fox, J.R. Tyeryar, Seam bonding of graphite reinforced composite panel using induction heating, *Industrial Heating,* p 32-4, March 1987.

86. Joining Composite Structures in Space, *Aerospace Eng.,* pp 9-11, July 1989.

87. S. Zelenak, D.W. Radford, M.W. Dean, The performance of carbon fiber reinforced PEEK subassemblies joined using a dual resin bonding approach, 37th Int. SAMPE Sym., pp 1,346-56, March 1992.

88. H.D. Swartz, J.L. Swartz, Focused infrared melt fusion: Another option for welding thermoplastic composites, Joining Comp. Conf., March 1989, Garden Grove, CA, EM89-175.

89. Focused Infrared Bonds, Shapes TPs, *Plast. Tech.,* pp 32-41, Jan. 1989.

90. Robotic Infrared Heating Technique Used for Joining PEEK Composites, *Reinf. Plastics,* **33,** (2) 48 (1989).

91. M. Girardi, Plastics at TWI - An update, *TWI Bull,* **5,** 100 (1992).

92. H. Gumbleton, Hot gas welding of thermoplastics: An introduction, Joining and Materials, **5,** 215 (1989).

93. M.N. Watson, Hot gas welding of thermoplastics, *TWI Bull,* pp 77-80, May-June 1989.

94. Hot-Gas Welding System Consolidates Complex Thermoplastics Composites, *Adv. Mater.,* **11** (16) 4 (1989).

95. M. Watson, Plastics joining at TWI, *TWI Bull,* Jan-Feb 1990.

96. R.A Grim, Fusion welding techniques for plastics, *W. J.,* pp 23-8, March 1990.

97. D.M. Maguire, Joining thermoplastic composites, *SAMPE Journal,* **25** (1) 11 (1989).

98. N.S. Taylor, Ultrasonic welding of plastics, *TWI Bull.,* pp 159-162, July-Aug. 1988.

99. G. Kempe, H. Krauss, G. Korger-Roth, Adhesion and welding of continuous carbon-fiber reinforced polyetheretherketone (CF-PEEK/APC2), Fourth European Conf. on Comp. Materials, ECCM-4, Stuttgart, F.R.G., pp 105-12, Sept. 1990.

100. K. Stokes, 'Vibration welding of thermoplastics, Parts I, II, III, and IV, *Polymer and Eng. Sci.,* **28** (11) (1988).

101. W.H. Schwartz, Ultrasonic sound off, Assembly Eng., pp 24-7, Jan. 1990.

102. A. Benatar, T.G. Gutowski, Ultrasonic welding of advanced thermoplastic composites, MIT Ind. Liaison Prog. Rep. 6-41-88, March 1988.

103. J. Wolcott, Recent advances in ultrasonic technology, ANTEC '89, pp 502-05, 1989.

104. E.C. Eveno, Experimental investigation of resistance and ultrasonic welding of graphite reinforced polyetheretherketone composites, Univ. of Delaware, CCM88-30, Dec. 1988.

105. A. Leatherman, Induction bonding finds a niche in an evolving plastics industry, *Plast. Eng.,* **4,** 27 (1981).

106. A.J. Smiley, A new concept for fusion bonding thermoplastic composites, SME Joining of Composites, Garden Grove, CA, March 1989.

107. M.J. Day, M.F. Gittos, The application of microscopy to welded joints in thermoplastics, The Welding Institute, Abington, U.K., 390/1989, Jan. 1989.

108. R.C. Don et.al., Fusion bonding of thermoplastic composites by resistance heating, *SAMPE Journal,* pp 59-66, Jan-Feb 1990.

109. E.C. Eveno, J.W. Gillespie, Jr., Resistance welding of graphite polyetheretherketone composites; An experimental investigation, Univ. of Delaware, CCM88-35, Dec. 1988.

110. R.C. Don, J.W. Gillespie, Jr., C.L.T. Lambing, Experimental characterization of processing – performance relationships of resistance welded graphite/ polyetheretherketone composite joints, *Polymer Eng. and Sci.,* **32** (9) 620 (1992).

111. L.Bastien, I.Howie, R.C. Don, et.al, Manufacture and performance of resistance welded graphite reinforced thermoplastic composite structural elements, Univ. of Delaware, CCM 90-33, 22 p, Oct. 1990.

112. X.R. Xiao, S.V. Hoa, K.N. Street, Processing and modelling of resistance welding of APC-2 composite, *J. of Composite Materials,* **26** (7) 1,031, (1992).

113. W.J. Cantwell, P.Davies, P.E. Bourban, et.al, Thermal joining of carbon fibre reinforced PEEK laminates, *Composite Structures,* **16** (4) 305 (1990).

114. R.J. Wise, Developments in techniques and research into bonding new combinations, *TWI Bull,* **33** (1) 4 (1992).

115. J.H. Devletian, SiC/Al metal matrix composite welding by a capacitor discharge process, *W.J.,* pp 33-6, June 1987.

116. A. Cox, W.A. Baeslack, III, S. Zorko, et.al, Capacitor-discharge resistance-spot-welding of SiC fiber-reinforced Ti-6Al-4V, proposed Welding Journal Supplement Paper #91018.

117. M.J. Cola, G. Martin, C.E. Albright, Inertia friction welding of aluminum 6061/Al_2O_3 particulates reinforced metal matrix composites, *EWI Res. Rept. Summary,* p 7, 1989.

118. O.T. Midling, Friction welding of SiC particle reinforced aluminum alloy, Selskapet for Industriell Og Teknisk Forskning, Norway, STF-34-A-91030, 48 p, April 1991.

119. S. Brusethaug, Mechanical properties and composition in A-357 (Al-7-Si-Mg) PMMC, Hydro Aluminium N-Sunndalsra, 1990.

120. G.H. Reynolds, Fusion welding of Al/Al$_2$O$_3$ composites- Phase I, MSNW, Inc., MTL TR 90-38, Final Rept. 31 p, Aug. 1989-March 1990.

121. D. Bhattacharyya, M.E. Bowie, J.T. Gregory, The influence of alumina microsphere reinforcement on the mechanical behavior and weldability of a 6061 aluminum metal matrix composite, *Proc. of the Mach. of Comp. Mater. Sym.*, ASM Materials Week, Chicago, IL, pp 49-56, Nov. 1992.

122. M. Gittos, P. Threadgill, Joining aluminium alloy MMCs, *TWI Bull*, **5,** 9 (1992).

123. A.Hirose, Y. Matsuhiro, M. Kotoh, et.al, Laser-beam welding of SiC fibre-reinforced Ti-6Al-4V composite, *J. Mater. Sci.,* **28,** 349 (1993).

124. S.A. Gedeon, I. Tangerini, A new method for bonding metal matrix composite inserts during casting, *Matls. Sci. and Engr.,* **A144,** 237 (1991).

125. S.A. Gedeon, R. Guerriero, I. Tangerini, Process for obtaining a metallurgical bond between a metal material, or a composite material having a metal matrix and a metalcasting or a metal-alloy casting, Eur. Patent 89202324.3-, Oct. 1989.

126. S.N. Rosenwasser, A.J. Auvil, R.D. Stevenson, Development of low-temperature, solid-state bonding approach for metal matrix composite joints, NSWC TR 89-302, LJ 89-040-TR, N60921-86-C-0279, 72 p, Oct. 1989.

127. D.V. Dunford, S.M. Flitcroft, D.S. McDarmaid, et.al., Forming and diffusion bonding of Al-Li alloy containing 20 wt% particulate SiC, Aluminum-Lithium, ed. by M. Peters and P.-J.Winkler, Vol 2, DGM, Informations-gesellschaft-Verlag, pp 1,087-92.

128. P.G. Partridge, D.V. Dunford, Interlayers and interfaces in diffusion bonded joints in metal matrix composites, ibid, pp 1,145-50.

129. P.G. Partridge, M. Shepherd, D.V. Dunford, Statistical analysis of particulate interface lengths in diffusion bonded joints in a metal-matrix composite, *J. Mater. Sci.,* **26,** 4,953 (1991).

130. P.G. Partridge, D.V. Dunford, The role of interlayers in diffusion bonded joints in metal-matrix composites, *J.Mater. Sci.,* **26,** 2,255 (1991).

131. J.P. Sorensen, Titanium matrix composites- NASP Materials and Structures Augmentation Program, AIAA 90- 5207, 4 p, Oct. 1990.

132. NASP Materials and Structures Augmentation Program "Titanium Matrix Composites," McDonnell Douglas Corp., Contract No., F33657-86-C-2126.

133. R.Larker, A.Nissen, L. Pejryd, et.al, Diffusion bonding reactions between a SiC/SiC composite and two superalloys during joining by hot isostatic pressing, *Acta Metallurgica et Materialia,* **40** (11) 3,129 (1992).

134. R. Larker, B. Loberg, T. Johansson, Proc. 3rd Int. Sym. on Ceramic Materials and Components for Engines, Las Vegas, NV, ed. by V.J. Tennery, Amer. Ceram. Soc., pp 503-12, 1989.

135. J.J. Lannutti, E. Park, J.D. Cawley, Active metal brazing of ceramic matrix composites, EWI Brief B9201, 2 p.

136. J. Cao, D.D.L. Chung, Carbon fiber silver-copper brazing filler composites for brazing ceramics, *W.J.,* pp 21s-24s, Jan. 1992.

137. F. Wakai, S. Sakaguchi, Y. Masuno, , Superplasticity of yttria-stabilized tetragonal ZrO$_2$ polycrystals, *Adv. Ceram. Mater.,* **1** (3) 259 (1986).

138. F. Wakai, H.Kato, Superplasticity of TZP/Al$_2$O$_3$ composite, *Adv. Ceram. Mater.,* **3** (1) 71 (1988).

139. F.Wakai, Y. Kodama, S.Sakaguchi, et.al, Superplastic deformation of ZrO$_2$/Al$_2$O$_3$ duplex composite, in *Proc. of the MRS Int. Mtg. on Advanced Materials,* vol 7, Superplasticity, ed. by M. Kobayashi and F. Wakai, Mater. Res. Soc., Pittsburgh, PA, pp 259-66, 1989.

140. F. Wakai, S. Sakaguchi, K. Kanayama, et.al, Hot work of yttria-stabilized tetragonal ZrO polycrystals, in *Ceramic Materials and Components for Engines,* ed. by W. Bunk and H. Hausner, Deutsche Keramische Gesellschaft, Bad Honnef, FRG, pp 315-22, 1986.

141. E,D, Weisert, G.D. Stacher, Fabricating titanium parts with SPF/DB process, Met. Prog., **111** (3) 32 (1977).

142. O.D. Sherby, J. Wadsworth, L.E. Eiselstein, et.al, Superplastic bonding of ferrous laminates, *Sci. Metall.,* **13** (10) 941 (1979).

143. J. Pilling, D.W. Livesey, J.B. Hawkyard, et.al., Solid state bonding in superplastic Ti-6Al-4V, *Met. Sci.,* **18** (3) 117 (1984).

144. T. Nagano, H. Kato, Diffusion bonding of zirconia/alumina composites, *J. Am. Ceram. Soc.,* **73** (1) 3,476 (1990).

145. A.J. Caputo, W.J. Lackey, Fabrication of fiber-reinforced ceramic composites by chemical or infiltration, ORNL Tech. Memo 9235, 1984,

146. D.P Stinton, A.J. Caputo, R.A. Lowden, Synthesis of fiber-reinforced SiC composites by chemical vapor infiltration, *Am. Ceram. Soc. Bull.,* **65** (2) 347 1986.

147. A.J. Caputo, et.al, Fiber-reinforced SiC composites with improved mechanical properties, *Am. Ceram. Soc. Bull.,* **66** (2) 368 (1987).

148. B.H. Rabin, Joining of fiber-reinforced SiC composites by in situ reaction methods, *Mater. Sci. and Eng.,* **A130,** L1 (1990).

149. B.H. Rabin, Joining of silicon carbide/silicon carbide composites and dense silicon carbide using combustion reactions in the titanium-carbon-nickel system, *J. Am. Ceram. Soc.,* **75** (1) 131 (1992).

150. S.A. David, M.L. Santella, Welding behavior and welding metallurgy of ductile intermetallic alloys, Advances in Welding Metallurgy, *Proc. of the 1st U.S.-Japan Sym.,* AWS, Miami, FL, p 65, 1990.

151. S.A. David, W.A. Jemian, C. Liu, et.al, Welding and weldability of nickel-iron aluminides, *W.J.,* **64** (1) 22-s (1985).

152. M.L. Santella, S.A. David, Weldability of Ni$_3$Al type aluminide alloys, *W.J.,* **65** (5) 129-s (1986).

153. S.A. David, J.A. Horton, C.G. McKamey, et.al, Welding of iron aluminides, *W.J.,* **68** (9) 172-s (1989).

154. S.A. David, D.N. Braski, C.T. Liu, Structure and properties of welded long-range-ordered alloys, *W.J.,* **65,** (4) 93-s (1986).

155. S.A. David, T. Zacharia, Weldability of Fe$_3$Al-type aluminide, *W.J.,* **72** (5) 201-s (1993).

156. P. Dadras, G. Mehrotra, Solid-state diffusion bonding of carbon-carbon composites with borides and carbides, EWI Rept. MR 9206, 21 p, June 1992.

BIBLIOGRAPHY – CHAPTER II

A. Hirose, M. Kotoh, S. Fukumoto, et.al, Diffusion bonding of SiC fibre reinforced Ti-6Al-4V, *Mater. Sci. and Tech.*, **8** (9) 811 (1992).

B. Irving, Demand for plastic parts builds the welding industry, *W.J.*, **72** (2) 31 (1993).

B. Irving, What's being done to weld metal-matrix composites? *W.J.*, **70** (6) 65 (1991).

FIRE welding thermoplastics, Connect, TWI, p 8, Jan 1993.

Thermoplastics inch ahead, Plast. Tech., pp 44-5, March 1992,.

J. Powers, W. Trzaskos, Recent developments in adhesives for bonding advanced thermoplastic composites, *34th SAMPE Int.Sym. & Exhib.*, Reno, NV, May 1989.

N.B. Dahotre, M.H. McCay, T.D. McCay, et.al., Laser joining of metal matrix composite, Proc. of the Machining of Comp. Materials Sym., ASM Materials Week, Chicago, IL, Nov. 1992.

R. Schonholz, G. Kleer, W. Doll, Reliability of interfaces in newly designed ceramic-ceramic and metal-ceramic systems, Fraunhofer -Institut fur Werkstoffmechanik, Freiburg, Germany, p 34, July 1990.

D. Hughes, Textron applies automation to titanium matrix process, *Aviation Week & Space Tech.*, pp 87-8, June 1992.

H. Hamada, K. Haruna, Z. Maekawa, et.al, Comparison of mechanically fastened joint strength between quasi-isotropic carbon-epoxy and carbon-PEEK laminates, *25th Int. SAMPE Tech. Conf.*, Oct. 1993, Session 4B.

K.D. Tackitt, R.C. Don, S.T. Holmes, et.al., Fusion bonding of thermoplastic composites: On-line, non-intrusive sensing for process control, ibid, Session 3C.

J. Dixon, J. Lonergan, Repairs for hat stiffened composite skins, ibid, Session 1E.

R.V. Dompka, V-22 composite repair development program, ibid, Session 1E.

K.A. Chabot, J.A. Brescia, Evaluation of primers for aircraft repair, Volume II, ibid, Session 1E.

J. Mahon, Materials and processes for rapid repair of composites, ibid, Session 1E.

P.A. Mehrkam, Cocure of wet lay-up patch and toughen adhesive for composite repair applications, ibid, Session 1E.

J.J. Connolly, A single side access repair solution for highly curved BMI composite structure, ibid, Session 1E.

S.-I. Y. Wu, Adhesive bonding of thermoplastic composites, 35th Int. SAMPE Sym., pp 846-58, April 1990.

J.M. Marinelli, C.T. Lambing, A study of surface treatments for adhesive bonding of composite materials, 38th Int. SAMPE Sym. & Exhib., May 1993, Anaheim, CA, Session 4C.

P.F. Packman, H.J. Hietala, K.C. Schulz, Fastener and hole imperfection effects on the bolted joint strength of graphite/epoxy laminates, ibid, Session 1B.

J. Browne, Aerospace adhesives in the '90s, ibid, Session 3B.

R.J. Kuhbander, Evaluation of low VOC chromated and non-chromated primers for adhesive bonding, ibid, Session 3B.

R.F. Wegman, The evaluation of the properties in adhesive bonds by non-destructive evaluation, ibid, session 3B.

D. Ruffner, R. Patel, Evaluation of low temperature curing repair adhesives, ibid, Session 4F.

R. Fredell, A. Vlot, Inspection and repair of fiber metal laminate aircraft fuselage structures, ibid, Session 4F.

G.-Z. Zhan, Mid-temperature cure carbon/epoxy prepreg for repair, ibid, Session 4F.

C.-L. Ong, Y.-C. Wu, C.-L. Lee, Structural repair of graphite/epoxy horizontal stabilizer, ibid, Session 4F.

S.T. Holmes, R.C. Don, J.W. Gillespie, Jr., et.al., Sequential resistance welding of large-scale IM7/PEEK double lap joints, Rept.CCM-91-55, 93 p, Dec. 1991.

M.R. Bowditch, D.A. Moth, Bonding techniques with particular reinforced metal matrix composites, British Defense Res. Agency, Sym. on Bonding of Advanced Composites, Adh. Grp.of Brit. Insti. of Materials, London, March 1993.

N.S. Taylor, A study of the ultrasonic welding of short glass fibre reinforced polyphenylene sulphide, TWI Members Rep. 388/1989, Jan. 1989.

J.A. Vaccari, Processing metal matrix composites, Amer. Mach., pp 95-7, March 1985.

S.S. Volkov, Welding of plastics, AFSC, FTD ID(RS)T 0104-88, April 1988.

R. Walsh, M. Vedula, M.J. Koczak, Comparative assessment of bolted joints in a graphite re-inforced thermoset vs. thermoplastic, SAMPE Q., pp 15-9, July 1989.

M.N. Watson, R.M. Rivett, Plastics joining at the Welding Institute, TWI Res. Bull., pp 5-13, Jan. 1986.

F. Zawistowski, Investigation into operation of new special drills for drilling holes in thin components made of composite materials, Technische Hogeschool, Delft, Netherlands, LR519, April 1987.

K.H.G. Ashbee, Fundamental Principles of Fiber Reinforced Composites, Technomic, Lancaster, PA, 1989.

A. Benatar, R.V. Eswaran, S.K. Nayar, Ultrasonic welding of thermoplastics, Part I: Near-field, Edison Welding Inst., Columbus, OH, MR8902, Feb. 1989.

T.G. Gutowski, A review of methods for fusion bonding thermoplastic composites. SAMPE Journal, pp 33-9, Jan -Feb 1987.

Z. Cheng, Ultrasonic welding of thermoplastics, Part II: Far-field, Edison Welding Insti., Columbus, OH, MR8903, April 1989.

J.H. Brahney, Fasteners for composite structures examined, Aerospace Eng., pp 8-14, June 1985.

D.J. DeRenzo, Advanced Composite Materials: Products and Manufacturers, Noyes Data Corp., 1988.

Lockheed-Georgia Begins Design of Metal Matrix Composite Tail Fins, Av. Wk. & Space Technology, p 127, Nov. 1985.

J. Carroll, S. Moran, A Review of Lockheed-Georgia's R&D Programs in Aluminum Matrix Composites. 17th Nat. SAMPE Tech. Conf., Kiamesha Lake, NY., Oct. 1985.

A.M. Maffezzolli, J.M. Kenny, L. Nicolais, Welding of PEEK/carbon fiber composite laminates, SAMPE Journal, 25 (1) 35 (1989).

L. McCarthy, Part design is critical for good welds, Plast.World, pp 61-7, June 1990.

M.G. Murch, Hot plate welding injection moulded thermoplastics: The effect of process parameters, TWI, Abington,U.K., 389/1989, Jan. 1989.

R.B. Ostrom, S.B. Koch, D.L. Wirz-Safranek, Thermoplastic composite fighter forward fuselage, SAMPE Quarterly, 21 (1) 39 (1989).

R.J. Pilarski, M. Chookazian, Ford bonds thermoplastics with electromagnetic welding, Adhesives Age, pp 20-21, June 1986.

G.H. Reynolds, L. Yang, Plasma joining of metal matrix composites, Interim Report, MSNW, Inc., San Marcos, CA, Dec 1985, AD-A164095, NTIS HC A02/MF A01.

N.E. Rouse, Improved methods for thermoplastic bonding, *Mach. Des.*, pp 72-9, April 1985.

B.A. Stein, et.al., Rapid adhesive bonding concepts, NASA Langley Research Center, NASA TM 86256, June 1984.

M.R. Morel, D.A. Saravanos, C.C. Chamis, Tailored metal matrix composites for high-temperature performance, NASA TM -105816, E-7247, 22 p, March 1992.

P.K. Mallick, S. Newman, eds., Composite Materials Technology, Hanser Publishers, Munich, Vienna, NY, Section III, Chapter 6A, pp 179-210.

C.-C. Chen, Mechanical fasteners for thick composites, ASM Int. and ESD 4th Ann. Conf. on Adv. Comp., Dearborn, MI, pp 119-26, Sept. 1988.

T.J. Lienert, J.C. Lippold, E.D. Brandon, Electron-beam welding of SiC-reinforced aluminum A-356 metal matrix composites, Edison Welding Institute, Columbus, OH, 74th Ann. AWS Int. Convention, Houston, TX, Session 24, Paper A, April 1993.

E.K. Hoffman, R.K. Bird, D.L. Dicus, Brazed joint properties and microstructure of β 21S titanium matrix composites, ibid, Session B3, Paper B3A.

M. Zhu, D.D.L. Chung, Active brazing alloy containing carbon fibers for metal-ceramic joining, ibid, Session B3, Paper B3B.

Carbon-Fiber-Reinforced Solder Preform, Patent # 5,089,356. Research Foundation- New York State Univ., Albany, NY.

S.H. McKnight, S.T. Holmes, I. Howie, Resistance heated dual resin bonding, Univ. of Delaware, CCM-92-43, 167 p, July 1992.

S.M. Todd, Joining thermoplastic composites, *22nd Int. SAMPE Tech. Conf.*, **22**, 1,313 (1990).

D. Whittaker, ed., Advanced Materials Technology International1992, Published by Sterling Pub. Ltd., London, U.K., 80 p.

J. Border, R. Salas, Induction heated repair of structures, Third DOD/NASA Composites Repair Technology Workshop, T. Reinhart, ed., Workshop Rept for 14-17 Jan 1991, WL-TR-91-4054, pp 82-4.

R.L. Keller, W.R. Mull, Supportability improvements in advanced composites field repair, ibid, pp 13-25.

B.K. Fink, R.L. McCullough, J.W. Gillespie, Jr., Heating of continuous carbon fiber thermoplastic composites by magnetic induction, ibid, pp 64-72.

S.V. Hoa, X.R. Xiao, Effects of repair processing on the mechanical properties of thermoplastic matrix composites, Concordia Univ., DREP-92-02, CTN-92-60594, p 132, Feb. 1992.

I. Birkby, Ceramic Technology International 1992, Published by Sterling Pub. Ltd., London, U.K., 235 p.

Adhesive Bonding of Composite Materials, Nerac, Inc., Tolland, CT, PB93-864163/WMS, April 1993.

S. Zelenak, D.W. Radford, M.W. Dean, Dual resin bonded joints in polyetheretherketone (PEEK) matrix composites, *SAMPE Quarterly,* **24** (3) 38 (1993).

D.A. Steenkamer, D.J. Wilkins, V.M. Karbhari, et.al, Preform joining technology applied to a complex structural part, Adv. Comp. Conf., Proc. of the 7th Ann. ASM/ESD; ACCE '91,Detroit, MI, pp 443-46, 1991.

S.K. Malhotra, Report on design guidelines for mechanically fastened joints in composite laminates, Deutsche Forschungsanstalt fuer Luft- und Raumfahrt e.V., Brunswick (Germany, F.R.). Inst. fuer Strukturmechanik. DLR-IB-131-92/21, 20 p, July 1992.

K. Stellbrink, Preliminary design of composite joints, ibid, Stuttgart (Germany, F.R) Gruppe Konstruktionssystematik, DLR-M1TT-92-05, ETN-92-92407, p 53, April 1992,.

K. Stellbrink, Preliminary design of composite joints, ibid, Cologne (Germany, F.R.), DLR-MITT-92-05, 51 p, 1992.

J.E. O'Connor, A.Y. Lou, D.G. Brady, Polyarylene sulfide composites, *Proc. of the American Society for Composites,* **1,** 21 (1986).

J.E. O'Connor, W.H. Beever, Polyphenylene sulfide pultruded type composite structures, *SPI Composites Institute Proceedings,* **42,** p 1-D (1987).

A.J . Klein, Joining plastics', Plastics Design Forum, Sep/Oct, p 39, 1988.

J.R. Krone, T.P. Hurtha, J.A. Stirling, Bonding and joining techniques for Ryton polyphenylene sulfide composites, *Proc. of the 33rd Int. SAMPE Sym.and Exhib.,* p 829 (1988).

P.Davies, W.J. Cantwell, P.-J. JAR, et.al, Joining and repair of a carbon fibre-reinforced thermoplastic, *Composites,* **22** (6) 425 (1991).

J.G. Dillard, I. Spinu, Plasma treatment of composites for adhesive bonding, *Proc. of the 4th Ann. Conf. on Adv. Comp.,* Dearborn, MI, ASM/ESD, pp 199-208, Sept. 1988.

G.Y. Yee, An overview of structural adhesive bonding needs for composites in the automobile industry, *Proc. of the 2nd Conf. on Adv. Comp.,* Dearborn, MI, ASM/ESD, pp 217- 220, Nov. 1986.

S.T. Holmes, J.W. Gillespie, Jr., Thermal analysis for resistance welding of large-scale thermoplastic composite joints, Univ. of Delaware, *Proc. of the American Society for Composites,* **1,** 135 (1992).

S. Kawall, G. Viegelahn, R. Scheuerman, LASER beam welding of Al_2O_3 reinforced aluminum alloy composite, 7th Ann. ASM/ESD Adv. Comp. Conf. and Expo, Sep 30-Oct 3, 1991, Detroit, MI, pp 283-90.

M. Mabuchi, T. Imai, K.Kubo, et.al, New fabrication procedure for superplastic aluminum composites reinforced with Si_3N_4 whiskers or particulates, ibid, pp 275-82.

J. Loucks, Process tolerances in ultrasonic welding, ASM Int.Materials Week '91, Cincinnati, OH, 21-24 Oct 1991, Session 1. B. Grimm, Thermoplastic welding processes, ibid, Session 1. M. Benfeldt, Vibration welding of thermoplastics, ibid, Session 1. J. Magenson, Recent advances in ultrasonic plastic welding joint design and process control, ibid, Session 1. R. Allred, Effect of plasma treatment of carbon/BMI interfacial adhesion, *34th SAMPE Int. Sym. & Exhib.,* Reno, NV, May 8-11, 1989.

R.R. Schmidt, W.J. Horn, Viscoelastic relaxation in bolted thermoplastic composite joints, *35th Int. SAMPE Sym.,* April 1990.

M. Sasdelli, V.M. Karbhari, J.W. Gillespie, Jr., The design and use of molded-in metal inserts and attachments in resin transfer molding, *Proc. of the 8th Adv. Comp. Conf.,* Chicago, IL, pp 193-98, Nov. 1992.

T. Schneider, Reversible bonding of composite aircraft structures, Soc. of Allied Weight Engineers, *51st Ann. Conf.,* SAWE-2055, 13 p, May 1992.

F. Lee, S. Brinkerhoff, S. McKinney, From 49-1 to HX1567- Development of a low energy composite repair system, 35th Int. SAMPE Sym, Apr. 2-5, 1990, Anaheim, CA, pp 2,202-15.

M. Mehta, A.A. Soni, Investigation of optimum drilling conditions and hole quality in PRM-15 polyimide/graphite composite laminates, 6th Ann. ASM/ESD Adv. Comp. Cong., ACCE '90, Detroit, MI, Oct. 8-11, 1990, pp 633-5.

CHAPTER III – FORMING AND FINISHING

3.0 INTRODUCTION

Thermoforming is the process of shaping a heated thermoplastic sheet by applying a positive air pressure, a vacuum, mechanical drawing, or combinations of these operations. Some of the older types of thermoforming techniques developed include plug assist forming, drape forming, and matched mold forming.

Currently, processes for thermoforming thermoplastic matrix composites (TMCs) are in various stages of development. Some, such as matched die flow forming and stamping, have been proven capable of high-volume commercial production. Others, such as diaphragm forming, show promise of commercial success in the laboratory but have not been proven or optimized in large scale operations. A number of processes are being used in the aerospace industry for limited small-volume production. These are compression molding, diaphragm molding, the Therm-x (SM) process, and rubber block molding.

3.0.1 Superplasticity

"Superplasticity is the ability of a polycrystalline material to exhibit, in a generally isotropic manner, very high tensile elongations prior to failure." It is exhibited in metallic, ceramic, composite, or intermetallic multiphase materials with uniform or non-uniform relatively coarse (20 μm) to ultrafine (30 nm) grain size that have isotropic or anisotropic grain (phase) shape,size, or orientation.[1]

Since superplasticity is concerned with achieving high tensile ductility, there is an interest in the maximum elongations which may be attained in different types of materials. The achievements to date in composite materials are summarized in Table III-1.

3.1 THERMOFORMING PROCESSES

A contrast should be made between the thermoforming of unreinforced thermoplastic sheets and continuous fiber reinforced thermoplastic laminates. When a thermoplastic sheet is thermoformed, the melted sheet thins and stretches to conform to the contours of the mold. The initial skin thickness and the depth of the draw determine the final part thickness. The surface area of the molded part is usually much greater than the initial surface area of the resin sheet, which results from the material stretching to cover the mold.

The concept of the laminate thinning and stretching is not valid for continuous fiber reinforced composites. Before the tool closes, the laminate is released from the clamp frame and allowed to lie on the lower tool. As the tool closes, the laminate slips into the tool from the edges to cover the contours. The thickness of the laminate does not change during the forming operation. Drapability and conformability are the preferred terms to use when describing the ability of a fabric to form to a contoured surface. As the fabric is forced to conform to the mold contours, the weave pattern will distort slightly and allow the fabric to drape the surface. Movement and slippage of the warp and weft fibers relative to each other account for the ability of a fabric to conform to contours. The factors which determine a weave's ability to conform are characterized by the particular weave design.

The thermoforming process has three key elements:

1) A laminate support frame which carries the laminate into the heat source, supports the laminate during and after the matrix melts, rapidly transfers the melted laminate from the heat source to the forming tool, and, then, releases the laminate onto the lower tool.

2) A heat source capable of evenly heating the laminate to its processing temperature in a reasonable period of time.

3) A forming tool capable of rapid closing speeds with sufficient clamp pressure to form the laminate.

3.1.1 Heating

Heating is required in all thermoforming processes. A number of methods available are:

- Infrared (IR) heating is a good choice when heating thin, well compacted flat sheets or gentle contours in a continuous oven feeding a molding process. It is widely used in sheet thermoforming. Under the proper conditions, it is fast and effective, but it is difficult to get uniform heating of highly contoured parts.

- In most cases, IR heating occurs at the surface facing the IR source, and must be conducted away from the surface to prevent overheating. For this reason, it is not a good method for heating loosely compacted plies. In this case, convection heating is a more effective method for penetrating the plies and preventing overheating. Impingement heating is a variation of convection heating in which a multitude of high velocity jets of heated air impinge on the surface of the heated article, greatly increasing the rate of heat transfer and reducing the length of a continuous oven.

- Contact heating heats by intimate contact between the item being heated and the heating surface. It is capable of faster heating rates than convection or impingement heating, but is not as convenient to use in a continuous heating line.

- Radio frequency (RF) heating can be a very effective method of heating, but it is highly dependent upon part composition and geometry, requires specialized design and set-up experience, and must be shielded to protect workers from RF radiation and to prevent interference with radio communications. Induction and dielectric heating are very similar in operation to RF heating, but operate at lower electromagnetic frequencies.

3.1.2 Compression Molding

Compressing molding thermoforming processes have had a history of commercial success in the auto industry over the past decade or more, and probably will continue to be given first consideration in any new applications. There are at least three variations of compression molding being used to thermoform fiber-reinforced thermoplastics. These processes typically use a compression molding press and a matched pair of molds. Starting materials may be in the form of rigid consolidated or unconsolidated sheet or stacks of rigid plies. Mold halves are aligned and mounted on parallel upper and lower platens of a compression molding press. The blank (the starting sheet or stack of plies) is heated and positioned between the mold halves. The press closes the mold rapidly, forming the blank to the shape of the closed mold cavity before it is cooled below its forming temperature by the mold. The mold is then opened, the molded part removed, and trimmed if necessary.[2] The schematic shown in Figure III-1 illustrates the process with a matched mold in the pressure forming stage.

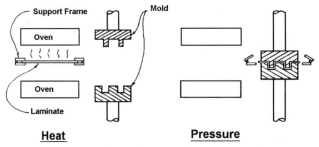

Heat   **Pressure**

Figure III-1. Matched mold for thermoforming.

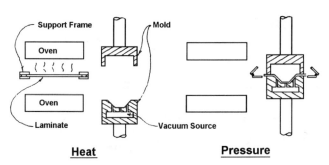

Heat   **Pressure**

Figure III-2. Vacuum forming.

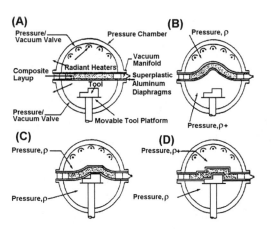

Figure III-3. Diaphragm forming in an autoclave. (Source: ICI, Harper). Reproduced from *Advanced Composites Magazine*, Sep/Oct 1988, p 49.

Stamping – Stamping is a variation of compression molding that is similar to the sheet metal forming process bearing the same name. The blank is a single sheet or a stack or plies cut to almost the exact shape of the cavity opening and to about 95-98% of the size of the cavity opening, so that virtually no flow is required to fill the cavity, and the blank and the final part are virtually the same thickness. This variation is like most thermoforming operations in the sense that the final part is basically the original sheet formed to a different shape with very little rearrangement of the material in the sheet. A principal advantage of this method is that highly ordered unidirectional fibers will tend to maintain their original spacing and direction in the final part. A disadvantage is that there is no built-in mechanism to prevent the heated sheet from buckling when it is draped on the mold and the mold is closed. As a result, this variation is limited with regard to the complexity of contours that can be formed. Compromises can be made to this variation by preforming the blank in some manner, adding flowable material in areas where increased thickness is required, etc.[2]

Flow Molding – Flow molding is a variation of compression molding that is similar to stamping. The principal difference is that the blank is a single sheet or a stack of plies much smaller in area and thicker than the molding and is often cut to simple shapes such as rectangles. Thermoplastic composites developed for this process, such as Azdel, can typically flow up to 50%.[3] Typical molding pressures are 10,300 to 17,200 kPa, and are reported as high as 48,300 kPa.[4] Flow molding is capable of forming complexly contoured parts and can accommodate, in a single part, large variations in part thickness such as molded-in ribs and bosses, as well as molded-in threaded holes, inserts, and the like.

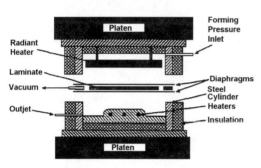

Figure III-4. Diaphragm forming in a press. (Courtesy of Univ. of Delaware.)

Stretch Compression Molding – In unreinforced thermoforming, buckling is normally prevented by clamping the heated sheet in a frame or ring, forcing the sheet to stretch as it changes shape. This same technique is being applied to reinforced thermoplastic stamping in an effort to achieve more complex contouring without buckling. Among the several methods mentioned in the literature[5], three are being studied actively.

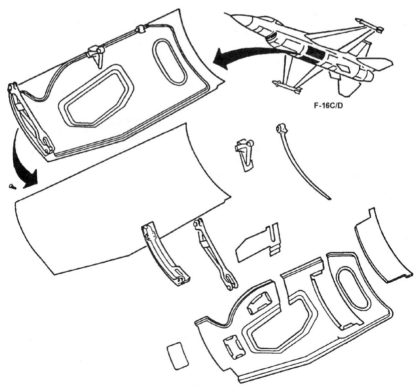

Figure III-5. Composite main landing gear.[21]

Clamping Ring with Stretchable Sheet – Clamping compression molding can provide a stretching action to the sheet being formed by the use of a stretchable form of reinforced thermoplastic composite sheet firmly clamped in a frame.[6] Useful forms of reinforcement include random chopped fiber mats, swirled continuous fiber mats, and DuPont's aligned discontinuous (or ordered staple) mats.[7]

Diaphragm Forming – A thermoforming process called diaphragm forming is being used with reasonable success with difficult-to-form composites, such as continuous carbon-fiber-reinforced PEEK. It is often considered a modification of the billowing/pressure forming process. The key step in this process is the placement of the sheet to be formed between two thermoformable or elastic diaphragms.[8-14]

This process can be carried out in an autoclave, (Figure III-3) or in a hydraulic molding press fitted with the mating halves of a two-piece pressure chamber mounted on its upper and lower platens, (Figure III-4). A sheet or stack of plies is sandwiched between two somewhat larger sheets of diaphragm material (e.g., superplastic aluminum), as shown in Figures III-3.

The diaphragms undergo a true stretching-type thermoforming action. However,the thermoplastic sheet, if reinforced with continuous fibers, cannot stretch and must slip with respect to the diaphragms. Therefore, the forming pressure is applied gradually, requiring possibly 10 minutes to reach final forming pressure, typically in the range of 344-689 kPa.[15]

With heating, gradual pressurization, and cooling of the formed part and the mold all taking place in the closed apparatus, cycle time might range from 20-100 minutes.[7-8] A patented diaphragm molding process for thermoforming contoured parts from sheets of carbon fiber/PEEK uses thin sheets of superplastic aluminum alloys, capable of deforming several hundred percent, as diaphragms.[7]

In ongoing investigations, Jar, Davies, Cantwell, et.al.,[16] studied autoclave forming of carbon/PEEK at 370°C and diaphragm forming at 385°C and that TMCs allows a wide range of manufacturing methods to be considered for the production of engineering components. They also found that the potential problems associated with the high forming temperatures and with the use of semi-crystalline matrix materials whose structures may be sensitive to processing history do not appear, regardless of short and long term properties. Pantelakis, Tsahalis, et.al.,[17] performed similar diaphragm forming techniques and arrived at similar conclusions.

Autoclave processed complex parts can be made from Avimid K® TPC material due to its unique tack and drape characteristics.[18] Laminate and sandwich skin panels, male and female C-channel, I-beams, and sine wave spars have been successfully fabricated using autoclave forming on low CTE tooling. Efforts by Baker and Gesell,[18] in the future will attempt to concept automated fabrication to eliminate hand layup, which is one of the two major cost drivers associated with composite structures.

Forsberg and Koch reported[19] on the design and manufacture of thermoplastic F-16 main landing gear doors. TPC left hand main landing gear doors for the F-16 were fabricated using T650-42/Radel-C, a 177°C service TPC material system. These doors were full form fit and function replacements for existing aluminum doors. The doors were essentially torque boxes formed by co-consolidating previously consolidated inner and outer skins. The inner skins were made using diaphragm forming. The outer skins were fabricated in an autoclave and co-consolidation of inner and outer accomplished using disposable mandrels. The finished assembly is 20% lighter than an aluminum door.

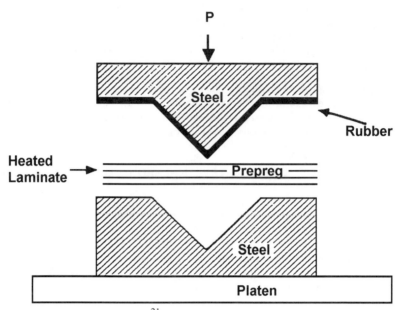

P

Steel

Rubber

Heated
Laminate →

Prepreg

Steel

Platen

Figure III-6. Press forming process.[21] The prepreg stack is clamped and heated in an infrared oven to the melt temperature of the material. This heated stack is then transferred into the mold cavity. The mold is closed, entrapping the stack between the rubber male and the heated (warm) steel female die. Pressure is applied and the laminate rapidly cools to below the melt. The mold is opened and the finished part removed.

Thermoforming of glass-reinforced composites is coming. G.E. Plastics' Polymer Processing Development Center is experimenting with thermoplastic sheet containing over 30% discontinuous glass fibers for an automotive application and is developing new thermoformable materials with even higher glass contents.[20] Meanwhile, Thermoforming Technologies Inc., is developing a single-station pressure former with profile heating capability to form reinforced thermoplastic sheet with up to 70% continuous fiber for automotive and aircraft applications.

Two design and manufacturing studies sponsored under government contracts conducted by Lockheed Aero. Systems Co.[21] and Northrop Co.[22] evaluated advanced TPCs. The first program conducted trade studies to evaluate thermoplastics versus bismaleimides for various generic fighter components and preliminary designs were developed followed by manufacturing verification and production demonstration tasks. These consisted of design, development fabrication and test of a thermoplastic F-16 main landing gear door component. The composite main landing gear door, Figure III-5, with thermoformed details was calculated to weigh 19.14 kg. This is a weight savings of 20.2% compared to the originally-designed aluminum door. The composite door was designed with manufacturability as one of the main criteria. Diaphragm forming was selected as a relatively mature process involving a short-cycle time, which has been successfully used for fabrication of high-temperature thermoplastic parts and assemblies. Diaphragm forming, Figure III-4, using superplastic aluminum as a diaphragm material was used in the fabrication of the complex shapes from the APC-2 thermoplastic. Matched rubber-metal press forming of the inner skin, Figure III-6, was also selected and the lower side bay bulkhead design shown in Figure III-7, was also designed for fabrication using diaphragm

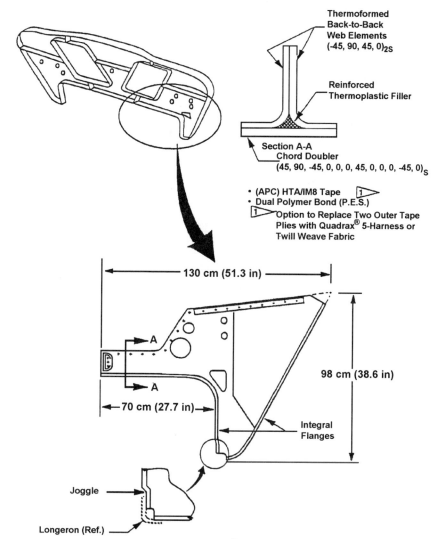

Thermoformed
Back-to-Back
Web Elements
$(-45, 90, 45, 0)_{2S}$

Reinforced
Thermoplastic Filler

Section A-A
Chord Doubler
$(45, 90, -45, 0, 0, 0, 45, 0, 0, 0, -45, 0)_S$

• (APC) HTA/IM8 Tape [1]▷
• Dual Polymer Bond (P.E.S.)
[1]▷Option to Replace Two Outer Tape
Plies with Quadrax® 5-Harness or
Twill Weave Fabric

← 130 cm (51.3 in) →

A

A

98 cm (38.6 in)

← 70 cm (27.7 in) →

Integral
Flanges

Joggle

Longeron (Ref.)

Figure III-7. Side bay bulkhead segment design.[21]

forming for the web and dual resin bonding for the web assembly. The selected fuel bay
bulkhead design is shown in Figure III-8.

The second program, DMATS,[22] was to design and develop a low cost manufacturing method
for thermoplastic structures for an F-15E forward engine access door (secondary structure)and
an advanced fighter fuselage structure as primary structure. The autoclave diaphragm forming
concept, Figure III-9, was used to produce bulkhead frames, webs, channels, upper/lower caps,
and amorphously bonded stiffeners for the primary fuselage structure shown in Figure III-10.
Other thermoforming studies and programs are in References 23 to 25.

119

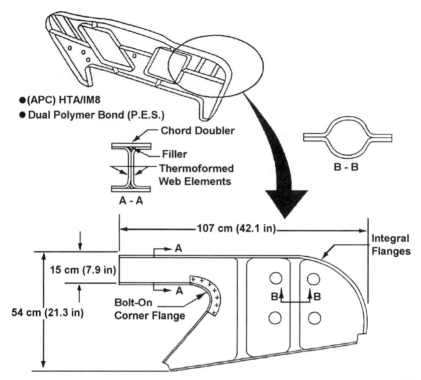

Figure III-8. Fuel bay bulkhead segment design using integrally formed web stiffeners.[21]

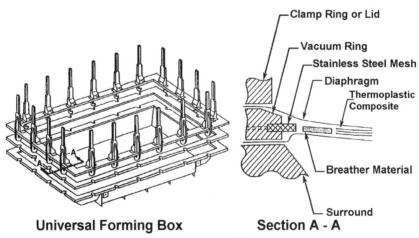

Figure III-9. Autoclave diaphragm forming concept.[22]

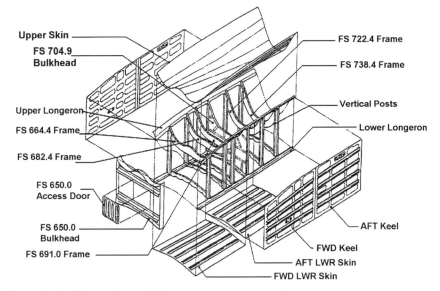

Figure III-10. Thermoplastic bulkhead thermoformed structure.

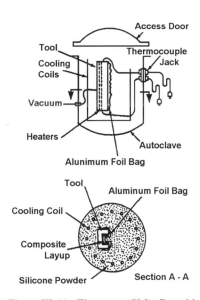

Figure III-11. Therm-x (SM) flowable elastomeric powder process. (Source: United Technologies. Reproduced from *Advanced Composites Magazine,* Sept/Oct 1988, p 54.)

Therm-x (SM) Process – The Therm-x (SM) process was developed and has been commercialized by the United Technologies Chemical Systems Division, and is described in detail in References 26 and 27. It is carried out in a Thermoclave™ pressure vessel.[28] It is an extension of the basic idea used in the "Trapped Rubber" molding process, in which molding pressure is applied to a prepreg preform on a mold-half by the thermal expansion of a rubber pad, confined in a pressure chamber with the mold and preform. In the Therm-x (SM) process, the thermally expandable process medium is a flowable elastomeric powder, rather than a rubber block. This eliminates severe pressure gradients experienced in the trapped rubber process and allows the safe generation of high pressures at high temperatures. The Thermoclave™ vessel is a specially designed autoclave, shown in Figure III-11. It contains a means to support the mold-half, a set of tubing coils to heat and expand the powdered silicone rubber process medium, and localized electric heaters to generate the cure temperature in the immediate vicinity of the part being molded. The Therm-x (SM) process is useful primarily in the compaction and curing of difficult-to-cure, contoured thermoset advanced com-

121

posites, but the process is capable of high temperature compaction and forming of high temperature thermoplastic advanced composites such as carbon fiber/ PEEK. A process capability of 816°C and 20,700 kPa pressure has been planned, along with vessels up to 5.1 m diameter for the future.

Rubber Block Molding, Hydroforming[29] – Rubber block molding is another process similar in some respects to the trapped rubber process. However, rather than using thermal expansion of the rubber block to pressurize the rubber block against the molded part on the mold-half, pressure is applied via a compression molding press. A rubber block, gradually spreading its area of contact with the sheet/mold-half combination as the press closes, generates a forming action with a reduced tendency to wrinkle areas of the sheet that must be formed by lateral compression, rather than by stretching. In another variation, called hydroforming, the rubber block is replaced by a heavy rubber diaphragm backed by a hydraulic fluid.[29-30]

Other Processes – There are a variety of other processes in use or being studied that are more specialized or not publicized for one reason or another. ICI, for example, uses a roll forming process, very similar to a hot ingot metal roll forming process, to make long, straight, uniform cross-section shapes such as channels and stringers.[31] Basically, roll forming contours the sheet by simple bending and does not form compound contours.

Incremental forming,[32] is a process being studied that can create large thermoplastic composite parts with relatively small equipment. It is compression molding, utilizing an oven, a press, matched molds, and a transfer system. A long laminate, such as a wing skin, is molded incrementally in short lengths as the sheet being formed is indexed lengthwise through the oven and the press. The oven and press are only the size of the section to be formed instead of the size of the entire part; thus, part length is not limited by equipment size. Modular molds can change shape quickly between sections,eliminating time-consuming mold changes. Five non-identical sections have been molded in 30 minutes. The potential advantages are particularly in the areas of equipment size and capital cost, labor and operating costs, and maximum part length. It has been proven on a prototype scale producing relatively small parts.

Vacuum forming (Figure III-2), which is conventionally used in thermoforming unreinforced thermoplastics, is also applicable to TPCs. When vacuum thermoforming, there are limitations to the types of parts which can be formed. Usually, this technique is applicable to gentle curvatures with shallow draws. With either pressure application technique, plug assist or vacuum, minimizing the time from when the laminate leaves the heat source until the press is fully closed is critical. The laminate begins to cool as soon as it leaves the heat source, and if not formed quickly, it will begin to stiffen and lose its drapability and formability. This transfer

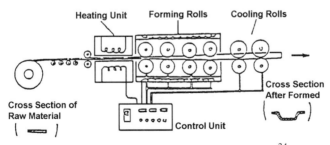

Figure III-12. Basic concept of continuous bend forming process.[34]

time is especially important with thin laminates of only one or two plies. Specific times depend on the laminate thickness; however, most thermoforming equipment can make the transfer and close the tool in 5-10 sec., which has been found to be adequate for the PPS-based material.[33]

Yamaguchi, Sakatani, and Yoshida,[34] examined press molding of AS4/PPS and IM7/PEEK materials and thermoforming (continuous bend forming, Figure III-12) of HM/PEEK composite beams for future large space structures. The process forms composite beams from thin flat sheet to hat-sections continuously with multi-step section rolls.

A new Inner Pressure molding technology,[35] method has been developed for producing pipes with excellent workability from a carbon fiber-reinforced composite prepreg using TPC and carbon fiber, while retaining the characteristics of the raw materials. Forming methods for these materials include the roll forming method, and curved forming methods, but there was no forming method for producing pipes accurately. The new molding technology was developed by employing a carbon fiber-reinforced composite prepreg using PEEK as the matrix. In principle, pipes were molded using other thermoplastic resins such as PPS and nylon 66 as the matrix. Carbon fiber or glass fiber were used as the reinforcing composite, and the technology is also applicable to producing woven and knit fabrics for use in molding.

Since the pipes are produced in the mold, they have excellent outside diameter dimensional accuracy, surface smoothness, and virtually no voids. The pipes are remarkably strong, reliable, and can have special sectional areas.

This new technology uses a thermosetting plastic that greatly improves the formability and eases forming control, and enables pipes of complicated shapes to be molded uniformly. The most important characteristics are that the pipes produced have 1) excellent toughness and strength provided by the raw materials and compatible properties, 2) excellent compressive fatigue strength and great chemical stability, 3) excellent mechanical properties at low temperatures, net shape molding, and excellent dimensional accuracy, and uniform consolidation, and 4) excellent productivity in mass production and a substantial reduction of manufacturing costs is possible.

These pipes have a broad range of applications including space development (space station truss structures, manipulator arms, etc.), aeronautics (energy absorption structural materials, wheel legs, torque tubes, etc.), leisure and sporting goods, and vehicles (sports cars and linear motorcar structural materials).

Another Japanese firm,[36] has developed a fiber-reinforced composite materials that can be melted by heating and can produce products with complicated shapes.

Conventional composite materials usually use a thermosetting resin and are inconvenient as reshaping and forming into complicated shapes are quite difficult since the material is brittle. Asahi Chemical Industry Co., Ltd. has mixed a fibrous heat melting resin with reinforcing fibers such as carbon and glass fiber, and produced a fiber-reinforced composite material that is flexible and can be bent in any direction. This material is called Web Interlaced Prepreg (WIP) and uses carbon fiber or glass fiber as the reinforcing material. These fibers are aligned in parallel and interlaced at the individual filament level with short fiber resin, such as nylon 6, nylon 66, PPS, PEEK, depending on the specific use. Shaping WIP at high temperature melts the resin and causes it to wrap the reinforcing material (such as carbon fiber) to provide a very strong composite material. PEEK and other engineering plastics are ideal for large products such as aircraft structural materials. Finally, thermofolding is a technology which implies the heating of a laminate along a line, along which the product is folded subsequently.[37] The

123

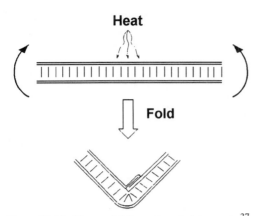

Figure III-13. Thermofolding of sandwich panels.[37]

thermofolding technique can be applied to both solid laminates and sandwiches. The solid laminate folded part has been applied to a prototype electronics housing box. The consolidated laminate consisted of glass fabric layers with a PEI matrix.

With thermoplastics, a fold can be obtained without milling or bonding. Heat is applied to one of the skins, Figure III-13. After heating sufficiently, the panel can be folded. The forming operation is very quick, limited only by the heating and cooling rate of the material. The strength of a thermal fold is sufficient for non-structural applications such as aircraft interior panels or non-structural aircraft exterior parts.

Prototype aircraft interior panels have been made by thermofolding the sandwich along all four edges, the maximum length being 1 m. For aircraft interior panels, fire safety is of prime importance. For post-crash fires, thermoplastics with continuous fiber reinforcements such as the glass/PEI laminate or sandwich with this material are very compliant to federal requirements for heat release in cabin fires.

3.2 SUPERPLASTICITY

There are two well-established types of superplastic behavior in polycrystalline solids.[38] The first type of superplastic behavior is known as fine structure superplasticity (FSS); the second type is known as internal stress superplasticity (ISS). In the case of FSS materials, a strain-rate-sensitivity exponent equal to about 0.5 is usually found and the materials deform principally by a grain boundary sliding mechanism. In the case of ISS materials, however, the strain rate sensitivity exponent is usually unity, i.e., they exhibit Newtonian viscous flow behavior. These ISS materials need not be fine-grained, and generally deform by a slip deformation mechanism. The concepts and principles described in FSS and ISS superplasticity have been applied to enhanced powder consolidation through superplastic flow and the development of superplasticity in laminated composites containing at least one superplastic component.[38]

3.2.1 Types of Superplasticity

Superplastic materials can be categorized into two groups. The first group is isothermal superplasticity and the other is ISS. Recent studies have demonstrated that many discontinuously reinforced aluminum MMCs tend to behave superplastically.[31-47] Comparative superplasticity data, representative of some of these studies, are summarized in Table III-2. All the composites listed in Table III-2 were made via conventional powder metallurgy techniques except $7475/SiC/15_w$, which was manufactured using SiC whiskers layered between specially prepared foils of superplastic aluminum alloy 7475.[42] Despite the fact that the reported strain rate sensitivities vary significantly, similar elongations (about 300%) were measured in each of the studies in Table III-2, with the exception of one of the thermally cycled materials

Table III-1 Achievements in Composite Materials in Superplasticity[1]

Property	Level of Achievement	Material or Contributor
Maximum Superplastic Elongation in an Intermetallic	> 800%	Ti_3Al (Super Alpha-2)
Maximum Superplastic Elongation in a Metallic	1,400%	SiC_w/Al6061 (Thermal Cycling)
aximum Superplastic Elongation in a Ceramic Composite	625%	Y-TZP/Al_2O_3
Finest Grain Size in a Consolidated Material	30 - 40 nm	Al_2O_3/SiO_2 (9:1) &in ZrO_2 (Y)

Table III-2 Superplastic Properties of Al/SiC Composites[49]

Material	Temp. (°C)	Max Elong. (%)	Strain Rate Sensitivity	Strain Rate (s^{-1})	Stress (MPa)	Ref.
2124/SiC/20w	475 - 550	-300	-0.33	3.3×10^{-4}	-10	1
2024/SiC/20w	100 ↔ 450*	-300	1.0	5×10^{-4}	-15	2
6061/SiC/20w	100 ↔ 450*	-1,400	1.0	10^{-4}	-7	3
7475/SiC/15w	-520	350	≥ 0.5	2×10^{-4}	-7	4
PM 64/SiC/10p+	-516	-250	≥ 0.5	2×10^{-4}	-1.4	5

*Thermal Cycling
+Back Pressure of 4.14 MPa Applied

Table III-3 HSRS Test Results and Conditions for Various Materials[49]

Material	Test Temp. (°C)	Solidus (°C)	Strain Rate (s^{-1})	Stress (MPa)	Elong. (%)
2124/β-SiCw	525	502	0.3	-10	-300
2124/β-Si3N4(w)	525	502	0.2	-10	-250
7064/α-Si3N4(w)	525	-525	0.2	-15	-250
6061/β-Si3N4(w)	545	582	0.5	-20	-450
MA IN9021/-SiCp	550	495	10.0	-7	-500

(6061/SiC/20w), in which a large elongation of 1,400% was found. The matrix grain sizes for the 7475/SiC/15w and PM 64/SiC/10p alloys were about 6 μm, whereas the grain sizes for the 2024/SiC/20w and 2124/SiC/20w were about 1 μm. Although not reported, the matrix grain size for 6061/SiC/20w was expected to be also about 1 μm.

Thermal-Cycling Superplasticity – As noted in Table III-2, for the cases of 2024/SiC/20w and 6061/SiC/20w, super-plasticity was observed under nonisothermal test conditions. In these cases,[40-41] the materials were demonstrated to be superplastic under conditions of thermal cycling between 100°C and 450°C (at 100 seconds per cycle). This resulted in the elongation of about 300% in 2024/SiC/20w and an exceptionally large elongation of 1,400% in the 6061/SiC/20w. In contrast, the MMC exhibits only 12% elongation under isothermal creep deformation at 450°C. Pickard and Derby[48] also demonstrated that an Al/SiC composite, thermally cycled between approximately 130°C and 450°C, exhibited a tensile elongation of 150%, whereas the normal tensile elongation is less than 10%.

Isothermal Superplasticity – In the case of isothermal superplasticity, the results can be further categorized into two subgroups, according to the strain rates at which superplasticity occurs.

As shown in Table III-2, the strain rate at which superplasticity takes place in the 2124/SiC/20$_w$ composite is about three orders of magnitude higher than that observed in the other Al/SiC composites which is about 10^{-4} - $10^{-5} s^{-1}$. Again, the latter strain rates are those usually observed in conventional superplastic aluminum alloys.

Conventional Superplasticity – As indicated in Table III-2, Mahoney and Ghosh[43] demonstrated that 7475/SiC/15w, after thermomechanical processing, behaved superplastically (350%) at conventional superplastic strain rates of 2 x 10^{-4} s^{-1}; this is also the strain rate at which superplasticity is observed in the monolithic matrix alloy 7475. These results are similar to those observed in the PM 64/SiC/10$_p$ composite systems.[43] (Alloy PM 64 is essentially a powder metallurgy version of the 7064 ingot alloy.) The above results suggest that superplastic deformation in PM 64/SiC/10$_p$ and 7475/SiC/15$_w$ composites is dominated by the behavior of the matrix. This is a quite different result from that observed in the reinforced 2124 and 2024 alloys. Specifically, whereas both the 7475 and the PM 64 unreinforced matrix alloys are superplastic, the unreinforced 2124 and 2024 matrix alloys are not. Despite this fact, both 2124/SiC/20$_w$ and 2024/SiC/20$_w$ composites behave superplastically and, furthermore, 2124/SiC/20$_w$ exhibits superplasticity at exceptionally high strain rates.

High-Strain-Rate Superplasticity (HSRS)

Among the studies listed in Table III-2, it is particularly noted that 2124/SiC/20$_w$ exhibited superplastic properties at extremely high strain rates. This HSRS has not only been observed in 2124/βSiC$_{w'}$, but also in 2124/β-Si$_3$N$_{4(w)'}$,[44-45] 7064/α -Si$_3$N$_{4(w)'}$,[45] 6061/β-Si$_3$N$_{4(w)'}$[46] and mechanically alloyed (MA) IN 9021/SiC$_p$.[47] These materials are shown in Table III-3 along with some of the test conditions and results. The HSRS results represent an important breakthrough, since a major limitation of current superplastic forming technology is the slow forming rate.[48-49]

3.3 APPLICATIONS OF SUPERPLASTICITY

Thin Sheet Forming – The origin of superplasticity in crystalline materials lies in their fine-grained nature. The most common superplastically-formed products are made from fine-grained sheets. The principal method is by blow forming, in which gas pressure is applied on one side of a sheet which then superplastically flows into a die of predetermined shape and complexity. Most superplastic components made to date have utilized the SUPRAL aluminum alloys, Al7475, Ti-6Al-4V, or stainless steel.[50-51]

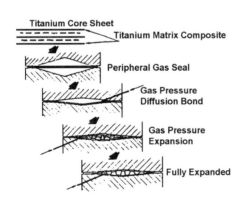

Figure III-14. SPF/DB process for forming hollow Ti MMC structures.[52]

A variation of gas pressure forming process is the combined superplastic-forming and diffusion bonding (SPF/DB). Many sandwich structure products are made by this process with fine-grained titanium alloys, usually the Ti-6Al-4Valloy.[38] Diffusion bonding is readily achieved because of the fine-grain size and because the oxide diffusion barrier in titanium dissolves at the temperature of processing, i.e., about 925°C.

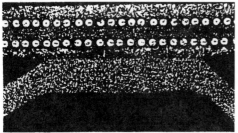

Figure III-15. Typical diffusion bond between core and MMC sheet showing the absence of sharp corners.

Table III-4 Ti Matrices
Used in Production of
SiC/Ti Alloy Composites

CP Ti
Ti-6Al-4V
Ti-3Al-2.5V
Ti-15V-3Cr-3Sn-3Al
Beta C Ti 3Al-8V-6Cr-4Mo-4Zr
IMI 829 Ti-5.5Al-3.5Sn-3Zr-1Nb
Titanium Aluminides (Alpha 2 Type)

Wedge shaped hollow MMC panel configurations have been prepared with SCS-6 fiber-reinforced titanium sheets. The process has been used to develop and process MMC blades, Figure III-14. Figure III-15 shows a typical section through a panel showing the bond between the MMC skin and core.

The demonstrated high strength and high stiffness of continuous SiC fiber reinforced titanium alloy with retention of usable properties up to 982°C, makes it a very attractive material for use in high temperature applications.[53] Titanium alloy matrices that have been successfully composited with continuous SiC are shown in Table III-4.

A prime example of the application of the SPF/DB approach to a high speed missile inlet is shown in Figure III-16. The reinforcement material is applied to the cowl lip and the main air duct, locally stiffening these areas with minimal thickness build-up and eliminating the requirement for splitters.

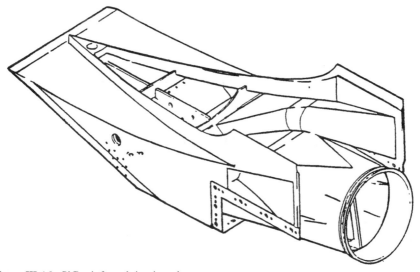

Figure III-16. SiC reinforced titanium duct.

Bulk Forming – Another superplastic method that is less well-developed, but of potentially great commercial interest, is that of superplastically forming parts by press-forging of bulk material into complex dies. Such processing of ultra-fine grained materials is still in the early stages of development. The best example of a product made in this manner is a disk with turbine blades made from a fine-grained nickel base alloy.[38] Another possible application of press forging superplastic metals is the production of gears. A bevel gear, for example, has been warm forged from a fine-grained ultrahigh carbon steel (UHCS).[38]

3.4 SP FORGING

Superplastic forging generally involves making a ceramic preform from a starting powder by wet- or dry- mixing a composition that can be from 70 to 99%Si_3N_4, up to 30% stabilizing additive, and up to about 7% free Si. Researchers cold-press the mixture into a preform and partially sinter it at atmospheric pressure in nitrogen at between 1,400 and 2,200°C. By isothermally pressing the preforms into near-net shape (NNS) at the proper temperature and strain rate, researchers avoid the formation of surface cracks. Scientists claim that the forging that takes place during superplastics processing densifies the end product more uniformly than conventional hot pressing does. In hot pressing, loose powder often doesn't densify uniformly when intricate shapes are formed. This causes the products to be weak. But in superplastic forging, the preform can be designed to have a shape similar to that of the final product, which allows the flow of material during processing to be controlled more uniformly. In addition to Si_3N_4, Al_2O_3, ZnS and some glasses have also been superplastically forged. At Japan's Govt. Ind. Res. Institute, Nagoya, Si_3N_4/SiC ceramic components have been superplastically forged. Potential applications for the technology are for wear- and heat- resistant mechanical components as well as piston rings for automotive engines.

3.5 MATERIALS

Ceramic base materials made superplastic to date are based on fine-grained materials. An example of such a ceramic is a 3 mol% yttria-stabilized tetragonal zirconia polycrystal (3Y-TZP) with up to 200% elongation and a strain-rate-sensitivity exponent of 0.5. More recently, studies by Nieh and Wadsworth[54] on this material have shown elongations of up to 800% in 3Y-TZP and 625% in a ceramic composite of 3Y-TZP containing 20% Al_2O_3.

The study of superplastic Al_2O_3 was first carried out by Carry and his group.[55] As a result of this study, a fine grain size (~ 0.66 μm) Al_2O_3 exhibited a rupture elongation of 65% at 1,450°C under an applied stress of 20 MPa. Additions of ZrO_2 have been shown to greatly stabilize the

Table III-5 Superplasticity of Al/SiC$_W$ and Al/Si$_3$N$_{4(w)}$ Composites[49]

Matrix	Whisker	Superplastic?
2124 Al	β-Si_3N_4	Yes
6061 Al	β-Si_3N_4	Yes
7064 Al	α-Si_3N_4	Yes
2124 Al	α-Si_3N_4	No
2124 Al	β-SiC	Yes
6061 Al	β-SiC	No

	Tensile Modulus (GPa)	Percent Improvement	Ultimate Tensile (MPa)	Percent Improvement
2124 Alloy	72	—	460	—
2124 + 20% SiC	100	38	620	35
8090 Alloy	80	—	450	—
8090 + 20% Sic	105	35	600	36
6061 Alloy	69	—	350	—
6061 + 40% Sic (Unidirectional Monofilament)	200	194	1,400	300

grain structure of Al_2O_3 during superplastic deformation. A 10 vol% ZrO_2-containing Al_2O_3 (with a grain size ~ 0.5 μm) was successfully stretched under biaxial tension at 1,400°C. Wakai et.al.,[56] also first showed that ceramic composites based on fine-grained zirconia, e.g., 20% Al_2O_3/Y-TZP, were superplastic. Other Wakai[57-58] work has shown that a fine-grained SiC/Si_3N_4 composite prepared initially from an amorphous Si-C-N powder which was produced by vapor phase reaction of $[Si(CH_3)_3]_2 NH/NH_3/N_2$ at 1,000°C has exhibited superplastic behavior (150%). As with Al/SiC composites, Tables III-2 and 3, not all Al/Si_3N_4 composites exhibit the superplasticity phenomenon. Table III-5 lists the $Si_3N_{4(w)}$ reinforced composite combinations that are superplastic and nonsuperplastic. Results from $2124/SiC_w$ and $6061/SiC_{w'}$ are included for comparison. As shown in Table III-5, specific combinations of matrix and whiskers in an MMC are critical to the development of HSRS. A simple criterion for superplasticity based on the individual types of whisker or alloy matrix appears to be impossible. This view-point is supported by the observation that $2124/\beta-Si_3N_{4(w)}$ and $6061/\beta-Si_3N_4$ are superplastic, but $2124/\alpha-Si_3N_{4(w)}$ and $6061/\beta-Si_3N_4$ are not.

Table III-6 illustrates typical elastic moduli and tensile strengths of particle and matrix reinforced composites in comparison with the matrix alloys. Clearly, the MMCs display higher strength than the matrix alloys.[59-60]

3.6 APPLICATIONS FOR SPF COMPOSITE MATERIALS

The effort and work program initiated by the U.S. govt. and General Electric,[52] to design, develop, fabricate, and demonstrate the structural and environmental suitability of reinforced titanium fan and compressor blades was successful. With SCS-6 fiber as the reinforcing material and the design of an SPF/DB reinforced titanium MMC advanced engine fan blade that meets mechanical requirements, the viability for advanced blading for military engines was shown. An $8090/SiC/16.6_p$ hatch cover was produced by SPF for a British Aerospace aircraft and is now in production.[59] Finally, subcomponents which are representative of NASP structural types including the fuselage and wing in order to demonstrate the material and structural performance of titanium matrix composites (TMC), are being fabricated by several methods including SPF and SPF/DB, Figure III-17. One of the most complex sub-components being made is the Lightly Loaded Splice Subcomponent, Figure III-18. It is representative of a lightly loaded, hat-stiffened structure such as a fuselage. Both SCS-6 and SCS-9 fibers and Beta 21S matrix were used. It includes a complex splice joint with manufacturing details such as joggles, build-ups, and single-sided assembly. Two of the largest remaining components are a stabilator torque box and a fuselage section. The stabilator which is 1.8 m x 2.4 m overall and consists

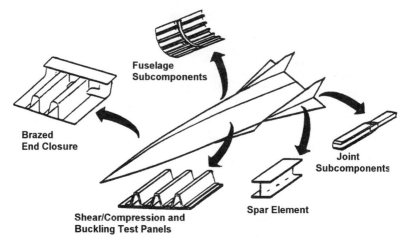

**Fuselage
Subcomponents**

**Brazed
End Closure**

**Joint
Subcomponents**

**Shear/Compression and
Buckling Test Panels**

Spar Element

Figure III-17. NASP subcomponents.

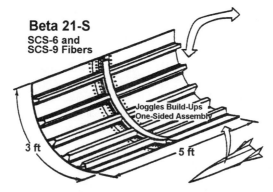

Beta 21-S
**SCS-6 and
SCS-9 Fibers**

Joggles Build-Ups
One-Sided Assembly

3 ft

5 ft

Figure III-19. NASP lightly loaded splice subcomponent.
3 ft = 0.9 m, 5 ft = 1.43 m.

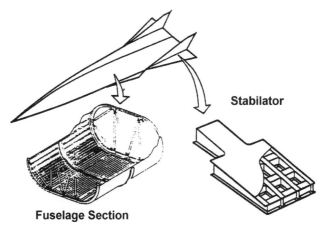

Stabilator

Fuselage Section

Figure III-18. Stabilator and fuselage section for proposed NASP aircraft.

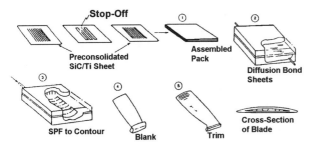

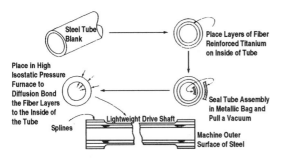

Figure III-20. Stages in SPF/DB of SiC/Ti alloy hollow fan blade.[53]

Figure III-21. Stages in the production of SiC/Ti alloy drive shaft.[53]

of large, flat skins, I-beam spars and intercostals of which some are SPF/DB, Figure III-19.

Another SPF/DB application mentioned earlier in the chapter is shown in Figure III-20. The sketch shows the sequence for producing hollow fan blades of Ti-alloy selectively reinforced with SiC fiber. A 15.24 mm dia. x 1.8 m long SiC fiber reinforced Ti-6Al-4V tube has been designed for 190,909 kg compression, fabricated and successfully tested to 206,818 kg. Being developed in primary and auxiliary engine power take-off shafts, the SiC/Ti alloy material provides increased axial specific stiffness, thereby allowing either an increase in the critical rotational speed of the shaft or a reduction in the length between support bearings. In production of the SiC/Ti alloy drive shaft, Figure III-21, the fiber preform is placed onto a titanium foil and then spirally wrapped for insertion into the steel tube blank. This assembly is sealed in a steel pressure membrane (bag) and a vacuum is drawn. After off-gassing, the assembly is HIPped. Subsequently, the steel is thinned down and machined to form the " spline attachment" at each end.[53]

POST PROCESS FINISHING

3.7 CORROSION PROTECTION

MMCs fabricated with high-modulus graphite or SiC reinforcements offer excellent structural properties. However, serious corrosion problems can occur, especially for Al/Gr MMCs, which contain some of the most powerful couples in the galvanic series. For Al/Gr, accelerated corrosion is likely to occur when the metal foils that cover the surface of the MMCs are penetrated by pitting attack, which leads to the establishment of a graphite/metal couple.[61-64] For Al/SiC MMCs, this galvanic problem might be less severe because of the insulating nature of SiC.

Anodizing of Al alloys is an electrochemical method of converting aluminum metal into aluminum oxide by adding an external current in an acid electrolyte. The most widely used electrolyte is sulfuric acid. There are two types of sulfuric anodizing: conventional anodizing, which is performed at room temperature and provides about 7 to 15 μm of oxide thickness and a fairly hard surface, and hard coat anodizing, which is performed at around 0°C and provides about 50 μm of oxide thickness with extreme hardness. The oxide film consists of a thin, continuous barrier layer below a thick, porous layer. The structure of the porous layer was characterized by Keller et.al.,[65] as a closed-packed array of columnar hexagonal cells that contain a central pore normal to the substrate surface. The porous layer can be sealed in hot water or a dichromate solution to close these pores.

In a program developed by Lin, Greene, Shih, et.al.,[66] 6061/SiC/25$_p$ with 10 μm size particulate powder was studied. Anodic coatings were produced by anodic oxidation in an acid bath to form an oxide layer. The procedures used in the study consisted of the following steps:

Conventional Anodized Coatings

a) The part was wiped with hexane.

b) The part was immersed in hexane at 50°C for 15 min and then immersed in an alkaline solution at 66°C for 8 min.

c) The part was immersed in a deoxidizer at room temperature for 10 min.

d) The part was immersed in 10 vol% H_2SO_4 at room temperature and constant current density applied. The coating time was about 30 min. (SAA)

e) The anodized test part was immersed in hot water at 90 to 100°C for 20min.[67-68] (HWS)

Hard Anodizing Coatings

a) The same procedure as for conventional anodizing coatings was used for degreasing and deoxidizing.

b) The test part was immersed in 15 vol% H_2SO_4 at 0°C, and a constant current density applied. The coating time was about 1 hr.

c) The anodized test part was immersed in hot water at 90 to 100°C for 30 min.

The results showed that, for 6061 aluminum, conventional and hard-anodized coatings (SAA+ HWS) provide excellent corrosion resistance to 0.5 N NaCl (open to air). Anodized coatings (SAA+ HWS) on Al/SiC provide satisfactory corrosion protection, but they are not as effective as for 6061 because the structure of the anodized layer is affected by the SiC particulates. The corrosion resistance of hard-anodized Al/SiC is less than that of conventional-anodized Al/SiC because the area fraction of the continuous barrier layer for hard-anodized SiC/Al is less than that for conventionally anodized SiC/Al. A mechanism for the formation of anodized layers on Al/SiC have been proposed by Lin, et.al.[66]

3.7.1 External Coatings for Filament Wound Components

A wide range of external coatings with corresponding functions and application techniques are now available.

The coatings, as well as the techniques by which they were applied and some of the critical parameters are as follows:

- Epoxy coatings (clear and opaque)

- Polyurethane elastomers and organic coatings

- Metal films

Epoxy Coatings – A wide range of two part epoxy (resin and curing agent)coatings are available. Each particular formulation has its own advantages and disadvantages and can be custom designed or matched to the particular application.

Epoxy coatings have excellent physical properties including bonding and adhesion. With proper mix viscosities, application techniques, and starting with a smooth surface, nearly flawless exterior surfaces can be produced.

The primary drawback to epoxy coatings is in exterior use. Due to the ultra-violet rays exposure, epoxy coatings tend to become chalky or discolor over long-term exposure. Use of various additives and using clear or translucent finishes while pigmenting the part itself tends to minimize the effect.

Application techniques are also quite varied. For coating of filament wound tubing with 0.005-.013 cm of material, the key to achieving a flawless surface is starting with a nearly flawless surface. Therefore, the grinding, machining, sanding, or material removal step is critical. For some applications, final sanding of material is done with 15 μm finishing paper to remove even minor scratches in the part.

Urethane Elastomers and Organic Coatings – The use of urethanes has become popular due to the excellent weatherability and durability of the coating. As a result of continuing advances in urethane chemistry, a number of options are available for a wide range of applications. Elastomeric urethane coatings and application techniques for filament wound composites are as follows:

- Conventional air spray gun

- Spraying using RIM type mixing and dispensing equipment

- Use of a dip tank

- Impregnation

Various organic coatings are becoming available which have excellent UV, scratch, and abrasion resistance. These coatings can be applied by spray, dip, spin, or gravure printing methods. Typical coating thickness is between 6-10 μm. These coatings are typically radiation cured by either UV or electron beam. These coatings are being used in many industries including composite sporting goods and recreational products.

Metal Films – For certain applications including antennae and telescope components, a metal film is required on the outside surface of the composite. In the case of placing a thin (0.005-0.013 cm thick) film over Gr/Ep composite, care needs to be taken to put a non-conducting material such as glass between the two to eliminate the possibility of a galvanic couple and degrading of the film onto the tube as it is rotated. Depending on the type of adhesive, the part can then be cured if necessary to achieve final physical requirements of the adhesive.

Other methods for obtaining a metallic surface on a composite structure include:

- Electroplating

- Vapor deposition

- Plasma coating

3.7.2 Coatings Removal

Paint coatings are used on military aircraft for many reasons. They act as protective layers for the substrates to which they are applied and also serve as a means for visually camouflaging aircraft. For composite materials, paints act as barriers protecting against environmental conditions, and ultraviolet radiation.

Paint removal is a necessary part of aircraft maintenance and is required in order to check for corrosion on metal and to repair composite structures. In many cases, paint is removed for purely cosmetic purposes or during a change in camouflage schemes. In the past, paint removal was primarily accomplished using chemical stripping agents. This process resulted in health hazards, waste disposal problems, and incomplete stripping jobs. Furthermore, many chemicals were found to degrade the organic matrix present in Gr/Ep composites and weaken the material.

Plastic media blasting (PMB) for use on Gr/Ep composites was investigated and compared with chemical and "hand" sanding removal to determine the potential for damage when removing paint from these materials. The research was conducted by the Air Force. The potential for streamlining stripping operations, (i.e.,using PMB on metal and composite airframe surfaces), for reducing health hazards, and for eliminating the potential for chemical strippers to damage composite materials are key reasons behind the interest in a more complete understanding of the effects of PMB on composite materials.[69] PMB was found to be a viable and safe paint removal method for use on Gr/Ep composites provided that special precautions are taken.

The following results from the Butkus, Behme, Jr., Meuer, study[69] show that PMB is a less damaging method of paint removal for Gr/Ep composites than is"hand" sanding. The overall effect of the PMB paint removal methods on the mechanical and physical properties of Gr/Ep composites is negligible if carried out using the"primer as a flag" criteria.

Using the Primer as a Flag – The most important aspect of the approach taken during the PMB stripping tests was the use of the epoxy primer coat as a stopping point. Unless otherwise noted, the PMB operator used the epoxy primer coat as a visual cue to aim the blast nozzle away from an area which had been stripped and on to another area which required stripping. This resulted in a mottled appearance of the remaining primer. However, it also minimized the amount of time which the PMB blast was aimed at a single area of a surface and minimized the possibility of degrading the mechanical properties of the composite by preventing damage to the surface of the composite. This technique was described as " using the primer as a flag." No internal damage to the Gr/Ep test panels was caused by either the "hand"sanding or the PMB methods. However, considerable surface erosion and surface fiber damage was caused by the sanding methods investigated and can be caused by PMB if the "primer as a flag" criteria is not used. Surface damage does result in the loss of mechanical properties of stripped composite materials. Using the "primer as a flag" criteria will significantly reduce the potential for causing surface damage and subsequent degradation to the mechanical properties of composites during PMB stripping operations. No direct relationships were discovered between the amount of mechanical property degradation and specific PMB parameter combinations or between mechanical property degradation and specific PMB media types used in the stripping operations. The results and conclusions concerning the effect of "hand" sanding and PMB on Gr/Ep composites should be applicable to other types of organic-matrix composites.[69]

REFERENCES – CHAPTER III

1. J. Wadsworth, T.G. Nieh, Superplasticity in ceramics and ceramic composites, intermetallics, and metal matrix composites; Superplastic Forming and Bonding of Metallic Alloys Conference, SME-EM-92-193, 19 p, June 1992.

2. R.C. Harper, Thermoforming of thermoplastic matrix composites, *SAMPE Journal,* **28** (3) 9 (1992).

3. T.E. Chilva, F.S. Deans, *Proceedings, Advanced Composites-The Latest Developments,* ASM Int., p 249, 1986.

4. E. Galli, *Plastics Machinery and Equipment,* **17** (3) 27 (1988).

5. R.C. Harper, J.H. Pugh, Thermoforming of thermoplastic matrix composites, in *Int. Encyclopedia of Composites,* vol 5, S.M. Lee, ed., VCH Publishers Inc., New York, NY, 1991.

6. R.K. Okine, D.H. Edison, N.K. Little, *SAMPE Int. Sym.,* **32,** 1,418, (1987).

7. P.J. Mallon, C. O'Bradaigh, R.B. Pipes, Thermoforming of fiber reinforced thermoplastic matrix composites, Univ. of Delaware CCM Report 88-11, 1988.

8. R.K. Okine, *Int. SAMPE Tech. Conf.,* **20,** 149 (1988), *SAMPE Journal,* **25** (3) 10 (1989).

9. A.J. Smiley, R.B. Pipes, Diaphragm forming of carbon fiber reinforced thermoplastic composites materials, Univ. of Delaware CCM Report 88-11, 1988.

10. C. O'Bradaigh, P.J. Mallon, Polymeric diaphragm forming of continuous fiber reinforced thermoplastics, Univ. of Delaware CCM Report 88-08, 1988.

11. P.J. Mallon, C. O'Bradaigh, R.B. Pipes, *Composites,* **20** (1) 48 (1989).

12. P.J. Mallon, C.O'Bradaigh, *Composites,* **19** (1) 37 (1988).

13. R.K. Okine, Int. *SAMPE Tech. Conf.,* **20,** 148 (1988).

14. R.K. Okine, *SAMPE Journal,* **25** (3) 9 (1989).

15. S. Witzler, *Advanced Composites, p 50,* Sept/Oct, 1988.

16. P.-Y. Jar, P. Davies, W. Cantwell, et.al, Manufacturing engineering components with carbon fibre reinforced PEEK, *Fourth European Conf. on Comp.Materials, ECCM-4,* Stuttgart, F.R.G., pp 819-21, Sept. 1990.

17. S.G. Pantelakis, D.T. Tsahalis, T.B. Kermanidis, et.al, Experimental investigation of the superplastic forming technique using continuous carbon fiber reinforced PEEK, ibid, pp 79-85.

18. E.T. Baker, T.L. Gesell, Autoclave molding of Avimid K$^{®}$ composites, *35th Int. SAMPE Sym. and Exhib.,* ed. by G. Janicki, V. Bailey, H. Schjelderup, **35,** 979, (1990).

19. G. Forsberg, S. Koch, Design and manufacture of thermoplastic F-16 main landing gear doors, *34th SAMPE Int. Sym. and Exhib.,* Reno, NV, pp 9-19, May 1989.

20. *Plastics Technology,* p 13, May 1992.

21. R.H. Stone, et.al, Design and manufacture of advanced thermoplastic structures, Final Rept. 9-15-87 to 12-20-91, WL-TR-91-8058, 416 p.

22. R.L. Ramkumar, Design and manufacture of advanced thermoplastic structures (DMATS), PR 11 April-June 1990, F33615-87C5242, 141 p, June 1990.

23. B.R. Bonazza, Ryton® PPS conductive composites for EMI shielding applications, *34th SAMPE Int. Sym. and Exhib.*, Reno, NV, pp 20-34, May 1989.

24. M.R. Monaghan, C.M. O'Bradaigh, P.J. Mallon, et.al, The effect of diaphragm stiffness on the quality of diaphragm formed thermoplastic composite components, *35th SAMPE Int. Sym. and Exhib.*, pp 810-24, April 1990.

25. Process produces practically perfect preforms, *Mach. Des.*, p 18, March 1993.

26. Therm-x(SM) Process – A revolutionary new curing system for advanced composites, United Technologies Chemical Systems Publication CSD-V-45058, May 1988.

27. R.B. Kromrey, Proceedings, SME Tooling for Composites '88 Conference, pp. EM88-219-1 to -11, May, 1988.

28. R.B. Kromrey, in Reference 27, pp. EM88-219-3 to -4.

29. T.P. Kueterman, Advanced Composites - Conference Proceedings, ASM Engineering Society of Detroit, SAMPE, SPE, SPI, pp 147-53, 1985.

30. L.M.J. Robroek, Introduction to the rubber forming of thermoplastic composites, NTIS Alert, May 1993, Technische Univ. Delft, Netherlands, p 57, Feb. 1992, LR-683, ETN-92-92877.

31. J.J. Cattanach, F.N. Cogswell, Dev. Reinf. Plast. 5, Process and Fabr.,Elsevier Applied Science Publ., London and New York, pp 20-25, 1986.

32. A.B. Strong, P. Hauwiller, *SAMPE Int. Sym.*, **34,** 43 (1989).

33. J.R. Krone, J.H. Walker, Thermoforming woven fabric reinforced polyphenylenesulfide composites, *CoGSME Composites in Manufacturing 5 Conf.*, pp 112-24, Jan. 1986.

34. Y. Yamaguchi, Y. Sakatani, M. Yoshida, Fabrication studies on carbon/thermoplastic - matrix composites, ASM Int. and ESD, *4th Ann. Conf. on Advanced Composites,* Dearborn, MI, pp 415-24, Sept. 1988.

35. Technology for producing carbon fiber-reinforced composite pipes, *New Technology Japan,* JETRO, 92-01-005-07, **19** (10) 34 (1992).

36. Fibre-reinforced composite material for forming complicated shapes, *New Technology Japan,* JETRO, 91-09-001-01, **19** (6) 18 (1991).

37. A. Offringa, From thermosets to thermoplastics: a logical evolution, *SME Composites in Manufacturing '92Conf.,* SME-EM-92-113, 12 p, Jan 1992.

38. O.D. Sherby, J. Wadsworth, *Prog. Mater. Sci.,* **33,** 169 (1990).

39. T.G. Nieh, C.A. Henshall, J. Wadsworth, *Scripta Metall.,* **18,** 1,405 (1984).

40. M.-Y. Wu, O.D. Sherby, *Scripta Metall.,* **18,** 73 (1984).

41. G. Gonzalez, et.al, *Comp. Sci. Tech.,* **35,** 105 (1989).

42. M.W. Mahoney, A.K. Ghosh, AFWAL-TR-82-3051, 1982.

43. M.W. Mahoney, A.K. Ghosh, *Metall. Trans.,* **18A,** 653 (1987).

44. T. Imai, et.al, Metal and Ceramic Matrix Composites:Processing, Modeling & Mechanical Behavior, ed. R.B. Bhagat et.al, TMS, Warrendale, PA, pp 235-42, (1990).

45. T. Imai, M. Mabuchi, Y. Tozawa, et.al., *J. Mater. Sci. Lett.,* **9,** 255 (1990).

46. M.Mabuchi, T. Imai, *J. Mater. Sci. Lett.,* **9,** 763 (1990).

47. K. Higaschi, et.al, *Scripta Metall. Mater.*, **26** (2) 185 (1991).

48. S.M. Pickard, B. Derby, *Mater. Sci. Eng.*, **A135,** 213 (1992).

49. T.G. Nieh, J. Wadsworth, Superplasticity and super-plastic forming of aluminum metal-matrix composites, *JOM,* **144** (11) 46 (1992).

50. Superplastic Forming of Structural Alloys, ed. by N.E. Paton, C.H.Hamilton, The Metall. Soc. of AIME, Warrendale, PA, 1982.

51. Superplasticity and Superplastic Forming, ed. by C.H. Hamilton, N.E. Paton,The Metall. Soc. of AIME, Warrendale, PA, 1988.

52. R.Ravenhall, W.E. Koop, Metal matrix composite fan blade development, AIAA 90-2178, 10 p, July 1990.

53. S.B. Lasday, Production and properties of continuous silicon carbide reinforced titanium for high temperature applications, Industrial Heating, pp 19-23, Dec. 1990.

54. T.G. Nieh, J. Wadsworth, *Acta Metall. Mater.*, **38,** 1,121 (1990).

55. P. Gruffel, P. Carry, A. Mocellin, *Science of Ceramics-Vol 14,* ed. by D. Taylor, Insti. of Ceramics, Shelton, UK,1988, pp 587-92.

56. F. Wakai, S. Sakaguchi, Y. Matsuno, *Adv. Ceramic Mater.*, **1,** 259 (1986).

57. F. Wakai, Superplasticity in Metals, Ceramics, and Inter-metallics, *MRS Proceeding No. 196,* ed. by M.J. Mayo, M. Kobayashi, J. Wadsworth, Pittsburgh, PA, pp 349-58, 1990.

58. T.G. Nieh, J. Wadsworth, F. Wakai, Recent advances in superplastic ceramics and ceramic composites, *Int. Mater. Rev.*, **36** (4) 146 (1991).

59. S. Miller, et.al, Metallurgical design of novel metal matrix composites for aerospace applications, Brit. Petroleum Co. Ltd., p 39, 1989.

60. T. Imai, Y. Nishida, T. Tozawa, *Proceedings of 4th Japan- US Conf. on Composite Materials,* Wash. D.C., Technomic, p 109, June 1988.

61. D.M. Aylor, P.J. Moran, *J. Electrochem Soc.*, **132,** 1,277, (1985).

62. P.P. Trzaskoma, *Corrosion,* **42,** 609 (1986).

63. W.F. Czyrklis, *CORROSION/85,* paper no. 196, Houston, TX: NACE, 1985.

64. D.M. Aylor, R.J. Ferrara, R.M. Kain, MP 23, p 32, 1984.

65. F. Keller, M.S. Hunter, O.L. Robinson, *J. Electro-chem Soc.*, **100,** 411 (1983).

66. S. Lin, H. Greene, H. Shih, et.al, Corrosion protection of Al/SiC metal matrix composites by anodizing, *Corrosion,* **48** (1) 61 (1992).

67. P.P. Trzaskoma, E. McCafferty, *J. Electrochem Soc.*, **130,** 1,804 (1983).

68. H. Ackerman, et.al, *Metals Handbook,* 9th Ed., **13,** 396, (1987).

69. L.M. Butkus, A.K. Behme, Jr., G.D. Meuer, Plastic media blast (PMB) paint removal from composites, *35th Int. SAMPE Sym.,* Anaheim, CA, ed.by G. Janicki, V. Bailey, H. Schjeldrup, pp 1,385-97, April 1990.

BIBLIOGRAPHY- CHAPTER III

C.A. Henshall, J. Wadsworth, M.J. Reynolds, et.al, *Materials and Design,* **8** (6) 324 (1987).

H. Watanabe, K. Ohori, Y. Takeuchi, Trans. Iron and Steel Inst. of Japan, **27,** 730 (1987).

D. Shin, D.A. Selby, J. Belzunce, et.al, Research in Progress, Stanford Univ.,1989; O.D. Sherby, ISIJ Int., **29,** 698 (1989).

A. Salama, Ph.D. Dissertation, U.S. Naval Postgraduate School, Monterey, CA, 1987.

T.G. Nieh, P.S. Gilman, J. Wadsworth, *Scripta Metall.,* **19,** 1,375 (1985).

P.C. Panda, J.W. Wang, R. Raj, *J. Amer. Ceram. Soc.,* **71,** C-507 (1988).

T.G. Nieh, J. Wadsworth, MRS Int. Meeting on Advanced Materials Vol 7 (IMAM-7 Super- plasticity), ed. by M. Kaboyashi and F. Wakai, MRS, Pittsburgh, PA, pp 43-50, 1989.

T.R. Bieler, T.G. Nieh, J. Wadsworth, et.al, *Scripta Metall.,* **22,** 81 (1988).

J.K. Gregory, J.C. Gibeling, W.D. Nix, Metall. Trans., **16A,** 777 (1985).

L.A. Xue, X. Wu, I.W. Chen, *J. Amer. Ceram. Soc.,* **74,** (1993).

W.J. Kim, G. Frommeyer, O.A. Ruano, *Scripta Metall.,* **23,** 1,515 (1989).

C.K Yoon, I.W. Chen:, *J. Amer. Ceram. Soc.,* **73,** 1,555 (1990).

T.G. Nieh, W.C. Oliver, *Scripta Metall.,* **23,** 851 (1989).

A.K. Mukherjee, T.R. Bieler, A.H. Chokshi, Proc. the Tenth Riso Sym. on Materials Archi- tecture, Riso National Laboratory, Sept. 1989, Denmark.

M.S. Paterson, in Superplasticity in Metals. Ceramics. and Intermetallics, MRS Proc. No. 196, ed. by M.J. Mayo, M. Kaboyashi, J. Wadsworth, MRS, Pittsburgh,PA, pp 303-12, 1990.

R.L. Ramkumar, J. Ogonowski, D.Stobbe, et.al, Design and manufacture of advanced ther- moplastic structures (DMATS),Phase II, Final Report Jan. 1988 to May 1992, F33615-87-C- 5242, p 422, WL-TR-92-8066.

I.A. Akmoulin, M. Djahazi, N.D. Buravova, et.al, Superplastic forging procedures for manu- facture of ceramic yttria stabilized tetragonal zirconia products, *Mater. Sci. and Tech.,* **9** (1) 26 (1993).

K. Kouba, O. Bartos, J. Vlachopoulos, Computer simulation of thermoforming in complex shapes, *Polymer Engr. and Sci.,* **32** (10) 699 (1992).

Superplastic forging applied to ceramics, *R&D Magazine,* p 28, Feb. 1992.

CHAPTER IV – MACHINING COMPOSITES

4.0 INTRODUCTION

Machining composites is tricky. Their abrasiveness results in a high wear ratio on cutting tools. When in tension or compression, the parts do not break away easily. In fact, layered materials tend to delaminate around the edges, particularly when tools begin dulling, and graphite fibers bend if cutting is occurring perpendicular to the surface. In this case, the fibers do not shear off. With graphite, dust is a major irritant as well as a health hazard. It can get into the machine tool and cause maintenance problems. Because the fibers are conductive, every effort is expended not to cut graphite or carbon-fiber-reinforced materials on NC-controlled equipment. Coolants make a major mess and can be a source of problems, too. The combination of dust and coolant creates a heavy, sludge-like material that is difficult to clean up. Some materials tend to absorb coolant, which can cause delamination. For example, when a coolant is introduced while cutting Kevlar®, it tends to run down into the fibers and become entrapped. Care must be exercised to prevent freezing/cracking if this material is used in a flying aircraft.

4.1 CONVENTIONAL METALS TO ALLOYS VS COMPOSITES

Machining of fiber-reinforced composites differs significantly from machining of conventional metals and their alloys. In the former, the material's response depends on diverse fiber and matrix properties, fiber orientation, and relative volume of the matrix and the fibers. The tool continuously encounters alternate matrix and fiber materials, the response of which to machining can vary greatly. For example, in an Al-B composite, the tool encounters a soft aluminum matrix and hard boron fibers. Similarly, in a Gl-Ep composite, the tool encounters a low temperature soft epoxy matrix and brittle fibers. In the case of Kv-Ep, the fibers have to be preloaded in tension and then cut with a shearing action. It is this variation in the requirements of a cutting tool that makes composites difficult to machine.

In view of high tool wear and high costs of tooling with conventional machining, noncontact material removal processes offer an attractive alternative; they can also minimize dust and noise. In addition, extensive plastic deformation and consequent heat generation associated with conventional machining of fiber-reinforced composites, especially those with an epoxy matrix, can be minimized.

4.2 MACHINING METHODS

A wide variety of methods are currently being used, or have been proposed for machining of advanced composite materials. Some of these, which can be classified as abrasive, nonabrasive, and combined processes, are listed in Table IV-1. They are not unique to advanced composites and were developed initially for application to metals. Some of these methods such as grinding, honing, lapping, polishing, and ultrasonic machining are extensively used for ceramics. Turning and milling methods, although extensively used in metal machining, have only limited application to advanced composites.[1]

139

Table IIV-1. Machining Methods for Advanced Composites

Abrasive Methods	Non-Abrasive Methods	Combined Methods
Grinding	Electrical Discharge Machining	Electrochemical Grinding
Honing	Laser Beam Cutting	Thermally Assisted Turning
Lapping and Polishing	Electron Beam and Ion Beam Cutting	Mechanical-Electrical Discharge
Ultrasonic Machining	Friction Cutting and Microwave	Chemical-Electrical Discharge
Liquid Abrasive Jet Cutting	Cutting	
Single-Point Turning		

4.2.1 Abrasive Methods

4.2.1.1 Grinding

Many different modes of grinding can be identified.[2] However, they all have as a common element, the use of a circular wheel or tool that is rotated about its axis of symmetry with some part of the periphery sliding against the workpiece. The entire tool, a layer a few millimeters thick, or only a single layer (such as in plated tools), consists of bonded abrasive grains. The tool may range in size from a tiny submillimeter abrasive coated drill to massive grinding wheels more than a meter in diameter. A variety of different abrasives are used. For application to advanced composites, only the superabrasives, diamond and CBN, are used with diamond being the primary choice.

4.2.1.2 Honing

Honing, like grinding, also uses fixed abrasives; and, for application to advanced composites, diamond is the abrasive of choice. The main difference between honing and grinding is the much lower surface speeds employed in honing. Honing, in general, is not used to remove large amounts of material. Its primary application is in obtaining a desired surface finish and in correcting errors in dimensional tolerance left by other rapid machining methods.

4.2.1.3 Lapping and Polishing

Similar to honing, lapping is largely a finishing process employed on parts that have already been machined to near final dimensions. Lapping differs from honing in that it is a loose or free abrasive process. Lapping is conducted by pressing the workpiece against a rigid surface, often cast iron, covered with a slurry of abrasive particles. Some of these particles are embedded into the lapping tool surface and produce a cutting action on the workpiece. Other particles that roll between the two surfaces, are also involved in the removal process. A distinction that is sometimes, but not always, made between lapping and polishing is in the use of a soft and flexible surface for polishing and a harder and rigid surface for lapping.

4.2.1.4 Ultrasonic Machining (USM)

USM encompasses rotary USM and ultrasonic impact machining. In both methods, the tool is vibrated along its axis normal to the surface of the workpiece at a frequency of 20 kHz typically. Otherwise, the two methods differ substantially. Rotary USM utilizes diamond coated grinding tools, most often core drills, in a drilling or vertical milling configuration. It is distinguished from ordinary milling and drilling operations only in that the rotating tool is vibrated with an amplitude of 0.025 to 0.05 mm while being held against the workpiece. The vibration serves to reduce friction, assist in the access of cutting fluid, and facilitate the removal of swarf. The net result is an increase in machining rate. The tools are generally limited to a diameter no

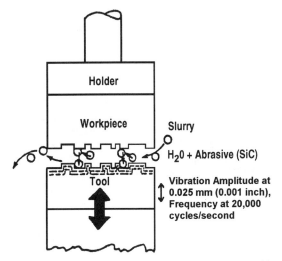

Holder

Workpiece   **Slurry**

H₂0 + Abrasive (SiC)

Tool

Vibration Amplitude at 0.025 mm (0.001 inch), Frequency at 20,000 cycles/second

Figure IV-1. Illustration of the ultrasonic impact machining, adapted from Rhoades.[2]

greater than 50 mm. In addition to core and solid drills, rotary USM has also employed small grinding and thread cutting wheels.

In ultrasonic impact machining, the tool itself does not incorporate abrasive particles and does not contact the workpiece. In this method, an abrasive containing slurry is circulated between the vibrating tool and the workpiece, Figure IV-1. The vibratory motion of the tool is imparted to the fluid and abrasive particles. Impact of the particles with the workpiece results in indentation and fracture with the consequent removal of material. The rate of removal is extremely sensitive to the gap between the workpiece and the vibrating tool.

Unlike thermal and chemical processes, USM is a mechanical material removal process applicable to both conductive and nonconductive materials and particularly suited to the machining of brittle materials such as graphite, glass, carbide, ceramics, and composites.

Because the abrasives contained in the slurry do the machining, they must be selected carefully based on the workpiece material and surface quality required. Typically, larger abrasive grain sizes yield higher cutting rates but, also, a higher surface roughness. Abrasive types commonly used in USM include Al_2O_3, SiC, B_4C, and diamond. CMCs may be machined with SiC or B_4C, but MMCs must be machined with B_4C or diamond. Since the abrasive cost increases dramatically from SiC to B_4C to diamond, the least expensive applicable abrasive is used.

The abrasive grain diameter affects machining rate, overcut, and surface roughness. When high surface quality is not required, 180 to 280 mesh abrasives provide high cutting rates. For finer finishes, 320 to 600 mesh abrasive is recommended and for super finishing, 1,000 mesh or finer. For cuts requiring large amounts of material removal and a good surface finish, it is possible to rough with a large grain abrasive and finish with a finer abrasive.

Abrasive wear is an important factor in material removal and overcut. Abrasive grains begin to wear as soon as machining begins. For example, machining CMCs with SiC 180 mesh abrasive, material removal rates with fresh abrasive require a feedrate of 15 mm/hr. However, after 10 hours, the feedrate drops to 5 mm/hr. MMCs tend to wear the abrasive at a higher rate. Therefore, when machining MMCs, B_4C should be used because it wears at a much slower rate than SiC.

4.2.1.5 Liquid Abrasive Jet Cutting

Jordan[3] has reviewed recent developments and applications of liquid jet cutting. Liquid jet systems are usually applied to cutting tasks, rather than to shaping or surface finishing.

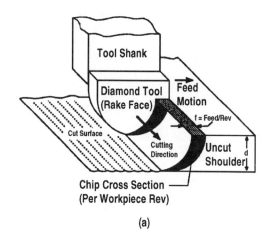

(a)

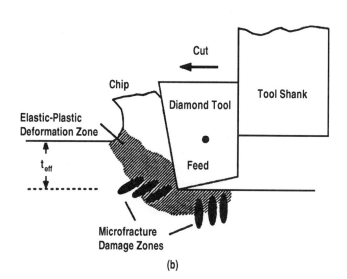

(b)

4.2.1.6
Single-Point
Turning

During the last decade, considerable attention has been focused on the single-point diamond turning process, which utilizes a polished single crystal diamond tool and requires a specially designed machine tool to allow very close control of the depth of cut and machine tool vibrations.[4-8] Single-point diamond turning was developed as a technique for producing highly smooth and precise surface contours for specialized applications. Since the process is inherently adaptable to computer control for producing curved surfaces in an automated manufacturing environment, it will likely see expanded use in the future.

Figure IV-2. a) Schematic of cutting geometry in single-point diamond turning, b) details of damage formed during cutting in schematic cross section.[10]

The machining geometry is shown schematically in Figure IV-2. In general, elastic-plastic deformation and microfracture (especially in brittle materials) occur in the workpiece ahead of the tool edge. If the conditions are established properly, most of the microfracture damage that occurs at a depth up to t_{eff} is subsequently removed as part of the chip. Single-point diamond turning has been successfully applied to a limited number of materials in producing very close tolerance surfaces of high smoothness and low levels of damage (Bifano et.al.,[9] and Blake and Scattergood[10].)

142

4.2.2 Nonabrasive Methods

4.2.2.1 Electrical Discharge Machining (EDM)

Many applications of EDM to advanced composites have been reported. EDM requires the workpieces to have an electrical resistivity less than 100 Ω-cm. Thus, EDM cannot he used for machining of glasses and some ceramic composites. Petrofes and Gadalla[11] have also described the application of EDM to composite ceramics, e.g., Si_3N_4-TiN, SiC-TiB_2 and cermet materials. Much effort appears to he underway in Japan exploring EDM for advanced composites. It is not clear, however, whether EDM can he regarded as a low damage machining method since it typically leaves a surface layer of melted or heat affected material containing a high level of residual stress and numerous microcracks.

4.2.2.2 Laser Beam Cutting (LBC)

Focused laser beams have been used to cut a large variety of composite materials. Certain cutting geometries may lend themselves well to such a technique, particularly when segments of workpiece material need to be removed. Recently, Taya and Ramulu[12] have reported laser machining of three-dimensional forms on CMCs.

4.2.2.3 Electron Beam and Ion Beam Cutting

Applications of both ion beams and electron beams in machining are only suitable for cutting and drilling thin sheets of composites.

4.2.2.4 Friction Cutting and Microwave Cutting

While these methods are suitable for cutting and slicing operations, they are not generally suitable for achieving a contoured surface with a high tolerance.

4.2.3 Combined Methods

4.2.3.1 Electrochemical Grinding (ECG)

The process consists of an electrolyte (water with dissolved sodium chloride, for example) is sprayed onto the grinding wheel and the surface being machined. A potential is applied between the metal-bond grinding wheel (cathode) and the workpiece (anode). Electrochemical action forms a reacted surface layer on the workpiece, which is removed by the grinding wheel. An advantage of this method lies in the anticipated low level of residual damage in the workpiece.

4.2.3.2 Thermally Assisted Turning

The technique involves plasma torch heating the workpiece material to temperatures as high as 1,000°C in front of a polycrystalline diamond compact (PDC) or CBN cutting tool. Where machinability is improved, it results from a transition to a more plastic type of deformation and removal processes at the elevated temperature.

4.2.3.3 Mechanical-Electrical Discharge

The combination of electrical discharge and ultrasonic machining methods has been reported[13] in experiments to machine TiB_2 with a metal bonded diamond tool. Applicability of this method to other materials of sufficient conductivity for EDM needs to be explored.

4.2.3.4 Chemical-Electrical Discharge

A combination of wire electrical discharge machining with electrochemical reaction in an electrolyte has been reported by Tsuchiya et.al.[14] The method permits cutting of surface contours in an effective manner without a direct contact between the tool and the workpiece.

4.3 CONVENTIONAL COMPOSITE MACHINING

Conventional machining of fiber-reinforced composites is difficult due to diverse fiber and matrix properties, fiber orientation, inhomogeneous nature of the material, and the presence of high-volume fraction (volume of fiber over total volume) of hard abrasive fibers in the matrix. Glass-, graphite-, and boron-reinforced composites (even polymer-based) are difficult to machine because of rapid tool wear. Since cemented carbide tools wear rapidly, diamond-impregnated tools are being chosen. Several advances have been made in the development of tool materials, including polycrystalline diamond (PCD) tools, diamond-plated tools, and diamond-impregnated tools in various forms, such as core drills, milling cutters, drills, and grinding wheels. Although cemented carbide and diamond are the most commonly used materials, high-speed steel (HSS) tools are used in some cases, but at the expense of rapid tool wear.

4.3.1 PCD

Traditionally, manufacturers have used carbide tools to machine composites, but tool life is disappointing. Carbide may start off turning or drilling cleanly, but it dulls fast and hairs and fibers are observed. As a result, manufacturers have chosen PCD. PCD lasts about 100 times longer than carbide, and it is three to five times harder. Using PCD, a smooth edge results and the heat dissipates, both in the chip and through the shank, leaving the tip cool. In spite of PCD's primary disadvantage (cost), in the proper applications, (where a good amount of machining is required), one PCD tool is cheaper than many carbide cutters.

When using PCD tools, it is preferable to clamp rather than braze the insert on the tool holder to avoid softening the braze material. The final choice of the tool material depends on the economics of the machining operation and the part requirements. Appropriate dust-collection and extraction systems should be in place when machining glass fiber-reinforced composites. Operator safety should be a prime consideration; use of masks, gloves, and aprons or lab coats is required to minimize the safety risks associated with loose glass fibers and dust. Cutting speeds range from 100 to 150 surface feet per minute for carbides and 500 to 1,500 surface feet per minute for PCDs.

In a recent study of machining graphite composites, Kohkonen[15] compared carbide and PCD end mills in terms of tool flank edge wear, work surface finish, and delamination. In the tests, fibers were laid out unidirectionally and cutting was done both parallel and perpendicular to them. The carbide wear ratio was much higher than that of the PCD tools. While surface finish was initially acceptable for both types of cutters, as the carbide dulled, finishes deteriorated rapidly. The PCD end mills continued producing excellent surface finishes as flank wear increased. (See MILLING below).

Another factor studied was tool geometry. Until recently, tool fabricators were limited to straight-flute end mills since it was thought PCD end mills would find greater application in cutting MMC and Gr/Ep materials if PCD could be configured as conventional helical end mills. Helical geometries reduce cutting forces, improve chip removal and surface finish, and allow higher cutting speeds than straight-flute geometries. The program also compared straight and helical-flute PCD end mills with tungsten carbide two-flute mills in cutting Gr/Ep. The results showed that the PCD straight-flute tools failed after cutting ~ 1.2 m of composite – the diamond broke away from both flutes. By contrast, the helical PCD end mill showed substantially less wear than tungsten carbide tools and produced superior surface finishes under all test cutting conditions. Surface finish values were consistent for each end mill type, apparently independent of fiber direction. Helical PCD tools manufactured to conventional end mill geometries clearly perform well under these conditions. More testing is required to establish actual tool life for a given set of machining conditions.

MMCs offer special machining challenges. Unlike resin-based composites, they have discontinuous reinforcements – particles or whiskers – and are more three dimensional. Additionally, they usually require considerable machining and employ traditional metalworking operations. Machinability trials show that the most cost-effective tool material by far for MMCs is PCD. High-speed steel dulls in seconds, and conventional and coated carbides last only a few minutes. Unfortunately, PCD tooling is not yet available for taps or very-small-diameter drills and reamers.

Abrasion is the primary mode of tool wear. Compared with a typical aluminum feed rate of 0.13 mm per tooth, heavy roughening feeds not only increase material removal rates but also remove up to five times as much composite per unit of tool wear. Tool life enhancement increases when cutting speed drops to 609.6 m/min or less, due probably to lower temperatures and micro-impact stresses. Grain size makes a significant difference in tool life: coarser grain diamond can increase it five fold. The theory currently holds that, when the diamond's grain boundary area is reduced, a site is established where aluminum tends to weld onto the diamond.

A study on the effect of coolant use on machining MMCs[15] involved machining automotive brake rotors made from Duralcan MMC aluminum with 20% volume SiC. The speed rate was 457 m/min, feed was 0.190 mm/rev, and depth of cut was 0.25 mm. A significant advantage was found when machining this material dry. One theory is that, when coolant is used, the SiC actually flushes into the cutting area as a slurry and adversely affects the cutting process.

The main problem in machining MMCs such as Duralcan is education. MMCs do not machine like aluminum. HSS and tungsten carbide tools do not last. Shops used to machining high-silicon aluminums do quite well because they are set up for diamond. Those that machine nothing but cast iron also do well because they are used to running at the right speed and feed routines. The result is conversion to PCD tools.

4.3.2 Diamond-Plated and Impregnated Tools

Boron is hard and abrasive while titanium is chemically more reactive with most tool materials. This dissimilar combination makes the task of machining a boron-titanium composite challenging. A variety of machining operations are performed on this material including drilling, reaming, countersinking, routing, milling, and sawing using diamond-impregnated or diamond-plated tools. Recommended drilling conditions are a cutting speed between 61-183 m/min and a feed rate between 0.012 and 0.126 mm/rev. To reduce heat buildup in both the tool and the part, cutting fluids are recommended, as well as protecting the machinery from the abrasive boron dust to prevent wear of machine elements. Since aramid-reinforced plastic composite is an inherently tough material, cutting tools should be sharp and clean. Tools should be cleaned frequently to remove buildup of partially cured resins, which can cause loss in cutting action of the tool. The requirements of tools for machining aramid-reinforced plastics are different from those for machining glass or carbon fibers. In many respects, aramid-reinforced polymer resembles wood; its structure is characterized by the presence of highly oriented fibrous material embedded in a matrix. The best results are obtained when machining occurs in a manner by which the fibers are preloaded in tension and then cut with a shearing action. Special cutters have been developed to handle this problem. Figure IV-3a illustrates a four-fluted spiral rotary carbide milling cutter with a unidirectional helix throughout much of its length and a reverse-directional helix adjacent to the cutting edge. Chip breakers are arranged along the lands with

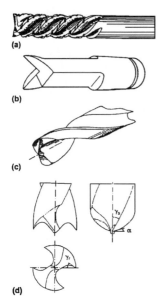

(a)

(b)

(c)

(d)

Figure IV-3. Special cutter designs for machining aramid-fiber-reinforced polymer composites: a) four-fluted spiral rotary carbide milling cutter; b-d) other designs and cutters for drilling.

the notches at alternate lands aligned and the notches of the other lands intermediately aligned. These cutters with angled chip breakers were designed to cleanly cut the aramid-fiber-reinforced plastic composites. The cutters operate at production speeds with minimal overheating. Figures IV-3b through IV-3d show some of the other designs of cutters specifically made for drilling aramid-fiber-reinforced resin composites. DuPont has developed guidelines for machining Kevlar® aramid composites and can be consulted for details.[16]

4.4 PMC MACHINING

Quality machining of composite materials must meet certain basic criteria, i.e., no fraying or delamination of cured composite edges. The following information applies to machining generic PMCs/RMCs (Gr/Ep, Kv/Ep, Gr/Gl/Ep, and B/Ep).

Standard machining equipment can be used with modifications. In general, spindle speeds and feeds depend upon the thickness of the laminate being machined and the type of cutting method. The cutting tools required to perform the machining operations include countersink, cutoff wheels, router bits, bandsaw blades, high-speed steel drills, and reamers. It is particularly important to keep tools sharp in order to provide quality cuts and minimize the possibility of delamination. Proper backup support of the work is required to eliminate delamination, along with cooling methods to control resin build-up on the tool caused by excessive frictional heat. Water and water-soluble coolants in a mist or flood application are satisfactory, as well as hydrocarbon fluids for B/Ep. If water-soluble coolants are used, the machined part must be rinsed thoroughly with water to remove excess coolant. When no liquid coolant is used with diamond cutting tools, the dust from machining operations must be collected in a vacuum system and the operator must wear a respirator. Some general rules are:

• Use diamond tools on all machining operations which require cutting B/Ep.

• Use solid backup for the workpiece, e.g., Masonite, to reduce chipping and delamination. Aluminum is a good backup material for drilling. It is advisable to have the backup cover the tool exit surface of the workpiece in the entire area to be cut. The backup must be held in intimate contact with the workpiece by clamping or otherwise.

• Extreme caution should be exercised to avoid overheating the resin when epoxy PMCs are being machined. Overheating destroys the part locally, and damage can be repaired only by removing the overheated area and filling the void with another material. Overheating is evidence by the machined surface turning brownish-black.

In view of the high hardness and abrasiveness of glass fibers in glass-fiber-reinforced plastic (GFRP) composites, cemented carbide and preferably single-crystal and PCD diamond tools are recommended for machining these materials. While HSS tools can be used to machine GFRPs, they wear rapidly and cutting speeds should be kept low enough to avoid overheating the tool. The machining of GFRPs can be considered similar to drilling a hole in a resin-bonded grinding wheel except that the abrasive in this case is glass. A dull tool can dissipate considerable heat into the workpiece and damage the resin-based composites. Although alumina-based tools can be used because of their hardness, the possibility of chemical reaction between alumina and glass should not be overlooked. PCD tools are preferred, particularly in the case of GFRP components with a high glass fiber content (about 60%), which have to be machined to tight tolerances and with good surface finish. Rigid machine tools are preferred when machining GFRPs with PCD in order to take advantage of PCD's superior cutting capability. Although carbon-fiber-reinforced plastic (CFRP) composites are generally fabricated to near-net-shape, additional machining operations such as drilling holes and trimming edges are needed. High tool wear and delamination of the composites are some of the concerns in machining. Numerous studies have been conducted by Koplev, et.al.[17] and Hasegawa, et.al.[18] evaluating the machining of CFRPs and GFRPs.

4.4.1 Laser

Laser machining (LBM) is based on the interaction of the work material with an intense highly directional and coherent monochromatic beam of light. Material is removed predominantly by melting and/or vaporization. In the case of resin matrix material, it is also removed by chemical degradation.

The physical processes involved in LBM are basically thermal in origin. When a laser beam impinges on a work material, several effects occur including reflection, absorption, and conduction of the laser beam.

Most metals absorb more readily at shorter wavelengths, hence, less power is required to machine these materials at these wavelengths. Therefore, Nd: YAG, with a wavelength of 1.06 micrometers, would be more suitable for machining MMCs than would CO_2. In contrast, some of the organic resins and other compounds have a higher percentage of absorption at higher wavelengths close to that of a CO_2 laser (10.6 micrometers), so that CO_2 would be more appropriate for machining such materials (e.g., aramid-resin composites). As melting begins or the material begins to interact with its atmosphere, the percentage of absorption may change as the process continues. For example, in drilling, the percentage of absorption in part of the hole drilled could be different from its initial value at the surface.

4.4.1.1 Types of Lasers

Several types of lasers are used for machining. Chief among them are gas (CO_2 and excimer) lasers and solid-state (ND:YAG and ND:glass) lasers. These lasers can be operated either in a CW mode or pulsed mode for machining. Benefits of LBM include minimum material waste (kerf width), minimum setup time, no tools (and thus no tool wear or replacement), smooth edge cuts, and low total heat input. As a result, there is low overall distortion or damage of the part. A possible limitation of LBM is the heat affected zone (HAZ), where high temperatures imparted to the workpiece at or near the last cut can cause metallurgical changes. This can reduce the fatigue properties of the work material and the quality of holes in deep hole drilling. Generally, the longer the wavelength of the laser beam, the higher the reflectivity of the metal workpieces. Similarly, the higher the thermal conductivity (thermal diffusivity), the higher the reflectivity. Nonmetals (e.g., plastics, glass, or ceramics) with low thermal conductivities are ideal candidates for CO_2 LBM (reflectivity is inversely proportional to the thermal conductivity). The amount of reflectivity can, however, be substantially reduced by modifying the surface conditions on the work materials.

147

The properties of aramid fibers are somewhat similar to those of resin with minor differences in magnitude. In contrast, the properties of carbon and glass fibers are different from those of the resin matrix material. As a result, large differences exist between the thermal properties of the resin matrix and glass or graphite fibers, while the difference is negligible with aramid. The laser power requirements, therefore, will depend on the fibers used and their volume fraction and not that of the matrix. However, too high a laser power may vaporize or chemically degrade the polymer matrix.

4.4.2 PMC Machining Behavior

The advanced composites CFRP, Kevlar® fiber reinforced plastics (KFRP), and GFRP are very similar in their forming process. But each material differs in its physical and mechanical properties and, hence, differs in machining behavior. The machining behavior of each composite varies widely depending on the fiber-matrix ratio, fiber orientation, and the form of the reinforcing fibers. A study by Santhanakrishnan and his associates[19] conducted in face turning trials on GFRP, CFRP and KFRP cylindrical tubes using HSS and sintered carbide tools. Apart from recording the surface roughness profiles of the machined surfaces, the morphology of the machined surfaces and the worn out tool portions were examined.

The results and conclusions found include:

- CFRP surface has the best finish value (R_a) in comparison with GFRP and KFRP surfaces.
- Due to the crushing and sharp fracturing of carbon fibers, the CFRP surface exhibits better surface texture than GFRP. The KFRP surface is poor due to the fuzziness caused by tough Kv fibers.
- HSS tools with sharp cutting edges appear to be well suited for machining KFRP, while the tool wear is extremely high for GFRP and CFRP composites.
- Sintered carbide tools used to machine GFRP exhibit strap wear on flank and secondary sides.
- Crater wear is predominant on sintered carbides while machined CFRP associated with uniform abrasion marks are seen along the primary and secondary edges.
- During HFRP machining, the sintered carbide tools fail by chip notching.

In another recently completed study, Textron Lycoming[20] studied the use of PMR-15 composite material for use as the accessory gearbox cover of a future helicopter. Comparisons were made to establish the composite covers vs. aluminum as a baseline. A comparison of cutting speeds, in sfm, between aluminum and PMR-15 is shown in Figure IV-4 for a variety of tool materials. The Figure is arranged so that choice of tool materials will reduce machining time as one moves from right to left. It is evident from the Figure that aluminum will machine more readily than PMR-15.

4.4.3 PMC Turning

PCD tooling has been used for diamond turning, boring, and milling tools. Some of the tools have brazed tips. Others accommodate replaceable inserts. For speeds and feeds, it has been recommended to begin with carbide guidelines and then, with experience, increase cutting rates to more optimum levels. A general range of cutting parameters for GFRP and CFRP materials can be found in Table IV-2.[21]

Kim and his associates[22-23] recently reported on a series of quantitative results for the cutting of CFRP materials and the turning of CFRP materials in order to obtain the Taylor tool-wear

Table IIV-2 Machining Parameters for PCD Tooling-Turning, Boring, Facing and Cutoff Operations[21]

Workpiece Material	Nose Radius, in.	Side Relief Angle, deg.	Positive Back Rake Angle, deg.	Speed, ft^2/min	Depth of Cut, in.	Feed Rate, in./rad
Glass-Fiber-Plastic Composites	0.030 - 0.090	5 - 20	0 - 5	400 - 3600	0.001 and up	0.001 - 0.010
Carbon-Plastic Composites	0.020 - 0.040	5 - 20	0 -5	500 - 2000	0.010 and up	0.005 - 0.015
1 in = 25.4 mm						

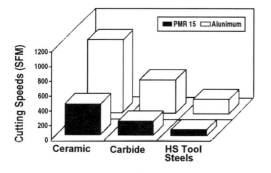

Cutting Tools

Figure IV-4. Cutting tool speeds of PMR15 and aluminum.[20]

constants and surface roughness as well as chip-formation mechanisms and the optimum cutting speed and feed in the turning of CFRP materials.

Conclusions from the study are:

- The flank wear of the tool increases with an increase in the fiber winding angle. Also, the dependency of the tool wear on the cutting speed was greater when the fiber winding angle was greater.

- The Taylor tool-wear constants n and C of the tungsten carbide tool (K10) were in the range of 0.45-1.12 and 55.4-241.5, respectively. The tungsten carbide tools in the machining of CFRP materials were uncoated. Multi-coated inserts and PCD tools might be more efficient in the machining of such materials.[23]

- The surface roughness was more dependent on the feed and the fiber-winding angle than on the cutting speed.

- The preferred cutting speed and the feed is 20-40 m/min and 0.2 mm/rev, respectively, from the point of view of the flank wear of the tungsten carbide tool (K10) and the surface roughness.

4.4.4 PMC Routing, Trimming, and Beveling

These operations are essentially equivalent, involving the use of hand routers, mechanical Marwin machine routers, and Roto-Recipro machines. Diamond-cut carbide and four-fluted milling cutters have been used to machine Gr/Ep and Gl/ Ep laminates, while carbide opposed-helical router bits have been used to machine Kv/Ep and Kv/Gr/Ep hybrids. Diamond-coated router bits have been used with the Roto-Recipro machine to rout and trim B/Ep and B/Gr/Ep hybrids. Speeds for these operations range from 3,600 to 45,000 r/mm.[24]

Routing – In manual routing, a higher-torque but lower-speed router (Buckeye) works best. Diamond-cut carbide bits attain higher feed rates at less operator effort than a six-flute configuration. Using a coolant tends to extend tool life and increase the cutting force (probably due to sludge formation) with no effect on cut edge quality. Based on operator effort, the maximum diametral tool wear would be about 0.04 mm before a tool change would be required. An opposed-helix carbide cutter has successfully routed Kv/Ep cured laminates. The opposed-helix cutter is especially useful for severing laminates into sections, cutting slots and notches, and trimming honeycomb-sandwich panels. The portable version of the cutter has performed better than diamond-cut carbide cutters with Kv/Ep and Kv/Gr/Ep hybrid laminates. Although cut quality is equivalent for both types of cutters, only the opposed-helix. carbide cutter can trim 6.4 mm material. For routing and trimming B/Ep and B/Ep and Gr/Ep hybrids, diamond-plated router bits are used. Spray mist or flood cooling is required for all power-feed routing of B/Ep; spray mist or air-blast cooling is recommended for hand routing. Power feed is recommended in routing B/Ep thicker than 1.3 mm.

Trimming and Beveling – Manual trimming and beveling with a Buckeye router against a guide has been successfully used on Gr/Ep, Gl/Ep, Kv/Ep, and Gr/Gl/Ep hybrid. The router bits were diamond-cut carbide types and depths of cuts were 1.5 or 3.3 mm and 45° by 6.4 mm. Good cuts have been made on thicknesses up to 6.9 mm.

4.4.5 PMC EDM

EDM can make complex shapes with high precision. It is a slow process, but automation can bring the cost of manufacturing down. The prerequisite for EDM is that the work material be electrically conductive. PMCs are, therefore, not candidate materials for this method of machining. They can be made conductive by being impregnated with metallic fillers (Cu, Al, or Ag powder), but that may defeat the purpose of composites for high-strength and lightweight applications.

4.4.6 PMC Sawing and Sanding

Reference 24 lists several critical points which must be remembered in sanding composite laminates for trim or fit and for bonding. Sanding Kv/Ep laminates requires wet paper for cooling and to prevent buildup of waste between abrasive particles. The preferred grit sizes are 120 and 240.

Sawing of cured laminates has been successfully performed with band, circular, or saber saws. The composite must be clamped to eliminate vibration, which can cause delamination. Cutting edges should be checked frequently to maintain sharpness.

Bandsawing – Normally, a fine offset high-strength steel staggertooth blade is used (8 to 12 teeth per in) and a surface speed of 22.9 to 33 m/s with 30.5 m/s preferred. This cutting method uses the heel rather than the hook of the cutting-tooth blade for cleaner cuts. The best feed rates are 3 to 5 m/s, and, as a general rule, the height of three teeth should match laminate thickness. Bandsaw cutting of Kv/Ep is usually performed with a fine-tooth-blade, preferably using water as a coolant. The band should be run in reverse, so that the heel of the tooth enters the composite first. B/Ep and B/Gl/Ep hybrids are bandsaw cut with diamond coated blades, 40 to 60 grit, and speeds of 10.2 to 30.5 m/s. Additional tool life has been obtained using carbide-coated blades, which operate at speeds of 7 and 15 m/s and feed rates at 8.5 to 16.9 mm/s.

150

Circular Sawing – By using a metal slitting saw, Kv/Ep has been successfully cut with circular saws, but diamond blades are preferred for most composite materials. Cutting speed for composites normally varies from 10.2 to 51 m/s; 6.4 mm thick Gr/Ep is best cut with a carbide-tipped blade and at a speed between 10.2 and 20 m/s.

Saber Sawing – Saber sawing normally cuts the outermost fibers of Kv/Ep on both sides of the laminate toward the interior. The blade has five alternating teeth in opposed directions; blade speeds of 2,500 strokes per minute are recommended, but blade speed and feed rates may vary with material thickness.

4.4.7 Countersinking

Conventional countersinking tools from 1,750 to 6,000 r/min have been used successfully on Gr/Ep, Gl/Ep, Kv/Ep, and Gr/Gl/ Ep hybrids while butterfly countersinking tools modified with serrations have also been used. Carbide countersinks have shown good results with Gr/Ep and Gl/Ep with optimum combinations of speed and relief angles.[25]

In countersinking B/Ep and Gr-Ep hybrids, the feed rate has a pronounced effect on diamond tool life. Low feed rates are recommended. Although the amount of wear on the countersink itself is slight, the angular change approaches the maximum allowable countersink-angle tolerance. The pilot on both types of countersinks wears rapidly when boron fibers are cut. The pilot in the plated tool is replaceable; the pilot in the sintered tool can be refurbished by nickel plating. Although plated tools wear more rapidly than sintered tools, their lower cost makes plated countersinks more cost-effective.

An approach[26] to tooling for drilling and countersinking Gr/ Ep composites has been introduced in the aircraft industry. This advanced tooling system (ATS) has replaced the manual approach using WC Daggar drills, where each drill lasted for 40 holes. The ATS consists of the following components:

- Specially shaped PCD drills
- A PCD countersink insert
- A drill countersink tool holder
- A hydraulic operated depth sensor

PCD was chosen for the cutting tool material in this system because of its outstanding wear resistance. PCD consists of fine grains of diamond directly bonded to each other and to a tungsten carbide substrate. PCD is 100 times more wear-resistant than WC.

4.4.8 Milling

To produce a good surface finish and good sharp corners without breaking the corners off, the edges of the composite block should be machined, leaving a step on the edges to reduce pressure on the laminate during milling. In milling from one side of the composite block to the other, it is always recommended that the block be precut. Doing this will prevent potential breakage problems in the lamination.

In milling composite materials such as unidirectional or cross-directional carbon fibers, it is always a good practice to make sure of the direction of layers. In milling composite materials, good results can be achieved by either plunging into or machining across the direction of the fibers, as well as by machining parallel to the fibers, according to Murray.[27]

End mills of WC with at least four flutes should have either a 45° angle or a generous radius. The four-fluted end mill will reduce cutting pressure on the laminate and keep it cooler. A sharp-tipped end mill will dull very quickly while machining composite materials. Tools should always be in good condition. A dull tool will cause overheating of the epoxy-resin and will cause delamination, not only on the surface but also within the layers, which cannot be seen. A dull tool requires more pressure against the composite material while machining, thus causing overheating. In milling operations, using a climbing cut works exceptionally well. This method will keep the fibers from separating. This method creates a chopping cut whereas machining in the opposite direction would create a pulling cut.

The use of fluorocarbon coolant is recommended because of its cooling efficiency. The coolant is applied as a spray mist during machining; the distance between spray applicator and cutter is adjusted so that frost forms on the cutter.

In milling Kv/Ep, conventional fluted cutters are satisfactorily operated at speeds of 0.4 m/s and feed rates of 5 mm/s. In milling Gr/Ep, HSS end mills or carbide cutters can be used provided they are multifluted. Four-flute end mills are recommended for efficiency and to reduce the cutting forces to a point where there is less chance of delamination. All cutters should be sharp; dull ones cause delamination. HSS cutters should always be of the four-flute positive-rake type; carbide milling cutters should also be of a positive-rake type with chip loads of 0.10 to 0.15 mm per tooth. Radiused-end mills last longer than square-cornered ones. Plunge-cut milling is not recommended unless there is sufficient backup support to prevent delamination.

Milling a complex shaped hole in graphite composites, AS4/3501-6 tape 0.152 mm thick, laid into 456 mm x 610 mm x 6.4 mm thick sheets with 12.7 mm, two-flute, PCD end mills using straight and 30° helical end mills, yielded some guidelines on machining parameters for this type of work.[15] For all tools tested, cutting parallel to the fibers produced greater wear than cutting perpendicular. Lower feedrates meant more cutter revolutions to remove the same amount of material, so the amount of wear increased, indeed doubled in some cases. But lower feedrates did produce a better surface finish. The effect of spindle speed on wear was much less significant than that of feedrate, but in the same direction. There is some decrease in wear at high rpms; 11,460 rpm was the maximum speed tested. Cutting speed also showed little effect on the surface finish which the researchers K.E. Kohkonen, S. Anderson, and A.B. Strong found surprising. The tools with the lowest wear were the 30° helical PCD end mills, because both types of end mills dulled much more slowly than the carbide end mills. The PCD end mills produced a superior surface finish over longer lengths of cut. Shallower depths of cut (1.27 mm), as expected, produced better surface finishes than greater depths of cut (2.54 mm). Interestingly, the same amount of flank wear (0.127 mm) produced more edge delamination with the carbide end mills than with the PCD end mills. New ceramic matrices, whisker-reinforced with SiC single crystals are being applied to materials as cutters in end milling, routing and drilling tools, Figure IV-5.

CVD and PVD – The chemical and physical vapor deposition coatings of carbide inserts have supplanted the dominance of uncoated carbides.[15,28-29] Currently, approximately two-thirds of all inserts sold are coated. New advanced material inserts which include ceramic, cermet, superhard and composite grades are projected to invade the insert market in the 90s. More later in this book on cutters.

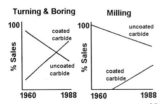

Figure IV-5. Application trends.[28]

Currently, Norton Diamond Film Co. has introduced CVD thin and thick-film diamond. CVD tools offer many advantages over carbide: lower friction, better heat removal, cleaner cuts, and they can run at high feeds and speeds because they tolerate higher temperatures. The advantages over PCD tools are less dramatic, but there are real behavior differences because PCD softens at a lower temperature and is subject to chemical reaction to some materials.

For the thin film, Si_3N_4 ceramic tools are normally coated with 5-15 μm of pure crystalline diamond with no porosity and no metal binder via a microwave plasma-assisted process. The coatings follow the tool contours. Also, diamond particles don't debond as they might with discrete particles in a matrix. Unfortunately, CVD is a moderately high-temperature process that doesn't work on WC. Most work has focused primarily on end mills and routers but now work has been initiated on drills and taps.[30] The thick film is produced by a proprietary DC arc jet plasma-assisted process and made into a wafer, 102 mm in diameter and 0.38-0.76 mm thick. This free-standing wafer is then laser-cut to desired shapes and brazed onto cutting tools. This process offers pure diamond at the cutting edge and good strength and toughness. It can also be used with WC tooling. The process has been applied to end mill cutters and routers. In a milling test on a multilayer CCC sandwich form, CVD thick-film-diamond coated end mills were compared with micrograin WC mills. The micrograin carbide tool showed 0.064 mm of wear after 9.85 m of travel and 41 cm^3 of material removed, and there was slight delamination. The thick-film diamond tool showed 0.127 mm of wear, with no delamination after 49.2 m of travel and 204.1 cm^3 of material removed. These results show that the diamond has an advantage of 1.5 x over the carbide before wear causes delamination. In effect, the diamond tool would show 0.38 mm of wear after 148 m of travel and 6,123 cm^3 of material removal.

4.4.9 Grinding

Kv/Ep composites can be ground or chamfered by strict adherence to specific conditions. A glazed work surface with a minimum of uncut fibers is obtained. Coolant is necessary to prevent wheel loading Al_2O_3 or SiC grinding wheels are satisfactory.

4.4.10 Shearing and Punching

Shearing can be performed, but the edge will be coarse or ragged. Recommended practice is to shear 0.25 to 0.50 mm outside of the final trim line and then sand or mill to the net dimension. Fiber tearing also occurs in punching composite materials, so that a reaming operation has been developed to minimize ragged edges. In most types of machining operations, it is good shop practice for an operator to wear a face-mask respirator to prevent inhaling the exceedingly fine but stiff particles that are given off.

4.5 MMC MACHINING

The machining of MMCs is not difficult but presents different machining requirements than for other materials. The use of the HSS tooling for MMCs is totally useless. Carbide tooling, either plain or coated, has limited capabilities. In general, carbide tooling will suffice if PCD or CVD tooling is not available, but only for very limited usage.

PCD or CVD are the most cost-effective and, by far, the best choice for machining MMCs: the larger the grain size, the greater the resistance to wear. Experimentation with 5 μm, 17 μm, and 25 μm grain size has proved that the cut edge of the tool exhibited less wear on the larger grain size. GE 1500 was the grade of the PCD insert that had the 25 μm grain size and it outperformed the smaller grain size.[31]

The machining characteristics of SiC-reinforced aluminum MMCs are different than those of unreinforced aluminum alloys. Very hard SiC ceramic particles in a relatively soft matrix are abrasive, creating higher temperatures between the tool and workpiece; this produces faster tool wear. Modified machining parameters or special tool materials (e.g., compacted diamond or carbide), however, produce close-tolerance, intricate parts with a very good finish, and surface integrity.

4.5.1 MMC Machining Behavior

Numerous studies, analyses, investigations, tests and examinations have been conducted and reported on the costs, economics, benefits, etc., in machining fiber, particulate and whisker MMCs on a prototype/production basis and compared to the machining of steels, aluminum and magnesium alloys, ceramics and titanium.[15,31-48] MMCs are considerably more difficult to machine than most conventional metals and alloys.

In an economic study for machining MMC parts, Schmenk and Zdeblick[32] found that laboratory turning and milling data for MMCs containing up to 15% volume of dispersed Al_2O_3 particles or up to 20% volume of dispersed SiC strongly suggested that the machinability of these MMC materials was quite similar to that of A390 high silicon aluminum. A390 aluminum and the MMC material tested contain comparable quantities of very hard, severely abrasive, dispersed particles. The oxide and carbide MMC particles and the somewhat larger, silicon particles in hypereutectic A390 rapidly abrade HSS and, even, carbide cutting tools. Rapid tool wear, however, is much less of a problem if PCD tools can be used to machine these materials. Comparisons of machining data for these MMC materials from in-house and outside sources with the extensive data available on A390[33,35,36] indicate that the machining behavior of these MMC materials was quite comparable to that of A390, This finding is important since there are a number of well-documented cases where A390 is economically viable for the production of a high volume component such as engine blocks and cylinder heads.

MMC disk brake rotors offer unique possibilities compared to conventional cast iron disk brake rotors. From the view of product cost, MMC rotors may well prove to be less expensive to produce than cast iron rotors as shown below:

- The estimated machining cost for MMC rotors is $1.57 per rotor compared to $1.72 for a cast iron rotor. This conservatively estimated saving of $0.15 per rotor (9%) for the MMC material does not consider the expected casting cost savings for MMC materials.

- The estimated machining production rate for MMC rotors is 3,131 per day compared to 1,900 per day for cast iron rotors, an increase of ~ 65%.

- Tooling expenses dominate the machining cost issue for MMC materials. Expected improvements in the performance versus cost ratio of PCD tooling would result in additional machining cost savings for MMC.

4.5.2 MMC Turning

Norton Diamond Film Corp.[15] performed turning tests on 6061-T6 aluminum composite reinforced with 15 v/o Al_2O_3 particulates, comparing wear resistance of two PCD grades and CVD thick-film diamond. Cutting speed was 680 m/min; feed rate, 0.12 mm/rev; depth of cut,

1 mm; tool entrance angle, 60°; and rake angle, 6° positive. The tools were inspected at five-minute intervals during machining and it was found that crater wear on the rake face was considerably less for the CVD tool which may be due to the higher thermal stability and hardness of the CVD diamond. Turning a negative or positive rake on the PCD insert seems similar in cutting ability. The positive rake, however, produces a smaller chip, better surface finish, and requires less horsepower. As a rule in turning, generally rough machining occurs at 2.2 mm depth of cut and 5.1 m/s feed rate and finished at 0.305 to 0.76 mm depth of cut and 0.09 to 0.13 mm feed rate depending on the micro finish requirement. Micro finishes of 16-30 are routinely maintained. It should be noted that the finish will appear much better to the naked eye than readings taken by instruments.

Finally, it should also be noted that it is sometimes virtually impossible to run the turning equipment at the recommended surface finish per minute. This is especially true in turning smaller diameters because conventional toolroom lathes cannot obtain the higher RPM that is needed. In roughening operations, avoid exceeding 1/2 of the PCD length or fracturing of the cutting edge may occur. It is crucial that adequate shielding and chip guards be in place while machining because of the higher chuck revolutions that are required.[31]

A machining cost comparison of 15 vol % SiC_w/SXA-2124 which included turning was conducted by Berlin.[37] The study produced a cost model which indicated SXA® is a bit more than half the cost to machine than titanium and slightly more than twice that of aluminum. The only major difference between the machining cost of SXA® and aluminum is the raw material cost.

PCD tooling has been effectively used to drill, bore, and turn 2080/15 vol% SiC_p material used for disc-brake rotors for automobiles. It appears that machining discontinuous-reinforced aluminum (DRA) composite rotors is cost competitive with cast-iron rotors. In part, this stems from the high productivity rates and the fact that balancing and gauging are not required for the lighter weight DRA rotors. The DRA rotors can also be cast closer to net shape, thus reducing the total amount of material to be machined.[43]

Looney, Monaghan, Reilly, et.al., conducted a series of turning tests in which a number of different cutting-tool materials were used to machine an Al/25 vol% SiC MMC, Table IV-3. The influence of the cutting speed on the tool wear, the surface finish, and the cutting forces was established for each tool material. It was found that carbide tools, both coated and uncoated, sustained significant levels of tool wear after a very short period of machining. The best overall performance was achieved using a solid cubic boron nitride, CBN, insert while the worst was encountered using a solid silicon nitride, SiAlON, tool.[47]

Table IV-3 Tool Materials[47]

Tool Type	Manufacturer	Tool Material/Coating
IC20	Iscar	Uncoated Carbide
GC3015	Sandvik	Carbide/Two Coatings: TiC and Al_2O_3
GC415	Sandvik	Carbide/Three Coatings: TiCn, Al_2O_3, and TiN
TX10	Seco	Carbide/Three Coatings: TiC, TiCn and Al_2O_3
Kyon	Kennametal	Ceramic: Silicon Nitride
Amborite	De Beers	Cubic Boron Nitride

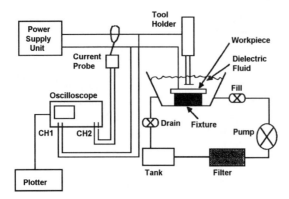

Figure IV-6. Schematic of EDM process setup.

Table IV-4 EDM Machining Characteristics of SiC$_w$/Al and TiB$_2$/SiC Using Brass Electrodes[50]

Material	Process Conditions	Relative Electrode Wear (%)
15% SiC w/Al	Fine	28
	Medium	29
	Coarse	32
25% SiC w/Al	Fine	44.5
	Medium	30.47
	Coarse	39.10
TiB$_2$/SiC	Fine	42.90
	Medium	82.00
	Coarse	98.00

A new whiskered ceramic cutting tool has now moved from research and development in the cutting-tool industry's most advanced metallurgical laboratories and is now demonstrating practical performance benefits in widespread production use on a range of difficult-to-machine materials.

One successful grade now on the market, GTE Valenite's QUANTUM 10 (Q-10), uses reinforcing SiC whiskers. The reinforced SiC whiskers add properties of high edge strength, increased fracture toughness, good thermal shock resistance, and excellent hot hardness to an abrasion-resistant oxide base.

In an effort to reduce machining time, a die manufacturer substituted Q-10, the SiC$_w$ cutter, for carbide in rough and finish turning operations on a heavily-scaled flow turn die forging of D-2 tool steel with a hardness of 60 Rc. When eight dies were completed, three weeks of machining time had been saved.[49] Other rough and finish turning applications on landing gears, artillery shells, and print rollers including materials, their forms, hardness, and other properties are detailed and described in Reference 49.

4.5.3 MMC EDM

MMCs are ideal candidates for EDM, especially where complicated shapes and high accuracy are required. Only a few CMCs that are electrically conductive can be shaped by EDM. However, recent improvements in the mechanical properties of CMCs – especially the fracture toughness and strength of whisker-reinforced ceramics through better processing technology and starting materials – make them ideally suited for high-temperature and fatigue-resistant applications.

The machinability of SiC$_w$/ 2124 Al and TiB$_2$/SiC by EDM was investigated by Ramulu and Taya.[50] The machined surfaces of these two high-temperature composites were examined by SEM and surface profilometry to determine the surface finish. Machinability was evaluated in terms of material removal rates, tool wear, and surface finish. They found that EDM was one of the most versatile and useful technological processes for machining intricate and complex shapes in various materials, including high-strength, temperature-resistant alloys.[51] The MMCs

156

Table IV-5 The Surface Finishes of SiC_W/Al and TiB₂/SiC Composites Cut By EDM[50]

Materials	Process	Rate of Cutting	Surface Finish (μm)
15% SiC w/Al	EDM	Fine Medium Coarse	3.56 3.68 6.99
25% SiC w/Al	EDM	Fine Medium Coarse	3.05 3.73 6.99
TiB₂/SiC	EDM	Fine Medium Coarse	1.75 2.60 3.80
*Electrode: Brass			

used in this study were 15% vol. and 25% vol. SiC$_W$/2124 aluminum matrix composites, in plate form with a thickness of 6.3 mm. Brass and copper were used in the EDM process as electrodes and cut the MMC and CMC at coarse, medium, and fine cutting conditions, Figure IV-6. As a result of the preliminary investigations, the following conclusions were made:

- Material removal rate in EDM processing increased with power to the electrode in both the TiB₂/SiC and SiC$_W$/Al composite material.

- Electrode wear rate increases with the increase in process conditions in the EDM process regardless of the workpiece material. Electrode wear rate in the machining of TiB₂/SiC is greater than in the SiC$_W$/Al composite material, Table IV-4.

- Surface quality is affected by the choice of machining process, and the brass electrode in the EDM process appears to give minimal surface damage. At higher cutting speeds, EDM causes severe microdamage to the surface. Surface roughness in TiB₂/SiC is found to be finer than in SiC$_W$/Al composite, Table IV-5.

- Microhardness tests on 25% SiC$_W$/Al composite revealed that the EDM process appears to cause surface softening at slower cutting speeds.

4.5.4 Milling

In milling MMCs, CVD thick-film-diamond end mills were compared to micrograin WC mills by Norton Diamond Film Corp. The end mills, 9.52 mm in diameter and 63.5 mm long, had three straight flutes. Cutting speed was 359 m/min; feed rate, 0.038 mm/tooth; axial DOC, 6.35 mm; and radial DOC, 1 mm. With the carbide, there was 0.79 mm of wear after 2 m of travel and removal of 13.1 cm^3 of material; for the diamond film, researchers recorded a wear land of 0.19 mm after 157.5 m of travel and removal of 655.5 cm^3. By extrapolating, it was determined that the CVD diamond tool would travel 327.5 m and remove 2112.8 cm^3 of material before the tool expired (0.38 mm of flank wear). This is more than 300 times better tool life for the CVD diamond end mill.

In another milling test,[15] micrograin carbide was compared to a Si₃N₄ end mill coated with thin-film CVD diamond. Cutting speed was 762 m/min; feed rate, 0.076 mm; axial DOC, 6.35 mm; and radial DOC, 2.54 mm. Geometry of the end mills was identical: three-flute, 60° spiral, 15.88 mm diameter, right-hand cut, right-hand spiral, 44.5 mm flute length, and 101.6 mm overall length. In this test, the micrograin carbide end mill showed 0.14 mm of flank wear after machining 104.9 cm^3; the diamond film had no measurable wear and no film chipping or spalling after 209.5 cm^3.

As tests have confirmed, HSS will not do the job for milling MMCs. Carbide inserts in face milling cutters had marginal success. Uncoated inserts are extremely limited with great wear apparent after removing very small amounts of stock. The TiC coated carbide inserts show removal of 3 to 4 times more material than the uncoated inserts but are still not acceptable because of the small amount of stock removal. Once the coating fails, the ability to machine deteriorates very rapidly. As is the case for turning, PCD inserts are far superior to carbide for milling.

In end milling, the wear of the solid carbide mills was significant, although, they do remove 10 times as much material as HSS end mills. PCD-inserted end mills remove material quite easily but deep milling as for pockets and slots must be done in increments of ~ 3.1 mm depth of cuts. This generally will leave slight steps in the side walls of the areas being milled. A CVD coated end mill could work well as an alternative.

4.5.5 Tapping

The most challenging machine operation for MMCs is tapping. Therefore, if machining operations were ranked on a scale of 1 to 10 with easiest being a one, then turning would be a 1, drilling would be a 2, face milling a 3, end milling would be a 4, but tapping would top the scale and be a number 12.[31]

Carbide plug taps with as many flutes as possible have been proven to work the best. The greater number of flutes is advantageous because the land area is smaller and less resistance to assist in the removal of the tap after threading. If a 2 or 3 flute tap must be used, then the back of the flutes must be backed off manually to reduce the land. Breakage of taps larger than a # 10 size is minimal. Number 8 and smaller will break quite readily although this can be minimized by use of a tapping head wherein the clutch has been properly set. Synthetic materials have been successfully used for tap lubricants and ease of removal (Trefolen).[31]

The tapping of MMC requires a constant feed and this is not possible when hand tapping. Each stoppage of the tap constitutes a dwelling in the hole allowing the tap to back off slightly because the torque has been removed. This will cause rapid breakdown of the cutting edges of the flute. Less than 10 holes can be tapped by hand, whereas, 100 holes have been tapped using a properly adjusted tapping head. The speeds used should be identical to tapping any heat-treated aluminum. Finally, roll form taps do not work very well in MMC. They will gall and seize up very quickly. Diamond-coated taps are in the development stage and should be very effective and give excellent results.

4.5.6 Deburring

Conventional methods for removing burrs work equally well for MMC. Carbide rotary deburring tools in an air-grinder can be used to remove heavy burrs resulting from milling and drilling operations. Holes can be chamfered with standard carbide countersink tooling. Vibratory deburring is highly recommended after removal of heavy burrs to obtain a constant radius on all outside corners and a uniform texture over the entire part. The medium selection does not seem to be a factor in obtaining results. Plastic impregnated medium, Al_2O_3, and ceramic have been used with equally comparable results. The time needed to obtain the desirable finish is a matter of personal preference, but, generally, 15 to 20 minutes is sufficient.

4.6 CMC MACHINING

Because of their hardness, ceramics and CMCs were produced to final size and shape without machining. However, new applications require precision, which cannot be obtained without machining. The hardness of ceramics and CMCs makes traditional abrasive machining slow and expensive; complex shapes may be impossible to machine. In order to achieve the finishes for today's and tomorrow's products, various processes are used including grinding, EDM, drilling, and cutting.[52-55]

4.6.1 EDM

One of the recently and successfully applied nontraditional machining techniques that utilizes electrothermal energy is EDM. Unlike traditional techniques, EDM does not depend on material hardness but requires that resistivity of the workpiece be lower than 100 Ω-cm.[56] The tool does not touch the workpiece. They are connected to electrodes separated by a dielectric fluid. There are two principal types of machines, the ram-type, also known as die-sinking, and the wire-type. For ram-type machines, the workpiece is generally the cathode and the shaping tool is the anode. However, for the wire-cut EDM, the reverse is applied.[56] Addition of computer numerical control provides nearly rectangular-shaped current pulses where the rise in voltage is held constant for a period of time known as "on-time". During such on-times, sparks are generated, and electrical energy is converted to thermal energy. This thermal energy can remove material from the surface of the workpiece by two erosion mechanisms: melting and spalling. According to a melt mechanism, the material is removed when the melted region is swept by the dielectric fluid.[57] According to the spalling mechanism, a steep temperature gradient exists and creates internal stresses high enough to overcome the bond strength causing cleavage of crystals on the surface.

Gadalla and Bedi initiated a study[58] on TiB_2-BN composites, materials used in the vacuum metallizing industry. Although TiB_2-BN can be readily machined with traditional tools, EDM is more attractive since it allows intricate shaping and is expected to yield higher material removal rates since the composite hardness does not effect machining in the latter case.

TiB_2 conducts current and forms a liquid phase at the interface with BN. Neighboring crystals of BN and some TiB_2 spall due to thermal shock. Composites rich in TiB_2 or with fine TiB_2 grains gave high material removal rates. Increasing the amount of the conducting phase by 10 x is as effective as decreasing the grain size from 11 to 7 μm. They found that coarse TiB_2 could withstand high pulse durations before wire breaks. Material removal rate increased with pulse duration, frequency, and current. For the same composition and grain size, increasing the pulse duration or current increased the crater depth (the roughness) up to a certain value, beyond which increasing these parameters yielded a smoother surface. The conductivity of the dielectric was effective only for compositions rich in TiB_2 content. In such cases, higher water conductivity lowered the energy required for material removal.

Generally, most of the properties of TiB_2 are similar to those of SiC. TiB_2 (16 vol%) reinforcement of SIC exhibits a 30% higher flexural strength (478 MPa) than SiC at 20°C. This strength is unaffected by temperatures up to at least 1,200°C.[59]

Addition of TiB_2 to the SiC matrix significantly reduces the grain size of the matrix in the sintered microstructure. TiB_2 particulates reportedly pin grain boundaries and inhibit grain growth. Wlectrical resistivity is lower for the SiC/TiB_2 composite because of the good conductance of TiB_2.

With appropriate composite ratios, SiC/TiB$_2$ can be EDM. For example, an EDM rocker arm insert made from a TiB$_2$ particulate toughened SiC composite is 50 to 75% tougher than direct sintered SiC.[59]

In laboratory tests, Calés, Martin, and Vivier[60] found that new conductive Si$_3$N$_4$ and Al$_2$O$_3$ based composites with high mechanical properties could be developed by adding amounts of TiC and/or TiN particles in the ceramic matrix. They concluded that they had developed special grades of conductive ceramic composites, with either a Si$_3$N$_4$ or an Al$_2$O$_3$ matrix for EDM. These composites exhibit high fracture strength, at least equal to that of the matrix, and high electrical discharge machinability. They are more easily machined than WC and the tool consumption is quite lower. Machined surfaces with low roughness and high material removal rate have been obtained after optimization of the EDM conditions. In the future, they expect to replace conventional Si$_3$N$_4$ and whisker reinforced Al$_2$O$_3$ ceramics in a number of applications where complex machined parts are required.

The fracture toughness of SiC whisker-reinforced Al$_2$O$_3$ is nearly double that of the material without the fibers. The same is true with Si$_3$N$_4$-based composites. The particle size and percentage of TiC or TiN to be added to the matrix can be adjusted to make it electrically conductive enough to carry out the EDM process without significantly compromising the ultimate properties and performance requirements of the material. Tests have shown that, for Si$_3$N$_4$-based composites with TiN added, EDM can be performed at a conductivity higher than $2 \times 10^2 \ \Omega^{-1}$ per centimeter (and preferably $5 \times 10^3 \ \Omega^{-1}$ per centimeter). In contrast, for wire EDM of SiC whisker-reinforced zirconia-toughened Al$_2$O$_3$ with TiC added, the minimum conductivity value was found to be $1 \Omega^{-1}$ per centimeter. With a 30 vol% TiC addition to the Al$_2$O$_3$ based composite, the bend strength was reported to be 860 MPa, while up to 50 vol% of TiN particles could be added to the Si$_3$N$_4$ matrix without reducing its fracture toughness. This is one example where difficult-to-machine materials such as ceramic composites can be tamed by making them electrically conductive and able to be processed by EDM.

An electro-conductive ceramic composite has been developed and tested for high temperature extrusion dies, cutting tools, and wear parts. By providing an electroconductive ceramic, the ceramic can be EDM to virtually exact shape subsequent to furnace sintering, hot pressing, or hot isostatic pressing to the final density and hardness of the ceramic. Thus, complex geometries and features such as holes, chamfers, slots, angles, changing radii, and other features can be EDM into the composite that could not previously be either economically provided or produced. After the EDM, minimal diamond grinding for final dimensionality or surface condition may be provided, if desired. Although the ceramic composite is specifically directed to die construction, tooling and wear parts, the properties of the ceramic allow other applications, in particular where intricate shapes are required.

The surface quality obtainable with EDM of the new composite is comparable with the best conventional grinding methods used to obtain a high quality surface finish. For example, a billet of the new EDM'able whisker ceramic composite was cut by a traveling wire electro-discharge machine and, upon, examination minimal removal of material, on the order of 0.01 mm, produced a surface finish of 16 micro-inches. This surface finish is comparable to surface finishes obtained by costly diamond grinding processes.

The electro-discharge cutting speeds of the new ceramic composite compare with the electro-discharge cutting speeds of typical tool steels and exceed the cutting speeds typically used to cut carbide materials. Moreover, the new ceramic composite does not exhibit edge chipping or a limitation with respect to minimum EDM section thickness, which is indicative of minimal reduction in strength associated with the thermally affected surface material.

4.6.2 Grinding

Rezaei, Suto, Waida, et.al.[61] conducted an experimental investigation into the creep feed grinding of CMCs. They believe that creep feed grinding with a diamond wheel has been shown to be the most efficient method of stock removal.[62] The main problem with grinding of ceramic is the rapid wear of the diamond abrasives which makes the process very costly. To increase the efficiency of the grinding process, novel grinding wheels have been developed which have superior cooling and lubrication ability.[63] Dressing diamond wheels has also been the subject of other research and unique dressing methods have been reported.[64] Their investigation compared Si_3N_4, Al_2O_3, SiC, ZrO_2, aluminum-reinforced Al_3O_3, SiC whisker-reinforced Si_3N_4 and C fiber-reinforced Si_3N_4. The latter material was ground in three directions: longitudinal, transverse, and normal to the fiber direction. All specimens were carefully prepared by grinding all sides. They were all 100 mm long except for the fiber-reinforced ceramics, which were 70 mm long. The grinding depth of cut was set at 2 mm and the table speed at 30 mm/min. For the case of whisker-reinforced Si_3N_4, the table speed of 18 mm/min was selected as this composite was found to be more difficult to grind. Their conclusions were that:

• Among the ceramic materials, SiC requires the highest grinding power, but Si_3N_4 induces the highest normal grinding force. The latter was found to be the most difficult ceramic to grind.

• Aluminum-reinforced Al_2O_3 is slightly easier to grind than Al_2O_3 and, consequently, the surface finish produced is rougher than that of Al_2O_3.

• Carbon fiber-reinforced Si_3N_4 was ground with a tenth of the power required for Si_3N_4. This is for the case when the direction of grinding was along the length of the fiber. When the grinding direction was normal to the fiber, the grinding power was half that for Si_3N_4.

• In contrast to other ceramic composites investigated, the SiC whisker-reinforced composite was more difficult to grind than the pure Si_3N_4. Consequently, the surface roughness of the ground composite decreased.

• The surface roughnesses of the fiber-reinforced Si_3N_4 specimen and aluminum-reinforced Al_2O_3 generally increased by a factor of 1.4 as compared with those for the pure ceramics.

Table IV-6 Machining Data for Gr/Ep Composites[65]

Tig Material	Material Thickness (mm)	Hole Diameter (mm)	Speed (m/s)		Feed Rate (in/rad)
			*	+	
Carbide	Up to 12.7	4.8 - 8.1	0.70	1	0.001 - 0.002
	12.7 - 19	4.8 - 8.1	0.55	0.7	0.001
PCD	Up to 12.7	4.8 - 8.1	1.625	1.625	0.002 - 0.0835
	12.7 - 19	4.8 - 8.1	1.625	1.625	0.002 - 0.0035
*Unidirectional Fibers +Multidirectional Fibers					

4.7 DRILLING

One of the first questions asked when attention began to focus on composites was why it takes so long to drill holes in these materials. The answer, of course, is that the available tooling was not designed for cutting composites. Gl/ Ep and Gr/Ep are so abrasive that, initially, only WC was used; however, PCD has been very successful, Table IV-6. Drill tips were designed for metalworking, the tip heating the metal to provide the plastic flow needed for efficient cutting. Because composites cannot tolerate this heat, production must be slowed down to reduce the heat. Drill designers had to abandon cutting tips with neutral and negative rakes and wide chisel points because a drill with a neutral rake scrapes the material and causes it to resist penetration by the drill tip. The operator must exert pressure to drill the hole, and pressure causes the heat buildup.

A neutral rake also tends to push the reinforcing fibers out in front, requiring a great deal of pressure to penetrate the piece. This pressure causes the fibers to bend, resulting in furry, undersized holes. The pressure also produces excessive heat, which causes galling and chip clogging in the resin. The release of pressure as the tool bit breaks through the part causes a sudden and momentary increase in feed rate. As the tool plunges through the last few fibers, the cutter shaft, not the cutting edge, removes the remaining material. The result is chipping and cracking.[24]

The best way to analyze a drilling operation is to examine the chips. The ideal chip form for composites is a dry, easily moved chip that looks like confectioner's sugar. If the speed of the cutting tool is too high, heat will make the resin sticky and produce a lumpy chip; if the cutting edge is scraping and not cutting the plastic, the chips will be large and flaky. Either type will eventually clog any removal system.

Good tool geometry for both resin and MMCs starts with a positive rake. The reinforcing fibers are pulled into the workpiece and sheared or broken between the cutting edge and the uncut material. A positive rake on the cutting edge removes more material per unit of time and per unit of pressure than a negative rake, but the more positive the rake, the more sensitive and fragile the cutting edge becomes. A small chisel edge, the second element of good tool geometry, improves the penetration rate, which translates into more pieces per hour. The optimum chisel edge for composites is as close to a point as possible. Good geometry also means a cutting-tool shape that facilitates chip handling, so that the chips are produced and then removed immediately above the entrance to the hole.

In the last few years, tooling has been developed that has greatly improved drilling operations in PMCs as well as MMCs. Some tool bits are made of particles of WC smaller than 1 mm. Another technique developed incorporates two different disciplines. Ultrasonic-assisted drilling involves the use of a rotary tool on which is superimposed an axial vibratory motion at high frequency. A special adapter is required to transmit the vibration from a piezoelectric transducer to the tool. Ultrasonic vibration can reduce friction, break chips, and reduce tool wear. It is a particularly useful technique when the matrix or reinforcing fibers are hard brittle materials. Use of a core drill permits cutting fluids to pass through its center. Ultrasonic machining, though slow, can result in high finish and accuracy of intricate parts. Hence, it is recommended for applications in which intricate shapes of high accuracy and finish are required. This technique has been used to drill and countersink Gr/Ep and B/Gr/Ep holes in prototype aircraft structures such as the B-1 horizontal stabilizer. The drills are water-cooled. Tools used are all-diamond types, sintered and coated. The machine is versatile in that it can drill, countersink, ream, and counterbore. The ultrasonic drill-countersink is fitted with a sintered-diamond core drill plus an electroplated nickel or diamond sizing band behind the tip to maintain size and concentricity. The countersink surface is also plated because it can be stripped chemically and replated at low cost to extend tool life. Depending on the application, specially designed, core-drill-countersink combination tools ranging in diameter from 4.8 to 12.7 mm have been used.[24] The application of ultrasonic energy to diamond core drills when drilling either B/Gl/Ep or B/Ep hybrids increases drill life. For example, 50 holes were drilled by a portable drill and diamond-core drill without ultrasonics in 10.2 mm thick Gr/B/Ep hybrid material. When ultrasonic energy was applied, a 100% increase occurred (200 holes in 5.6 mm thick hybrid).[24]

4.7.1 PMC Drilling

4.7.1.1 Kv/Ep

Spade drills are best for drilling Kv/Ep, leaving very little fuzz and fraying on hole edges. Like carbide drills, these drills have a tendency to burn if they are used too long. Conventional drills can also be used, but a firm sacrificial backing must be provided at the exit surface. Twist and flat-ended HSS drills perform quite well on Kv/Ep, especially with a firm backup of the composite to eliminate fuzzing and delamination at the hole exit. The 0.08-mm layer of fiber glass on the top and bottom surfaces of the Kv/Ep composite produces the best holes, leaving clean entrance and exit surfaces. Drill speeds range from 25,000 to 35,000 r/min. The HSS drill is shaped like a twist drill fluted on the end. In a production environment, 45 to 50 holes have been made before resharpening was required although signs of wear have been found after drilling five holes. The best lubricant is water.

4.7.1.2 B/Ep, Gl/Ep and Hybrid Drilling

Drills with titanium diboride coating improve the life of drills by 877% when used on Gl/Ep.[17] The best way to drill B/Ep and B/Gl/Ep hyhrids is with diamond-impregnated core drills and an ultrasonic machine as a drilling-assist tool; however, the cores have a tendency to get stuck in the drills. The best combination is a diamond-impregnated core, reamer drills, and counter-sink tools. Although the cores sometimes get stuck, similar to ultrasonically powered drills, they can be removed by stopping the drill spindle and leaving the ultrasonic unit on. However, this is a slow process for production applications. The best speed for drilling B/Gl/Ep laminates is 2,000 to 3,000 r/min. The best machine spindle feed rates for drilling B/Gl/Ep laminates are 0.14 to 0.08 mm/s. A 5% commercial surfactant in water is the best coolant when drilling B/Ep and B/Gl/Ep hybrids.[24]

4.7.1.3 Gr/Ep Drilling

WC drills have been successfully used in drilling Gr/Ep using a backing plate. Various coolants have been tried to prevent drills from breaking, dulling, or burning up. A lubricant called Boelube has worked well, but water is the best. To reduce surface delamination during drilling, a layer of woven glass (0.08 mm) on both sides of the laminate is recommended.

Solid carbide Daggar drills have also proved quite successful in drilling Gr/Ep, producing a clean hole with little or no breakout.[66] Drilling has been accomplished without using backup material. Since the Daggar drill will maintain hole tolerance, no reaming is necessary. The overheating problem can still occur in drilling other composite materials such as carbon chopped fibers and glass-filled plastics; using a double-angled drill has worked best.

A new approach to produce good holes when drilling Gr/Ep composites is to add an abrasive slurry. Conventional drill bits alone are blunted very quickly and produce holes with cracked edges when drilling Gr/Ep composites, but, when an abrasive slurry is added, good holes are produced inexpensively and efficiently.[24] The technique was developed for preparing specimens of Gr/Ep for shear testing. A slurry of SiC powder in water is fed onto the drill. It contains 60% SiC by weight. The powder particles become trapped between the cutting edges and the composite and, apparently, remove the composite material by abrasion. The drill is also dulled, but researchers who developed the process report that the cutting is unaffected.[67] With the slurry, the dull drill cuts as fast as or faster than the sharp ones. With the slurry, the holes could be drilled efficiently regardless of the ply orientation.

In the search for an optimal cutting tool material for machining Gr/Ep, PCD inserts have been shown to be more abrasive resistant than carbide (C6 grade) inserts.[68-69] In a recent study, Wern, Ramulu, and Colligan[70] showed that Gr/Ep test panels were drilled using two different PCD tipped drills at a constant speed of 4,550 rpm, for a feed range from 0.0254 to 0.254 mm/rev in increments of 0.0254 mm/rev. Drill A was 12.7 mm in diameter with an axial rake angle of 27°, while drill B had a diameter of 13.9 mm and an axial rake angle of 7°. Apart from the difference in axial rake angle and drill size, the two drills have essentially the same geometry. The surface produced by drill geometry A was superior to that produced by drill geometry B for three reasons. First, when the feed rate is low, the surface produced by drill B is almost four times rougher than that produced by drill A. Drill B also produces a surface with a wider surface damage zone. Lastly, the probability of the occurrence of deep valleys in surfaces produced by drill B is greater than that in surfaces produced by drill A. Feed rates have a definite effect on the machined surfaces. Between 0.0254 and 0.1778 mm/rev, the surface roughness decreased with an increase in feed rate. At feed rates above 0.1778 mm/rev, the surface roughness increases with an increased feed.[70]

A study on tool wear thrust and torque using HSS and WC coated drills by Malhotra[71] evaluated CFRP and GFRP materials. He found, in general, that tool wear increases with the number of holes and flank wear is higher than chisel edge wear. Carbide drills perform much better than HSS drills with both the materials. Tool wear, thrust, and torque are much higher in drilling of CFRP as compared to GFRP and this is due mainly to the higher abrasiveness of carbon fibers compared to glass fibers.

4.7.1.4 Other Composite Materials

Hocheng and Puw[72] found that chips from drilling carbon fiber-reinforced ABS indicated that considerable plastic deformation had occurred. Compared to the chip formation of thermoset plastics, it contributes to the improved edge quality in drilling. The edge quality is generally fine except in the case of concentrated heat accumulation at the tool lips, which is generated by high cutting speed and low feed rate. Plastics tend to be extruded out of the edge rather than neatly cut. The average surface roughness along hole walls is commonly below 1 μm for all sets of cutting conditions; values between 0.3 and 0.6 microns are typical. The HSS drill presents only minor tool wear during the tests. Based on these results, they concluded that the carbon fiber-reinforced ABS demonstrated good machinability in drilling.[72]

4.7.1.5 PCD Drills

Fully fluted drills with PCD tips on solid carbide provide the strength of carbide and hardness of diamond while allowing innovative geometries. Users must be careful, however, to adjust feeds and speeds for each layer of composite. Use of a torque sensor to find the layers is very helpful.[73]

For drilling a graphite composite with an aluminum matrix backing, the PCD double angle drill is recommended. With an ordinary carbide drill technology, the hole finish quality of the graphite layer would be damaged with chips of aluminum being pulled into contact with the

Table IV-7 Hole Drilling Characteristics of Three Materials[43]

Material	Optimum Tool Material	Machining Rate (cm/min)	Drilling Operations Per Hole	Reaming Operations Per Hole	Tool Life (No. of Holes)
2080/SiC/15p	Diamond	63.2	1	0	10,000
Steel*	Steel	11.7	2	2	2,000
Ti-6Al-4V	Carbide	11.7	2	2	1,450
*Conventional powder metallurgy connecting rod steel (0.5C, 2Cu)					

graphite. The double angle drill allows the shallow center section of the tool to center the tool and prevent wandering, while the larger angle eases the tool into the hole up to its diameter, creating less heat and eliminating interface bellmouthing.

The life expectancy of PCD drills is now well-documented. Even without resharpening, which can be accomplished at least two or three times per tool, diamond drills have increased holes in composites per tool by a factor of one hundred or more compared to ordinary carbide drills. Even though diamond drills are more expensive than carbide drills, they offer the same flexibility of design. The diamond drill often offers manufacturers the most cost-effective method of optimizing production rates without the need for a large investment in new equipment.[74]

4.7.2 MMC Drilling

Ricci[75] evaluated the following MMCs:

- $SiC_f/6061$ - T4Al
- $SiC_p/7079$ - T6Al
- $B_4C_p/AZ61$ A - Mg

and used the following drills and processes:

- PCD-tipped twist and spade
- Diamond-plated twist and core
- Abrasive water-jet hole-cutting

The diamond-tipped twist drills outperformed all the other drills. Core drills were found to be viable alternatives for the production of larger holes in high-volume fraction composites. Plated twist drills were viable alternatives for low-volume fraction particulate composites. Spade drills failed due to low edge strength. Abrasive water-jet hole cutting was successful for rough, large-diameter holes.[76]

Allison and Cole[43] in their study of opportunities and challenges for MMCs in the automotive industry compared hole drilling machining rates of $2080/SiC/15_p$ (2080 aluminum alloy reinforced with 15 vol% SiC particulate), a conventional PM steel connecting rod (0.5 C, 2 Cu), and Ti-6Al-4V, Table IV-7. Diamond tooling is more expensive than traditional HSS or carbide tooling but tool life can be quite long (see Table IV-7). Therefore, these high tooling costs can be amortized over many parts. Because of the higher productivity rates, the capital required for machining lines can be reduced. Moreover, machining tolerances are greatly improved using diamond tooling, so subsequent finishing operations can be reduced or even eliminated. Again, for hole drilling (Table IV-7) in discontinuous-reinforced aluminum composites, holes are finished with such high tolerances that subsequent reaming operations can be eliminated.

Monaghan and O'Reilly[77] confirmed the conclusions of Allison and Cole and others. They found that, on a 1050 series aluminum alloy (12 % Si-rem Al)/SiC/25$_p$, tool hardness plays a significant role in the efficient and cost-effective drilling of MMCs. It was found that the PCD-tipped tools, while they lasted, were superior to the carbide tools, and they, in turn, were much better than the coated and uncoated HSS drills. This ranking applied to all of the recorded values for flank and rake-face wear, torque, thrust-force, and surface finish. The effect of the ceramic TiN coating on the HSS drill did not improve its performance significantly compared to the uncoated high-cobalt HSS drills. It was found that, in each case, there was a similarity between the trend of the flank- and rake-wear results with speed and feed rate and the variation of the drill thrust-force and torque, respectively.

The four vertical tail stabilizers discussed in Chapter III as one of the largest MMC primary structures built confirmed that existing manufacturing technology for mass production (such as cutting and drilling) was possible with the MMC components. The program[78] proved the viability of using conventional shop procedures for drilling and fastening these MMC components. It also tested innovative manufacturing equipment such as PCD tooling for all the drilled and countersunk holes, and abrasive water-jet cutting. Diamond-tipped drills were used because the MMC was considerably more abrasive than aluminum due to the SiC content. The tips were refurbished and reused after making 75 holes. The abrasive water-jet cutter used a thin stream of high-pressure (45,000 psi) water containing powdered garnet.

B.R. Berlin[37] made a machining cost comparison of SiC discontinuously reinforced aluminum, unreinforced aluminum, and titanium and included drilling. He found that drilling was the least economical operation in all materials. While representing relatively small portions of machining times, the relative tool costs were quite high. The SXA®(2124/SiC/15$_w$). in particular, was extremely expensive to drill. Representing only 6.3% of total machining time, it required 39.0% of the total machining cost. Therefore, the average machining cost per minute was extremely high (over $2.02 per minute). Aluminum and titanium follow similarly, but not nearly to the extremes of the MMC This high cost was due to the relatively short lives drills typically exhibit, especially in small diameters, and the cost of PCD drills. PCD drills averaged approximately $285 a piece.[37]

Finally, B. Dwenger,[31] in his prototype MMC machining study, concluded that HSS drills, both coated and uncoated, absolutely would not work with MMCs. All drilling in his work had been done with carbide drills. Hole sizes ranged from 2.3 mm through 25.4 mm diameter. High helix drills were also used, but, for normal depths of 3 to 4 times diameter, there was no proven advantage. For the larger hole sizes, the carbide tipped drills ar generally used. They worked equally well and had less initial cost than the solid carbide drills. Speed and feed rates were the same as if drilling standard aluminum. The use of a coolant during drilling was mandatory, preferably, a water-soluble oil. Coolant fed drills worked quite well for 9.5 mm diameter and larger, especially if the hole was quite deep, for clearing chips from the flutes.

One point to note is never allow the drill to dwell in the hole This will dull the cutting edge immediately and drill failure will occur leading to out-of-tolerance hole sizes or breakage of the drill.

166

With carbide drills, upwards of 200 holes were drilled between resharpenings. However, if diamond-tipped drills were substituted it is believed that results would range from 2,000 to 3,000 holes between resharpenings.

Other studies conducted in various laboratories on $2014/SiC/15_p$ with a WC drill have concluded that:

- Tool wear is accelerated due to the presence of SiC particles.

- Use of a coolant is essential.

- Chisel edge, margin, lip crater and lip flank wear were observed under most cutting conditions. Wear of the outer corner of the lip flank was found to have the most significant effect on performance.

- The generation of burrs on the exit side of the holes related to the rounding of the outer corner of the lip flank. Hence, higher speeds and feeds result in the earlier appearance of such burrs.

- The most evident mechanism of wear observed on the drills was that of abrasion by the SiC particles present in the aluminum.

4.7.4 Future Systems

Expanding from the manual machining and drilling of today, future systems will be tailored to several manufacturing approaches. One is the FAS (Flexible Assembly Subsystem), an Air Force-generated program in pilot production at Grumman.[79] The FAS program's overall objective was to develop and demonstrate key enabling technologies that are critical for achieving both flexible robotic assembly systems and the integration of these systems into the existing factory environment. To demonstrate its reliability and repeatability, FAS was introduced to industry in 1991 via successful fabrication of an EA-6B type bulkhead and five flightworthy, prototype production EA-6B pylons. Today, the FAS is meeting the production schedule needs for two planar bulkhead assemblies, as well as two other complex closed-contoured pylon assemblies.

Other techniques being implemented in 1993 include the drilling and fastening of advanced composites. The FAS has been recognized by the DOD, industry, and numerous universities as being "state-of-the-art" in complex assembly robot systems.

LTV Aerospace and Defense has recognized that assembly of aircraft components typically accounts for more than half of the total labor cost of an aircraft, making assembly automation one of their top priorities. As a result, they have developed one of the world's largest computer-controlled robotic systems for automated drilling and fastening of composite/metallic aerostructures.

The Robotics for Major Assembly (RMA) developed by LTV is an automated, five-axis robotic system that performs the drilling and fastening operations of large, highly contoured aircraft parts that combine composite materials with titanium or aluminum substructures. Aircraft sections are rigidly fixed in the RMA cell while the computer-controlled robot, moves over the assembly, drilling holes, installing fasteners, and applying sealant – all in one pass. The new drills allow one-shot drilling, producing a finished hole in one pass. Instead of the traditional liquid coolant, 5.2-MPA gaseous nitrogen passes through the coolant feeding drill to cool the drill and flush away the chips. The system produces precise holes and eliminates the need for a liquid coolant recovery system. Finally, a U.S. based machinery manufacturer has introduced a series of cantilevered-arm robots as a low-cost alternative to CNC drilling and routing of composites. The robots can cut or drill any plastic material as well as glass-reinforced composites, at speeds of 0.008 - 0.021 m/sec. The company claims that the machine is accurate to 1 mil at the highest speed.

4.8 CUTTING

Cutting is necessary at one or more points in the manufacture of most structural fiber composite parts. Cutting operations can be divided into those that occur before cure and those that occur after cure.

Materials cut before cure include various mats, fabrics, and other dry reinforcements, as well as prepreg layers. Cutting tools range from manual scissors to automated laser or water- jet devices. Between these extremes are power shears, rotary power cutters, reciprocating knives, and steel-rule dies for blanking. Which tool is used depends on the material to be cut and the speed and accuracy desired.

As for post-cure cutting, traditional tools include various bandsaws, hand routers, shears, and sanders. These tools are highly labor-intensive, noisy, and create serious dust problems. While they are still very widely used, they are steadily being replaced by more advanced cutters. The most important of these are the high-pressure water jet, abrasive water jet, and the laser.

A set of criteria has been established in order to evaluate cutting operations. The major areas of concern are:

Part Contamination – Contamination arising from cutting operations can be a serious problem with laminates. A number of impurities may be driven or absorbed into the laminate during cutting. These include cutting fluids, lubricants, and solids from the base or backing materials or from the cutting machine itself.

Cut Quality – A high-quality cut is a smooth, clean cut without fiber fraying or cracking between plies.

Stress on the Workpiece – Too much stress placed on a laminate by the cutting tool can cause ply separation. It also can force movement of the cutting surface, giving a misplaced or otherwise bad cut.

Local Heating Effects – The main problem here is heat distortion in the area being cut. This is of special concern in laser cutting, but can also arise from the friction of traditional tools.

Resolution – This term refers to the tool's ability to cut small-radius curves with minimal error. A related term is flexibility. This refers to the tool's ability to change directions without interrupting the cut.

Postcut Operations – A good cut should leave as little as possible to be done in follow-up operations. Such operations increase both costs and the chance of damaging errors.

Freedom From Wear – Cutting tools or machinery that wear easily raise costs in several ways. Output is reduced because of machine downtime. Inspection requirements are increased. Part rejection becomes more likely toward the end of the machinery's wear life.

Versatility – This is the ability to cut different materials without major loss of effectiveness.

Material	Max No. of Plies	Configuration	Cut Quality
Gr-Ep	18	Circles: 2.5 in (64 mm) Diameter with 0.44 in (11 mm) Diameter Holes	Excellent
		Triangles: 3 in (76 mm) with 0.68 in (17 mm) Diameter Holes	Excellent
B-Ep	18	Triangles: 3 in (76 mm)	Excellent
		Same with 0.68 in (17 mm) Diameter Holes	Excellent
Kv-Ep	12	Triangles: 3 in (76 mm)	Good
Gl-Ep	27	Triangles: 3 in (76 mm)	Excellent

*All blanking was done with top and bottom polyethylene cover sheets.

4.8.1 Uncured Composites

4.8.1.1 Reciprocating-Knife

There are two reciprocating-knife cutting machines, both of which incorporate high-speed reciprocating knives that are driven through the material to be cut by a minicomputer-controlled, xyzc positioning system. In one system, the cutting knife penetrates through the material into closely packed plastic bristles that constitute the surface of the cutting table. The surface is nondegradable and does not require periodic removal.[80] The cutting knife ranges in width from 1.3 mm for the diamond cutter up to 4.4 mm for the carbide cutter. The system cuts in a chopping mode; i.e., the knife rises above and plunges through the material onto the table.

The second system[25] can cut desired patterns in a continuous line at high speed. Curves, sharp corners. and notches can also be cut without lifting the knife from the material. The knife can be lifted. as required, to start new cutting lines, to pass over sections without cutting, or to cut

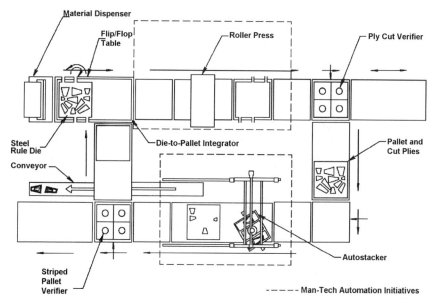

Figure IV-7. Integrated, automated, low-tack, prepreg cutting, and laminating system.[81]

169

holes of any diameter. The system uses a blade 6.4 mm wide and cuts either by chopping or slicing. In the slicing mode, the knife remains buried in the material after the first stroke (each stroke is 19 mm) and is always at least 3.2 mm below the material being cut. Computer-controlled rotation of the knife about the c axis keeps the blade properly positioned at all times.

Cutting test results for both systems indicate that the slicing and chopping system can cut a greater number of Gl-Ep and Gr-Ep plies at twice the feed rate of the chopping system. Visually, the quality of the edges of laminates cut by both systems is about equivalent.[25,80]

4.8.1.2 Steel-Rule-Die Blanking

Steel-rule blanking dies are an economical method for cutting relatively small quantities of composite prepreg materials. If a die does require reconditioning, only a minor expenditure is involved. Normally, the steel-rule dies are positioned above a flat mild-steel plate to permit blanking on the downstroke. . Each die consists of 3-mm one-side-beveled, hardened-steel strap embedded in a wooden base with a cork stripper plate. This cutting-edge configuration gives a higher quality than other standard configurations. Single-ply laminates of all composite materials except Kevlar® cut cleanly and easily, requiring only minor die-position and pressure changes. Kv/Ep requires more buildup with paper and metal for an additional pressure and a better impression than the other materials to achieve a clean separation of the blanked configuration. Multiple cutting is clean and easily performed for all materials except Kevlar® prepreg. The criterion used for selecting the maximum number of plies is squareness of the cut. As the number of plies increases, edge squareness decreases, Kv/Ep requires significantly higher pressures since the fibers are difficult to sever, limiting the maximum number of plies cut (Table IV-8).

A recently completed "Manufacturing Technology Advanced Materials Program for Composite Vanes" evaluated[81] automating the cutting and stacking of 29 pieces of Gr/PI composite prepreg used for the outer shell of the engine inlet guide vane. The program examined various methods of cutting the prepreg with the need for an automated system. A steel rule die with elastic foam ejector pads that cuts 20 plies of prepreg simultaneously was selected as the most appropriate, flexible, automated system for cutting these unique composite airfoil components, Figure IV-7.

4.8.1.3 Laser-Beam cutting

Continuous-wave 250-W CO_2 lasers have produced cutting speeds up to 169 mm/s in single ply Kevlar® prepregs and up to 127 mm/s in single ply B/Ep. Although B/Ep laminates, up to four plies thick, were also laser-cut at a feed rate of 25.4 mm/s, the slow feed rate resulted in a marginal cut with a significant amount of resin retreat. The best cuts have occurred in two ply B/Ep laminates cut at a feed rate of 51 mm/s. A feed rate of 127 mm/s on single-ply Gr/Ep tape is the most effective cutting rate; for two ply laminates, 63.5 mm/s is recommended. Although Gr/Ep tape laminates up to three plies thick have been successfully laser-cut, an excessive edge bead of resin develops at 25.4 mm/s. In uncured Gl/Ep laminates, laser cutting has been accomplished at 127 mm/s for single-ply thicknesses and at 38 mm/s for thicknesses for up to three plies. Higher feed rates reduced cut quality, apparently because the resin could not vaporize quickly enough.

Woven graphite broad goods are readily cut at 127 mm/s with nitrogen-assist gas pressure. No evidence of heat damage to the fibers was visible. A Gr/PS with its thermoplastic matrix exhibits the greatest matrix damage. In tests, as much as 0.6 to 0.9 mm of the matrix material has been removed. Cutting rates up to 114.3 mm/s are easily obtained with the laser.[24]

4.8.1.4 Ultrasonic Cutting

Ultrasonic cutters are being used on most natural- and synthetic-fiber materials. With materials such as Gl, aramid, and C fibers, ultrasonic machines make clean cuts without displacing fabric or warp ends. On thermoplastic materials, both separation and selvedge sealing are done to prevent the edges from unraveling. And, unlike hot cutting, the selvedged edges are free of welding beads. The ultrasonic process also prevents yellowing because heat, generated by mechanical vibrations between a sonotrode and countertool, occurs on the inside of the material rather than the outside.

The ultrasonic cutters, vibrating at 20,000 strokes per second, use TiC blades at the end of a "horn" to reduce frayed edges in composite materials. The machine can cut polyimide-based resins, epoxies, and adhesive films.

4.8.2 Cured Composites

4.8.2.1 Sawing

"Flash" is excess material that must be removed in order to bring the composite part to its finished size. Where the flash is more than a mm thick or contains a high proportion of reinforcement, a robust cutting tool is required. The most suitable hand tool is a standard hacksaw with a metal-cutting blade, and this can also be used for general cutting operations on most composites.

To speed up the rate of cutting, some form of power tool is necessary. The jigsaw is a versatile and effective power tool but relatively slow. To achieve a high rate of cut on thermoset composites, the high speed abrasive disc cutter should be used. A diamond-impregnated cutting wheel is essential to achieve a reasonable wheel life. Circular saws can be used to cut composites, but both saw and workpiece must be firmly held to avoid binding. This means that the saw needs to be bench-mounted and the workpiece held in a traversing table or vice versa and a diamond-impregnated saw blade is necessary. Most of the cutting methods discussed above involve high surface speed abrasion and the inevitable result of this is the generation of large volumes of very fine dust. Efficient dust extraction facilities are essential to provide tolerable and safe working conditions.

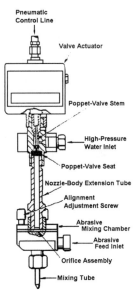

Pneumatic
Control Line

Valve Actuator

Poppet-Valve Stem

High-Pressure
Water Inlet

Poppet-Valve Seat

Nozzle-Body Extension Tube

Alignment
Adjustment Screw

Abrasive
Mixing Chamber

Abrasive
Feed Inlet

Orifice Assembly

Mixing Tube

Figure IV-8. AWJ cutting head.[24]

4.8.2.2 Water Jet (WJ) Cutting

In the past, cured composite parts were trimmed by bandsawing, hand routing, shearing, nibbling, and sanding. These processes were labor-intensive, noisy, and produced critical dust problems. WJ cutting has reduced the dust and noise problem and produced clean trimmed edges that eliminate hand sanding. Cured materials follow the rule of thumb; i.e., the cut quality improves with increasing nozzle pressure, increasing nozzle orifice diameter, decreasing traverse speed, and decreasing material thickness and hardness. Softer materials cut with a better edge quality than the harder ones. The order of decreasing hardness is B/Ep, Gr/Ep, Gl/Ep, and Kv/Ep, with hybrid materials occupying positions midway between the parent materials.[24,82]

Most of the water jets are based on the concept of a router walking around a part, pressing a template against it and trimming. Some system feature WJ cutting heads suspended in the air by an articulated, counterweighted arm. The user pushes against the template and walks around the part, trimming it as he goes. These systems cut quickly and leave an excellent edge. They do not have the tearing, fraying, or delamination of some router applications, and they are cleaner. Unfortunately, they are effective on material thicknesses of up to 2.54 mm.

A basic WJ cutting system consists of a water pressure booster and filtration system, hydraulic pump, nozzle, and catcher. The booster increases incoming water pressure to the level required by the intensifier; filters remove particles that would damage pump seals, check valves, and orifices. Usually, the unit runs at 345 MPa in conventional operation; the water is 12% compressible; for the first 12% of the stroke of the plunger, nothing will come out. To keep the jet running, compressed water is stored in the accumulator when the pump is making its initial compression (like a storage tank on an air compressor). Traveling at speeds up to 900 m/s, the water moves to the nozzle, where it passes through an orifice and emerges as a coherent cutting stream. At this point, the jet is a tight core surrounded by a shroud of light mist.

4.8.2.3 Abrasive Water Jet (AWJ) Cutting

The process (also known as hydroabrasive machining) is identical to waterjet cutting except for the injection of abrasive into the cutting stream.[83] To the above basic ultrahigh-pressure water-generating system is added recent developments that revolve around entraining abrasives (typically 80- to 100- mesh garnet) into the water stream. Part of the water's momentum is transferred to the abrasive particles which do most of the cutting. This cutting system adds an abrasive hopper, abrasive metering valve, and specially designed mixing chamber, and focusing nozzle,[24,84-87] Figure IV-8. Several attributes make AWJs superior to more conventional cutting methods in a variety of applications:

- Low operating temperature avoids the heat affected zones and distortion created by laser and plasma-arc systems. Thus, there is no need for expensive grinding, annealing, or secondary machining of heat affected surfaces to obtain final dimensions.

- The jet cuts with little lateral or vertical force and is virtually vibration-free. This minimizes mechanical stress at the material edge and eliminates the need for costly part-support fixtures.

- Abrasive water jets cut in any direction, and the small, lightweight cutting head allows AWJs to take full advantage of robotic manipulation that is often limited by the weight and geometry of traditional tools. Integrating AWJs with computer-numerical-control (CNC) manipulators combine omnidirectional cutting capability with the flexibility and repeatability of comput-erized cutting programs.

- Because the cutting head does not contact the workpiece, AWJs are highly maneuverable and excellent for cutting precise right angles and special shapes. In more conventional methods, the cutting path is restricted by blade or tool geometry.

- The 0.76 to 2.5 mm thick water stream minimizes width of cut, or kerf, which maximizes material use. AWJs cut small holes, narrow slots, and closely spaced patterns, and small parts can be nested within larger ones. Unlike a saw rough cut, which leaves material that must subsequently be machined away, an AWJ often requires only a single pass to produce final dimensions and finish.

Hand-held AWJ cutting is too dangerous, however. Entraining the garnet abrasive into the stream allows cutting much thicker materials than lasers can handle. AWJ cutting uses erosive action rather than friction and shearing for a finished edge. Low operating temperatures do not affect the material being cut at speeds of 381-762 mm/min and thicknesses of 0.5 to 25 mm in most materials and some difficult to cut materials up to 100 mm in Gr/Ep or MgB_4C. Moreover, there's less dust, especially in cutting Gr/Ep. While AWJ is not completely clean, it leaves no more than a small puddle on top of the part, whereas routers spray particles out at high velocity. Even the most elaborate vacuum systems have trouble keeping the dust down. With the abrasive cutter, a small point catcher follows the head, collecting residue, and dissipating kinetic energy. In addition, abrasive jets operate with low cutting forces. Typically, a workpiece would have about 4.4 N of force on it. This can be increased by cutting faster, but not above about 35.6 N on composites. At that level, the tooling will not handle any real weight. For example, on the 777, Boeing uses universal tooling for the composite tail sections it cuts with abrasive jets; with a router, it needs different tooling for every part.

Another advantage of waterjets is they can produce inside corners with 0.76-mm diameters. These tight corners still have some round to eliminate stress risers. Routers usually need tool bit diameters of 6.3 mm and that sacrifices speed. It is suggested to use the largest cutting head possible. The bigger the cutting head, the larger the stream, but you can cut up to 50% faster with a large head.[88-89] To reduce tooling and labor costs associated with net trimming of composite components, systems have been designed and implemented which combine the flexibility and accuracy of robotics with the productivity of AWJ cutting. The system is comprised of a large, six-axis gantry robot which uses specially developed abrasive water jet end effectors to trim the edge-of-panel (EOP) and integral stiffener blades. These end effectors employ compact catchers to contain the spent system and, thereby, eliminate the need for large catcher tanks commonly used in cutting. The robot is offline programmed to perform trimming on large, complex contoured panels while a vision system monitors kerf width.[90]

AWJ technology has become a much more reliable and proven technology over the past five years. On the cutting-equipment side, the work life of AWJ cutters has been extended, and the wearout rate been made more predictable.[91] Sapphire orifices now last 80 to 100 hours while their wear-out rate can now be predicted, making it possible to anticipate and compensate for changes in the size of the water/abrasive stream through appropriate computer programming. Nozzle life, too, has been increased to a maximum of 100 hours or more. On the robotics (or motion-equipment) side, improvements in programming have made the equipment far more accessible and user-friendly. The technology has gone from a very hands-on, operator-dependent technology to a hands-off, accepted manufacturing process.

Now that water-jet cutting systems can operate with as many as six axes of motion, they are capable of cutting three-dimensional, complex contoured parts, catching the kerf, water, and abrasive residue, and, in some cases, providing inspection via coordinate measuring machine (CMM) technology. For flat parts or parts with a slight contour, two- or three-axis machines can be used. A manufacturer reports that it can invert and mount the articulated arm that comprises the sixth axis of motion onto a vertical, servo-driven track positioned above the part to constitute a seventh axis.91

4.8.2.4 AWJ versus WJ

In several investigations and studies, Ramulu and Arola[92] found that the surface characteristics of WJ and AWJ machined unidirectional Gr/Ep composite show the following.

- The principal material removal mechanism present in WJ machining of unidirectional Gr/Ep composite is material failure associated with microbending-induced fracture and out-of-plane shear. WJ machining exploits the weakness of the composite mechanical properties.

- AWJ machining of unidirectional Gr/Ep composites consists of a combination of material removal mechanisms including shearing, micromachining and erosion. AWJ machining is found to be a more feasible machining process for unidirectional Gr/Ep due to its material removal mechanisms, and superior quality surface generation.

Geskin, Tisminetski, et.al.,[93] demonstrated that AWJ cutting of composites with continuous matter distribution is similar to cutting of conventional materials. The specific feature of composites machining is a narrow range of process variables which assures the optimal process results. AWJ can be used routinely for cutting of structural type composites which is possible only with reduced requirements to the surface waviness. However, the results of their study showed the feasibility of developing a reliable technology for shaping of these materials. The use of WJ was best suited for rough cutting of materials, not to generate finished surfaces. The jet based machining enables the generation of a wide variety of flawless surfaces with acceptable topography. The improvements of flow and robot control will result in the development of AWJ based technologies for manufacturing complex parts into the next decade.[93]

4.8.2.5 Abrasive Liquid Jet Machining (Cutting)

Machining with abrasive liquid jets entails the removal of material by the erosive action of impacting particles. Relatively high levels of particle kinetic power density are delivered to the workpiece. The key to obtaining these high levels of kinetic power is the use of liquid (as opposed to gas) jets to accelerate the particles.

There are two main methods of forming abrasive liquid jets:

- Entrainment: Abrasives are entrained and accelerated using a high-pressure liquid jet. Entrainment systems typically employ a high-velocity waterjet to entrain abrasives and form an AWJ to perform the cutting.

- Direct Pumping: Abrasives are premixed with a suspending liquid to form a slurry. The slurry is pumped and expelled through a nozzle to form an abrasive slurry jet (ASJ).[94]

Abrasive liquid jets can be used in many types of machining applications. Due to the large number of parameters involved in abrasive liquid jet machining, optimization is an important task. To help provide a basis for such optimization, a comparative study of entrainment AWJ systems and directly pumped ASJ systems was undertaken and the following conclusions were made by Hashish.[94]

- High-pressure entrainment AWJ systems (up to 345 MPa) are more effective in material removal than low pressure directly pumped ASJ systems (up to 70 MPa) in the use of hydraulic power and a consumption.

- Directly pumped ASJ systems are generally more efficient than entrainment AWJ systems and produce greater power density for impacting particles. At $r = 1$ and $k = 2$, directly pumped ASJs will be at least eight times as dense in power as entrainment AWJs. For $k = 10$, directly pumped ASJs will be approximately 24 times as dense in power as entrainment AWJs.

4.8.2.6 Water Jet/Laser Grooving Process

Chryssolouris, Sheng, and Anastasia[95] examined a method for reducing the thermal damage of composite materials during laser grooving. The method utilizes a water jet in tandem with the laser beam in an attempt to control the heat conducted into the workpiece.

Laser grooving is a steady-state process, since the erosion front is stationary with respect to the moving laser beam coordinates. It has a heat affected zone (HAZ) at both the walls and the bottom of the groove, and heat conduction occurs in three dimensions. It differs from laser cutting (LBC – more later) which is also a steady-state process, but, the HAZ is only formed at the walls of the kerf, since the kerf is "bottomless."

174

In subsequent tests by the Laboratory for Manufacturing and Productivity,[95] the following effects were found:

- Up to 70% reduction in heat affected zone was achieved by using a water jet, with corresponding groove depth reductions from 43 to 45%.

- The analysis for laser grooving without a water jet showed linear relationships (in log scale) between groove depth and damage width versus non-dimensional energy. The analysis for laser grooving with a water jet showed a linear relationship between damage width and NDE while the groove depth showed a relationship that increased non-linearly with increase in NDE.

- The results derived from theoretical analysis for the damage width predict an 80% reduction in the HAZ, and, therefore, were within the range of experimental measurements.

4.8.2.7 Laser Beam Cutting (LBC)

Laser cutting has become an industrially accepted production process mainly because of the flexibility of the technique and the high quality cuts which can be made. With the development of composite materials and the introduction of a reinforcement phase, engineers found that the fiber type had a significant influence on the cut quality, carbon fiber reinforced materials proving most difficult to cut. The main problem with the carbon fiber reinforced composites in laser cutting was fiber/resin separation.

For glass and aramid reinforced composites, less separation between the fiber and the resin has been observed. From this work at TWI,[96] it was concluded that CO_2 laser cutting is not capable of producing cuts in carbon PMCs without a degree of separation between the matrix and the fibers. If avoiding this separation is considered to be vital to the performance of the components, alternative cutting techniques, such as water jet, should be considered. Also, the laser cutting process may create hazardous by-products.[97] For example, aromatic compounds and polycyclic aromatic hydrocarbons have been identified from the degradation of aramid fiber composites.

Mukherjee and associates at Michigan State University[15] studied the three major PMCs using various combinations of laser input power (800-1,500 W) and cover gas pressure (0.07-0.55 MPa). For cutting, maximum speed was 0.17 m/sec. They also used several slower speeds for Gr/Ep composites. Cover gas pressure of 0.55 MPa produced the least charring and most efficient cutting.

Gl/Ep – Cutting speed was much higher than conventional machining and more complex cutting geometries could be programmed with the laser.

Kv/Ep – With this material, the extent of fiber damage and edge fraying was much lower than with conventional machining. There was no visible delamination with either straight or cross-cutting.

Gr/Ep – Unlike the other materials, Gr/Ep did not laser-machine well. Because the ablation temperature of graphite fibers is high, extensive ablation, burning, and thermal damage occur long before the graphite is cut. The graphite-reinforced composite also absorbs a lot of incident laser energy. Finally, it has high thermal anisotropy, which produces a preferential temperature gradient along the direction of easy conduction, regardless of cutting geometry or speed.

Another series of tests by Viegelahn, Kawall, Scheuerman, et.al.[98] were conducted to determine the basic cutting parameters for laser cutting of fiberglass composites of different thickness with basic tensile strength determinations made from the different thicknesses.[98]

Conclusions drawn were that:

- The laser could be used as a tool for quickly cutting out sections from fiberglass-polyester composites.

- Tension tests indicated no noticeable change in the strength characteristics due to laser cutting.

- Black sooty edges could be quickly cleaned by lightly sanding the surface.

- The cutting residue, in the form of smoke and vapor, can be handled quickly and easily by a ventilation system.

LBC has been applied to MMCs. However, as with PMCs, there are problems in removing the matrix and leaving the fiber materials protruding for contact techniques (tool wear) and non-contact techniques (e.g. laser and water jet cutting).[96]

Using a 1.5 kW CO_2 laser, the scientists at TWI[96] laser cut 8090 Al/SiC/20$_p$. A 2 mm thick sample showed, at the cut edge, only a small amount of dross and a cut surface where no fibers were protruding from the matrix. The cuts were similar in appearance to those achievable on a monolithic aluminum alloy.

A Ti-6Al-4V alloy with SiC fiber reinforcement unidirectionally aligned revealed the cut edges of fibers, 1 mm thick sample, protruding from the matrix. Two distinct modes of behavior in laser cutting of MMCs were identified. One appears to be where the composites behaved as a metal and the other where fiber and resin separation occurs in a similar way to carbon fiber reinforced polymer composites. However, at present, there is limited information available on laser cutting parameters for MMCs with different types of reinforcement.

Lau, Yue, Jiang, et.al.,[99] used a pulsed Nd:YAG laser to cut 8090 Al/SiC/20$_p$ to study the cutting behavior of the MMC. They found that it was possible to minimize the HAZ and to improve the quality of the machined surface by varying the laser operating parameters and, at the same time, to achieve the best possible machined surface quality.

In pulsed laser cutting of SiC/Al-Li MMC, the thermal effect will cause microstructural changes in the adjacent machined area. In practice, the HAZ width can be minimized mainly by using smaller pulse energy and shorter pulse duration. It is obvious that the higher feed rate will increase the speed of cutting. However, the pulse overlap condition must be fulfilled. In determining the cutting parameters, the depth of cut, the cutting efficiency, and the HAZ width should be considered simultaneously.[99]

Meinert, Martukanitz, Bhagat[100] studied laser cutting of MMCs with a 1.5 kW CO_2 laser. They demonstrated that the absorptivity of the composites is higher than that of the unreinforced alloy 6061/Al$_2$O$_3$/20$_p$-T6 composite compared to the matrix alloy. This is an inverse relationship to conventional machining practices where the increased reinforcement content leads to decreased processing efficiency. The use of a filler alloy containing titanium, to form TiC in preference to Al$_4$C$_3$, eliminated the formation of Al$_4$C$_3$ during welding of 6061/SiC/20$_w$ composites. Titanium additions result in the formation of an in-situ, TiAl$_3$ matrix having TiC reinforcement within the fusion zone of the weld. Further development of filler alloy additions and processing techniques are necessary to obtain a fusion zone having an aluminum alloy matrix with fine dispersed TiC dendrites.[100]

Table IV-9 Wide Range of Ceramic Cutting Tool Compositions

Category	Compositions	Comments
Carbides	WC, TiC, MoC, NbC	Cemented Carbides (WC) are Sintered with Metals SiC-Whisker Reinforcements in Ceramics
Nitrides	BN, Sialon, TiN	BN is Expensive Because of High Manufacturing Costs TiN is Used as a Coating for WC and High-Speed Steel
Oxides	Al_2O_3, ZrO_2	Pure Alumina has Been Replaced with Alumina Containing $\leq 15\%$ ZrO_2
Carboxides	TiC Dispersed in Al_2O_3	Called Black Hot-Pressed Ceramics, Contain $\leq 40\%$ TiC to Improve Hardness
Carbonitrides	TiN with Carbide Additions	More Commonly Known as Cermets Because of Metal Binder Phase

4.9 CUTTERS

4.9.1 Composites and Cutting Tools

Cutting tools exemplify the application of materials in several high-stress and high-temperature environments. In metal cutting, the cutting tool, acting as a blunt wedge, removes material from the workpiece by deforming a thin layer of the workpiece and separating the material near the cutting edge of the tool by a high-strain-rate plastic shear mechanism. A significant part of the mechanical energy supplied in the work of cutting is converted into thermal energy and, consequently, high temperatures are generated. Heat produced by the primary shearing process flows into both the workpiece and the chip, with the majority going into the chip, while the friction developed at the tool-chip interface and rubbing along the tool flank heats both the tool and the workpiece.

The primary mode of mechanical wear for ceramic cutting tools is abrasion, the removal of tool material by a scouring action of protruding asperities and by hard phase inclusions in the workpiece and chip. The two body abrasive wear resistance of metals that can accommodate large strains prior to fracture has been the subject of numerous investigations, and it has been shown that the wear resistance is determined by the hardness of the material.

There have been no clear-cut scientific selection criteria developed to match the specific workpiece with the optimum and most cost effective type of cutting tool. Most often users follow the recommendations of a vendor or the guidelines in handbooks, and they use their own trial-and-error experience to arrive at the proper choice of cutting tool.

The scientific approach would involve all of the measured data and the criteria developed as the fallout of all test results for the user to be able to systematically arrive at the best choice for a specific machining application.

4.9.2 Cutting Tool Materials

Because hard materials are used to cut softer materials and because ceramics are inherently hard, an interest in the use of ceramics for cutting tools developed in Germany around the beginning of the century (1905).[101] Alumina was the material considered for cutting tools. Patents were issued in 1912 and 1913 in the United Kingdom and the Federal Republic of Germany, respectively. Cemented carbides were introduced in Europe in 1926. During World War II, carbides were used to increase the cutting capability of the earlier HSS tools. This was followed by the introduction of Al_2O_3-coated WC. In the 1960s, ceramic cutting tools were

177

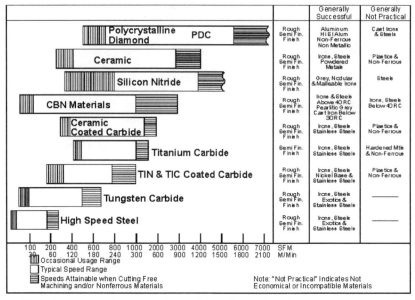

Figure IV-9. Spectrum of cutting tool materials and guidelines for use.

based on sintered or hot-pressed polycrystalline alumina. Two factors limited the widespread adoption of these Al_2O_3 tools by the metal cutting industries. The inherent low fracture toughness of the material limited the range of application, and the low thermal conductivity increased the susceptibility of the tool to damage by thermal shock.

It is apparent that, since the end of the 1960s, the strengths of ceramic tools have been improved through a better control of the microstructure and by the addition of additives. With a better understanding, improved mechanical, thermal, and chemical properties of ceramics were obtained and adapted to ceramic cutting tools. In addition, composition variables gave rise to the cermets, and additions of titanium, TiC, and WC were made to improve the strength of Al_2O_3 cutting tool, Table IV-9.

Harder materials such as PCD and cubic boron nitride (CBN) were developed for use during the 1972 to 1974 time period. CBN, Al_2O_3-TiC ceramics, and the newer CMCs based on Si_3N_4, SiC_w/Al_2O_3, Al_2O_3/ZrO_2, are very promising because of their wear resistance and effective application to the machining of nickel- and cobalt-based superalloys in the jet engine industry. Polycrystalline diamonds find excellent application in machining abrasive materials such as silicon-aluminum automotive alloys and glass- and carbon-fiber-reinforced composite materials.

The demand for CMCs will strengthen as the ceramic industry scales up manufacturing processes, costs become more competitive with conventional materials, and users gain confidence in the composites' properties and reliability. Major market segments are the cutting-tool inserts, TiC_p/Si_3N_4, TiC_p/Al_2O_3, and SiC_w/Al_2O_3. Therefore, the history of ceramic cutting tools has evolved from the early aluminas to the newly-tailored aluminas and other CMCs, Figure IV-9.

4.9.2.1 Alumina

Two innovations in alumina material technology, aimed at increasing the fracture toughness of Al_2O_3, have found commercial utilization as ceramic cutting tools. The first, transformation-toughened alumina, was introduced to the marketplace about 1980 and was targeted to expand the application range of ceramics for steel machining. This family of ceramics consists of particles dispersed in an alumina matrix. Under the application of load, the tensile stress field surrounding a crack tip causes the tetragonal ZrO_2 to transform to the stable monoclinic polymorph. This transformation with an accompanying volume expansion of the dispersoid, leads to an increase in fracture toughness and retards crack propagation.

The newest alumina-base materials, SiC-whisker-reinforced Al_2O_3, was added to the cutting tool arsenal only within the past five years. The whisker reinforcement produces a two fold increase in fracture toughness of the composite relative to monolithic alumina. SiC-whisker-reinforced Al_2O_3 has received widespread acceptance in the aerospace industry, where it is considered the state-of-the-art cutting tool material for rough machining of superalloys.

One of the first ceramic composite cutting tools was SiC_w/Al_2O_3.[102-106] Advanced Composite Materials Corp. (ACMC) grows SiC crystals from rice hulls, which are composed mostly of relatively pure silica. The silica reacts with carbon when heated to 1,800°C. Joining these SiC whiskers with Al_2O_3 produces ARtuff®, ACMCs high-performance ceramic composite. ARtuff® ceramic composites make high-performance cutting tools for superalloys, nozzle inserts for water-jet looms in the textile industry, cutting tools in the furniture industry, and can-forming tools in the beverage industry. As with most whisker-reinforced ceramics, ARtuff® exhibits several advantages, including strength, fracture toughness, and resistance to wear. and thermal shock.

Another firm, Greenleaf Corp., uses ACMCs SiC whiskers in their machine tool inserts. Their WG-300 Al_2O_3 with SiC_w forms dozens of cutting tools that make cuts on nickel-based alloys without smearing.

WG-300 has high thermal and mechanical shock resistance that holds up six to eight times better than cemented carbides. It has been tested at speeds as high as eight times carbide at 1.5 times carbide feed rates, and looks particularly good for machining forged high nickel alloys (e.g., Inconels and Hastelloys), where high thermal and mechanical shock are encountered.

Whisker reinforcement more than doubles the fracture toughness of oxide ceramics up to temperatures as high as 1,000°C; this makes them more resistant to catastrophic failure; and reduces their chances of failure with time. The composite's thermal conductivity is also about 40% better than the unreinforced matrix material.

Because of WG-300's increased toughness, cutting tools made with it have sharp rather than beveled edges. Cutting tools with beveled edges cause finished cuts in the hardest alloys to look smudged. Cutting and shaping tools using WG-300 inserts have 10 x the metal removal rate of those fitted with WC. Whiskers also greatly enhance the material's thermal conductivity and thermal shock resistance, permitting use with coolants.

Greenleaf Corp. also makes another whisker-reinforced cutting tool, WG-100, which forms drills, mills, routers, and other tools to machine advanced composites such as CCC, Kv/Ep, and nonmetallic substances.

The Al_2O_3 with 5% ZrO_2 has been used to machine gray cast iron for automotive components. Toughened Al_2O_3 with 20% ZrO_2, has been used at cutting speeds of up to 5,000 sfpm, and was adopted by the steel industry because of its capability for breaking steel chips. Al_2O_3-TiC composites, containing at least 20% TiC, formed by hot pressing and sintering in an inert atmosphere has been used for milling and rough turning applications of cast iron and high temperature nickel-based superalloys. These Al_2O_3-TiC carbide tools are cost competitive with traditional carbide tools, but can operate at significantly higher cutting speeds. A cutting tool insert of SiC-Al_2O_3, Quantum 10, is claimed to out-perform other ceramic tool inserts, such as coated carbides, cermets, and CBN, both in cutting speeds and tool life when cutting hardened steels and superalloys. Competitive materials are being produced in Japan where a ceramic composite ultrahard tool material based on SiC_w and ZrO_2 in an Al_2O_3 matrix is reported with a fracture toughness of 8 (MPa) $m^{1/2}$ and a three point bend strength of 1,600 MPa.

4.9.2.2 ZrO_2

Two new cutting tools have been introduced from toughened ZrO ceramic.[107] Corning claims that the material offers three times the ductility of a comparable ZrO_2 material, as well as metal-like, strain-to-failure characteristics normally found only in fiber-reinforced ceramic materials.

The toughened ZrO_2 combines with Al_2O_3 and has successfully machined ferrous-based alloys. Initial test results show that ceramic composite cutter offers up to six times the wear resistance of cutting tools made from coated carbides.

The Greenleaf Corp. claims that the new cutting tools have outperformed all traditional ceramic and non-ceramic products at both high and low cutting speeds in most applications. The ceramic material doesn't chip or fracture at the edges, as often occurs in conventional ceramic cutting tools. Greenleaf has found that the new products have twice the strength of the company's more traditional Al_2O_3 products.[108]

Norton Co. has developed Norzide® zirconia (TZP). The material belongs to a relatively new class of sintered ZrO_2-based ceramics characterized by high strength, high toughness, and superior wear resistance. One formulation of the material, Norzide® TZ-110HS, has been used as a bed knife in a granulator and lasted three times longer than the submicron grain size grade C-10 WC previously used, according to Norton.[107] In another knife application – a slitting operation – the same material's precision-honed cutting edge outlasted razor blade grade stainless steel by a factor of 24.

J. Withers[109] examined two composites, sol-gel derived Al_2O_3-ZrO_2-SiC_w and pressureless sintered ZrO_2-TiC composites which were determined to process the optimum properties for cutting tool applications. The ZrO_2-TiC composites, which were prepared and consolidated by novel processing, exhibited exceptional high temperature properties. Machining tests of HSS were performed with these composites. The SiC whisker-reinforced composites and the ZrO_2-TiC composites had a significantly improved volume of metal removal over the commercial cutting tools for machining steel. The ZrO_2-TiC composites have the advantage, in addition to superior cutting performance, of being produced via a cost effective method (pressureless sintering).

4.9.2.3 Si_3N_3

Ford Motor Co. introduced an S-8 ceramic cutting tool material consisting of Si_3N_4 with additions of Y_2O_3 and Al_2O_3. It has been used to machine production auto engines. Several General Motors divisions are also using these tools to bore cut 2.5-L engines, turning brake drums, and face-milling 4.1-2 V-8 engine blocks.[110] Norton Co. has developed a Si_3N_4-reinforced with SiC_w ceramic cutting material.

Table IV-10 Materials Comparison Chart

Material	Strengths	Weaknesses	Typical Applications
HSS	Superior Shock Resistance Versatility	Poor Speed Capabilities; Poor Wear Resistance	Screw Machine and Other Low-Speed Operations, Interrupted Cuts, Low-Horsepower Machining
Carbide	Most Versatile Cutting Material; High Shock Resistance	Limited Speed Capabilities	Finishing to Heavy Roughing of Most Materials, Including Irons, Steels, Exotics, and Plastics
Coated Carbide	High Versatility; Good Performance at Moderate Speeds	Limited to Moderate Speeds	Same as Carbide, Except with Higher Speed Capabilities
Cermet	High Versatility; Good Performance at Moderate Speeds	Low Shock Resistance; Limited to Moderate Speeds	Finishing Operations on Irons, Steels, Stainless Steels, and Aluminum Alloys
Ceramic Hot/Cold Pressed	High Abrasion Resistance; High Speed Capabilities; Versatility	Low Mechanical Shock Resistance; Low Thermal Shock Resistance	Steel Mill-Roll Resurfacing, Finishing Operations on Cast Irons and Steels
Ceramic Silicon Nitride	High Shock Resistance; High Thermal Shock Resistance	Very Limited Applications	Roughing and Finishing Operations on Cast Irons
Ceramic Whisker Reinforced	High Shock Resistance; High Thermal Shock Resistance	Limited Versatility	High-Speed Roughing and Finishing of Hardened Steels, Chilled Cast Iron, High-Nickel Superalloys
Cubic Boron Nitride	High Hot Hardness; High Strength; High Thermal Shock Resistance	Limited Performance on Materials Following: 38 RC; Limited Applications; High Cost	Hardened Work Materials in 45 - 70 Rockwell C Range
Polycrystalline Diamond	High Abrasion Resistance; High Speed Capabilities	Limited Applications; Low Mechanical Shock Resistance	Roughing and Finishing Operations on Abrasive Non-Ferrous or Non-Metallic Materials

4.9.2.4 Sialon (Si_3N_4-Al_2O_3)

Kennametal[110] has developed several sialon cutting tools. Kyon 2000 has an impact resistance approaching that of most ceramic-coated carbides and is capable of cutting at the high speeds normally associated with conventional ceramics. Kyon 2000 also offers excellent thermal-shock resistance; coolant can be used to improve both tool life and surface finish. The sialon cutting tools, such as Kyon 2000, increased metal-removal rates by 100-300% for cast iron and 400-800% for high-temperature nickel-based alloys when compared to carbides. Recently introduced was Kyon 3000 which has better performance than 2000

4.9.2.5 Titanium Carbide

GTE has developed a composite of Si_3N_4-TiC with a high fracture toughness and hardness better than sialon alloys. Originally introduced as Quantum 5000, the composite consists of a matrix phase (70%) and a dispersed phase of 30% TiC. The Si_3O_4 matrix phase also contains additions of Y_2O_3 and Al_2O_3. GTE claims higher productivity (higher feed rates and cutting speeds) than sialon; has introduced an improved version known as Quantum 6000 which it claims has superior performance.

4.9.2.6 Titanium Boride

In a series of developments in Japan, several laboratories have developed composites of 70 wt% Ti(C,N)-30 wt% TiB$_2$, a three-point flexural strength of over 800 MPa at room temperature, K_{1c} > 5 MPa $\sqrt{m}$,H > 2,500, α = 8 x 10^{-6} °C^{-1}, as well as excellent cutting tool behavior. The Ti(C,N)-Cr$_3$C$_2$ system is also now under investigation. Both Ti(C,N)-TiB$_2$ and Ti(C,N)-Cr$_3$C$_2$ system cutting tools have been verified to have a longer lifetime for the high speed (300 m/min) cutting of plain carbon steel than the lifetime for WC-Co and TiN cermet tools under comparable circumstances. These new composite materials are both now under evaluation for cutting heat-resistant and stainless steels.

4.9.3 Composite Cutter Applications

A new variety of ceramic cutting-tool materials which was introduced in recent years and is now used commercially in a wide range of metal-cutting applications, including turning and milling, is comprised of three main categories: metal oxides, metal nitrides, and metal carbides, Si$_3$N$_4$-based ceramics, generally used for machining gray cast irons, also have sufficient toughness for milling applications. However, the low toughness of the oxides and carboxides allows for interrupted cuts only at very low feed rates. Consequently, ZrO$_2$ is added to Al$_2$O$_3$ to improve toughness of this composite cutting tool.

A hot-pressed ZrO$_2$-Al$_2$O$_3$ ceramic cutting tool[111] insert is being used by an aircraft engine manufacturer to machine a variety of engine parts made of Inconel 718. Often, many of the company's parts must be production-machined in the fully heat-treated condition, some as hard as 48 HR$_c$. Cutting speeds were increased from 34 m/min for cemented carbide to 244 m/min, which is nearly double the speed of conventional hot-pressed ceramics and a seven-fold improvement over cemented carbide.

Higher cutting speeds now mean that operations involving long cuts do not have to be stopped for tool changes. Surface finishes are improved due to higher cutting speeds. Feed rates are comparable to rates formerly used with carbide cutting tools. The depth of cut is unchanged except for severe roughing cuts. Even chip removal is easier, because at higher cutting speeds chips are broken into flakes, reducing chip control problems.

Another category is the Al$_2$O$_3$-TiC material. The addition of up to 30 wt% of TiC to Al$_2$O$_3$-based cutting tools has resulted in what is often called a composite or black ceramic, and increased the transverse rupture strength (breaking strength) over the basic Al$_2$O$_3$ ceramic. These ceramic materials have been very successful in turning hard cast irons and heat-treated steels hardened to 64 to very fine surface finishes and tolerances normally obtained only by grinding. Parts machined with this category of ceramic cutting tool range from small gear parts for the automotive industry to large form rolls for steel mills. The cutting speed is 1.3 m/s, the feed rate is 0.25 mm/rev. Productivity can double with this type of material compared to cemented WC.

Ceramics which combine Al$_2$O$_3$ and TiC are known by many names in addition to hot-pressed and black ceramics. Other terms used to describe such ceramics include cermets, composite ceramics, and modified ceramics. Hot-pressed ceramics have found application in the replacement of conventional carbide in many jobs. For example, Al$_2$O$_3$-TiC hot-pressed ceramic inserts have replaced carbide tooling to effect a productivity increase and an improvement in surface finish, eliminating the need for a subsequent polishing operation.[112]

When using Si_3N_4-based composite tools which are applied correctly, there are multiple benefits:

- Increased machine tool capacity.
- Better quality parts.
- Reliable tool life without insert breakage.
- Fewer tool changes and less machine down time.
- Increased tool life.

These add up to improved productivity and reduced cost.[113]

The higher toughness and thermal-shock resistance of Si_3N_4-based ceramics, compared to Al_2O_3, permit the rough turning and milling of cast irons under severe conditions. These conditions include heavy interruptions, variations in depth of cut, rough scale, or the use of a coolant. High-temperature nickel-based alloys can also be turned quite efficiently with this type of cutting tool. However, steels are unsuitable for machining with Si_3N_4 ceramics. An example of an application with this material involves the milling of a housing made of gray cast iron. The cutting speed was 13 m/s and the feed rate 0.15 mm per tooth. This is the third group of ceramic cutting tools. Inconel 718 castings are machined using a Si_3N_4-based ceramic with a 90% reduction in cutting time and an 84% reduction in total machining cost. In another application using Si_3N_4 tools, a gray cast iron automotive brake disk increased feed rate over Al_2O_3 from 0.38 to 0.51 mm/rev and provided an average tool life of 400 pieces per cutting edge versus 150 for Al_2O_3.

Another class of cutters are the sialons.[114] Field tests have shown sialon cutting tools to be successful in the machining of cast iron, nickel-based superalloys, and aluminum-silicon alloys. In these tests, the sialon tools exhibited longer tool life (at both low and high cutting speeds) than the cutting tools traditionally used for these materials.

By using pressureless-sintered β-SiAlON, materials perform better than HIPed β-Si_3N_4 ceramics in machining nickel-based superalloys and the longest tool life has been obtained for (α + β)-SiAlON materials.[114]

Cutting tools are the leading edge of commercial ceramics, and breakthroughs have a way of working back to other applications where wear and heat resistance are important. It is heartening to see how researchers have solved sintering problems that occur when Si_3N_4 cutting tools are strengthened with TiC.

While TiC-Si_3N_4 tools do have excellent properties, the two materials react to form nitrogen gas when sintered. The released gas forms pores, which concentrate stresses and weaken the tool. The solution found was to coat TiC particles with thin TiN coatings. The TiN film keeps the TiC from reacting with Si_3N_4 during sintering, so no nitrogen gas or by-products are formed. The resulting tool is easier to sinter and performs better than TiC-Si_3N_4 tools.

The fourth and final grouping of tools are SiC whiskers, which are now commercially available from various manufacturers as reinforcements for cutting tools. The whiskers improve toughness, strength, thermal-shock resistance, and reliability. These reinforcements can be added to oxide, nitride, and carboxide ceramics. Commercially available grades contain fewer than 30% of whiskers, resulting in higher toughness and better thermal conductivity.

SiC whisker-reinforced tools have been applied mostly in the machining of high-temperature alloys, resulting in significant productivity increases. The rough turning of a turbine rotor made of Incoloy 901 (40 HRc) was machined using round inserts 12.7 mm in diameter at a cutting speed of 4.5 m/s and a feed rate of 0.15 mm/rev. Stock removal was about five times faster than with conventional cemented carbides.

SiC whisker-reinforced Al_2O_3 inserts have been evaluated in the machining of Inconel 718 also.[104] Thangaraj and Weinmann used 12.7 mm diameter round inserts at cutting speeds ranging from 6.0 to 13.0 m/s and the feed rates ranged from 0.13 to 0.51 mm/rev. They found that tool failure in the cutting of the relatively soft (220 HB) nickel-based superalloy was due to excessive wear. Flank wear played a larger role at the lower speeds, but depth-of-cut notch wear was significant at the higher speeds. Abrasion, adhesion, and chipping were found to be dominant wear mechanisms.

Al_2O_3-SiC composite inserts generally contain 30 to 45% SiC whiskers. Extensive physical, mechanical, and performance data on Al_2O_3-SiC whisker material are now available.[115-116] It has been reported[117] that the "characteristic strength" of hot-pressed Al_2O_3 bodies increased from about 504 to 690 MPa as the content of SiC whiskers increased from 0 to 30 wt%. Adding SiC whiskers, followed by hot pressing, can more than double Al_2O_s fracture toughness to a value of 8.7 MPa • $m^{1/2}$.[118] In actual metal cutting of alloys such as Inconel 718, composite tools have performed up to three times better than conventional ceramic tools and eight times that of carbide.[119-120]

SiC/Al_2O_3 whisker orientation, fracture behavior, and cutting performance have been studied at laboratories and universities throughout the world.[105-106,121]

4.10 CONCLUSIONS

Advanced ceramic and ceramic composite cutting tools offer improved productivity in metal-cutting compared to more traditional (WC) cutting tools. The advanced ceramic and ceramic composite tools can achieve longer lives at higher speeds, resulting in less downtime and less operation time which results in less required per part machined. These reductions in routine machining costs may be partially due to increases in non-routine machining costs due to the brittle nature of ceramic materials. The brittleness causes reliability problems that, in turn, may result in a greater likelihood of premature tool failure and lost production.

The competitiveness of U.S. industry in the field of advanced ceramic and ceramic composite cutting tools is of great concern. At stake is more than just the market for these tools. Because cutting tools represent an early commercial application for structural ceramic applications, a dominant technological and market position in cutting tools may have important synergistic effects on a country's future technological and market positions in these structural ceramic areas. To some extent, U.S. industry appears to trail European and Japanese industry in advanced ceramic and ceramic composite cutting tools. Japanese and European producers have had more production experience and greater commercial success with advanced ceramic and ceramic composite cutting tools than have their U.S. rivals. Indeed, approximately 50% of the U.S. consumption of advanced ceramic and ceramic composite cutting tools is accounted for by imports. The next decade will allow CMC materials and ceramic cutting tools to realize their full potential.[101,110,122]

184

REFERENCES – CHAPTER IV

1. S. Sahanmir, L.K. Ives, A.W. Ruff, Ceramic Machining: Assessment of current practice and research needs in U.S., National Insti. of Standards and Technology, NIST-SP-834, 102 p, June 1992.

2. L.J. Rhoades, Advances in some specialized grinding processes, Milton C. Shaw Grinding Symposium, R. Komanduri and D.Maas (Eds.), ASME Press, New York, NY, pp 107-26, 1985.

3. R. Jordan, Waterjets on the cutting edge of machining, Proceedings of the Conference on Nontraditional Machining, ASM Inter., Materials Park, OH, pp 13-22, 1986.

4. M.G. Schinker, Subsurface damage mechanisms at high-speed ductile machining of optical glasses, *Precision Engineering,* **13,** 208-18 (1991).

5. T. Sugita, K. Ueda, T. Hashimoto, Fracture mechanics and material removal mechanism in microcutting of sintered alumina, *Proceedings of the Inter. Symposium on Ceramic Components for Engine,* Tokyo, Japan, pp 690-99, 1983.

6. K. Ueda, T. Sugita, H. Hiraga, A J-integral approach to material removal mechanisms in microcutting of ceramics, Manufacturing Technology, *Annals of CIRP,* **40,** 61-64 (1991).

7. K. Ueda, T. Sugita, H. Tsuwa, Application fracture mechanics in microcutting of engineering ceramics, *Annals of CIRP,* **32,** 83-86 (1983).

8. N. Nakajima, Y. Uno, T. Fujiwara, Cutting mechanism of fine ceramics with a single-point diamond, *Precision Engineering,* **11,** 19-25 (1989).

9. T.G. Bifano, T.A. Dow, R.O. Scattergood, Ductileregime grinding of brittle materials, Ultraprecision in Manufacturing Engineering, M. Weck and R. Hartel (Eds.), Springer-Verlag, New York, NY, pp 22-40, 1988.

10. P.N. Blake, R.O. Scattergood, Ductile regime machining of germanium and silicon, *Journal of the Amer. Ceram. Soc.,* **73,** 949 (1990).

11. N.F. Petrofes, A.M. Gadalla, Electro-discharge machining of advanced ceramics, *Ceram. Bull.,* **67,** 1,048 (1988).

12. M. Taya, M. Ramulu, (Eds.), Conference on Machining Composites, ASME Press, New York, NY,1988.

13. T. Uematsu, K. Suzuki, T. Yanase, et.al., New complex grinding method for ceramic materials combined with ultrasonic vibration and electro-discharge machining, Machining of Advanced Ceramic Materials and Components, ASME Press, New York, NY, pp 135-40, 1988.

14. H. Tsuchiya, T.Inoue, M. Miyazaki, Wire electrochemical disc machining of glass and ceramics, *Proceedings of the Fifth Inter. on Production Engineering,* Tokyo, Japan, pp 413-17, 1984.

15. R.R. Schreiber, Composites: A mixed blessing, *Manufacturing Engineering,* **109** (1) 87 (1992).

16. Kevlar Cutting and Machining Handbook, E.I. du Pont de Nemours & Co.

17. A. Koplev, A. Lystrup, T. Vorn, The cutting process, chip and cutting forces in machining composites, *Composites,* **14** (4) 371 (1983).

18. Y. Hasegawa, S. Hanasaki, S. Satanaka, Characteristics of tool wear in cutting of GFRP, *Proceedings of the 5th Int'l. Conf. on Production Engineering,* Tokyo, Japan, pp 185-90, 1984.

19. G. Santhanakrishnan, R. Krishnamoorthy, S.K. Malhotra,. Machining of composites: Surface morphology and tool wear studies, pp 411-17, NASA CP3018, Metal Matrix, Carbon, and Ceramic Matrix Composites 1988, ed. by J.D. Buckley, Cocoa Beach, FL, Jan. 1988.

20. A.G. Meyer, Composite gearboxes, designing for the 90's, AIAA/SAE/ASME/ASEE 27th Joint Propulsion Conference, Sacramento, CA, AIAA 911910 CP, 8 p, June 1991.

21. T. Beard, Machining composites – New rules and tools, *Mod. Mach. Shop,* pp 7485, April 1989.

22. W. Konig, C. Wulf, P. Grab, et.al., Machining of fiber reinforced plastics, *Ann. CIRP,* **34** (2) 537 (1985).

23. K.S. Kim, D.G. Lee, Y.K. Kwak, et.al., Machinability of carbon fiber-epoxy composite materials in turning, *J. Mater. Proc. Technol.,* **32** (3) 553 (1992).

24. M.M. Schwartz, Composite Materials Handbook, 2nd edition, McGraw-Hill, Inc., New York, NY, 650 p, 1992.

25. W. Marx, S. Trink, Manufacturing methods for cutting, machining and drilling composites, Vol I, test results, and Vol II, Final Rep August 1976 - August 1978, Grumman Aerospace Corp., Bethpage N.Y., AFML-TR-78-103, Contr. F33615-76-C-5280.

26. T. Kuhl, G. Devlin, J. Bunting, The mechanics and economics of advanced tooling systems for drilling and countersinking composite materials, SME, AD-89-645, FASTEC'89, Oct. 1989.

27. T.P. Murray, Machining composites for medical devices, SME EM-89-104, *Comp. in Manufact.,* Oct 1989.

28. C.W. Beeghly, Get your shop out of the uncoated age, Proc. of SCTE '89 Conference, High-Speed Machining: Solutions for Productivity, ed. by G. Schneider, ASM Inter., pp 49-56, Nov. 1989.

29. M.C. Vagle, A.S. Gates, PVD coatings on carbide tools, ibid, pp 57-67.

30. J.H. Forrest, The use of titanium nitride coated tools, ibid, pp 69-72.

31. B. Dwenger, Prototype machining of MMCs, Machining Requirements for Metal Matrix Composites Conference, SME EM-92-251, p 10, Sept. 1992.

32. M.J. Schmenk, W.J. Zdeblick, An economic study for machining MMC parts, ibid, SME EM-92-253, p 9, Sept. 1992.

33. D. Stashko, A study of sintered diamond capabilities in machining 390 aluminum, *Cutting Tool Engineering Diamond/ Superabrasive Directory,* **30** (56) 59.

34. Machining Data Handbook,3rd edition, Institute of Advanced Manufacturing Sciences, Cincinnati, OH,1980.

35. J.A. Sacha, Machining 390 aluminum alloy, ASM Metals Congress, Paper No.8201-050. St. Louis, MO, 18 p, Oct 1982.

36. J.C. Miller, J.W. Sutherland, R.E. DeVor, Surface roughness characteristics for turning 380 and 390 aluminum casting alloys, 10th North American Manufacturing Research Conference Proceedings, NAMRI/SME, pp 282-88.

37. B.R. Berlin, Machining cost comparison of silicon carbide discontinuously reinforced aluminum, unreinforced aluminum, and titanium, ibid, SME EM-92-252, 15 p, Sept. 1992.

38. M.K. Brun, F. Gorsler, M. Lee, Wear characteristics of various hard materials for machining SiC reinforced aluminum alloy, ASME Inter. Conference, April 1985.

39. G.A. Chadwick, P.J. Heath, Machining metal matrix composites, Metals and Materials, pp 73-6, Feb. 1990.

40. D.M. Ginburg, Cost effective secondary fabrication processes for metal matrix composites, SME MF-87-570, 1987.

41. W.S. Ricci, Machining metal matrix composites, SME MR-87-827, 1987.

42. Guide to Machining SXA® Engineered Materials, Advanced Composite Materials Corp., 1988.

43. J.E. Allison, G.S. Cole, Metal-matrix composites in the automotive industry: Opportunities and challenges, *JOM, 45* (1) 19 (1993).

44. F. Folgar, J.E. Widrig, J.W. Hunt, Design, fabrication, and performance of Fiber FP/Metal matrix composite connecting rods, SAE Paper 870406, Warrendale, PA, 1987.

45. A.M. Urquhart, Advanced Materials and Processes, pp 25-9, July 1991.

46. C.T. Lane, Paper presented at the Inter. Seminar on Machine Tools and Advanced Materials, Oct 1992, Milan, Italy.

47. L.A. Looney, J.M. Monaghan, P. O'Reilly, et.al., The turning of an Al/SiC metal-matrix composite, *J. Mater. Proc. Technol,* 33 (4) 453 (1992).

48. G. Schneider, Highspeed machining: Solutions for productivity;, Proceedings of SCTE '89 Conference, Soc. of Carbide & Tool Engrs., ASM Inter., 163 p, 1990.

49. Whiskered Ceramic Cutting Grades, Matl. Engineering, p D, June 1989, GTE Valenite Corp, Troy, MI.

50. M. Ramulu, M. Taya, An investigation of machinability of high-temperature composites, pp 423-31, NASA CP3018, Metal Matrix, Carbon, and Ceramic Matrix Composites 1988, Ed. by J.D. Buckley Cocoa Beach, FL, Jan. 1988.

51. N. Saito, Recent electrical discharge machining (EDM) techniques in Japan, *Bull. Japan Soc. of Precision Engr.,* 18 (2) June 1994.

52. M. Ramulu, EDM sinker cutting of ceramic particulate composite, TiB_2/SiC, Advanced Ceramic Materials, 1988.

53. E. Savrun, M. Taya, Surface characterization of SiC whisker/2124 aluminum and Al_2O_3 composites, *J. of Mater. Sci.,* 1988.

54. M. Ramulu, M. Taya, EDM machinability of SiC_w/Al composites, *J. of Mater. Sci.,* Nov. 1988.

55. E. Savrun, M. Taya, M. Ramulu, A preliminary investigation of machinability of high temperature composites, Univ. of Washington, Dept. of Mechanical Engineering Technical Report, June 1987.

187

56. A.M. Gadalla, Thermal spalling during electro-discharge machining of advanced ceramics and ceramic-ceramic composites, *Proc. of the Machining of Composites Materials Symposium,* ASM, Chicago, IL, pp 151-57, Nov. 1992.

57. A.M. Gadalla, W. Tsai, *J. Am. Ceram. Soc.,* **72** (8) 1,396 (1989).

58. A.M. Gadalla, H.S. Bedi, Effect of composition and grain size on electrical discharge machining of BN-TiB$_2$ composites, *J. Mater. Res.,* **16** (11) 2,457 (1991).

59. C.H. McMurty, W.D.G. Boecker, S.G. Seshadri, et.al., *Ceramic Bull.,* **66** (2) Carborundum Co., 1987.

60. B. Calés, C. Martin, P. Vivier, High strength ceramic composites Symposium, Las Vegas, NV, Nov. 1988, Ceramic Materials & Components for Engines, V.J. Tennery, ed., pp 1,189-1,201.

61. S.M. Rezaei, T, Suto, T. Waida, et.al., Creep feed grinding of advanced ceramics, *Proc. Instn. Mech. Engrs.,* **206** (2) 93 (1992).

62. K. Suzuki, T. Uematsu, S. Asano, et.al., Advances in grinding ceramics and WC, *Ind. Diamond Rev.,* 1988, p 278, June 1988.

63. T. Suto, T. Waida, H. Noguchi, et.al., High performance creep feed grinding of difficult-to-machine materials with new-type wheels, *Bull. Japan Soc., Precision Eng.,* **24** (1) 39 (1990).

64. M. Rezaei, T. Suto, T. Waida, et.al., A novel dressing technique for diamond wheels, *Ind. Diamond Rev.,* **6** (89) 258 (1989).

65. J.D. Kellner, W.J. Croft, L.A. Shepard, Titanium diboride electro-deposited coatings, AMMRC Tech. Rep. 77-17, June 1977.

66. C.L. Stotler, S.A. Yokel, PMR graphite engine duct development final report, NAS 3-21854, NASA CR 182228, Aug. 1989.

67. *Amer. Mach.,* p 25, Aug. 1986.

68. M. Ramulu, M. Faridnia, J.L. Garbini, et.al., *J. Eng. Mater. Technol.,* **112,** 430 (1991).

69. K. Colligan, M. Ramulu, *Manuf. Rev.,* **5,** (1992).

70. C.W. Wern, M. Ramulu, K. Colligan, A study of the surface texture of composite drilled holes, *J. of Mater. Proc. Tech.,* **37** (14) 373 (1993).

71. S.K. Malhotra, Some studies on drilling of fibrous composites, *J. Mater. Proc. Technol.,* **24,** 291 (1990).

72. H. Hocheng, H.Y. Puw, Machinability of fiber-reinforced thermoplastics in drilling, ASME Transactions, *J. of Engr. Mater. and Technol.,* **115** (1) 146 (1993).

73. R.G. Kinkaid, PCD drills for composite and metal matrix materials, SME Superabrasives '91 Conference, SME MR-91-174, 9 p, June 1991.

74. Machining of graphite/epoxy materials with PCD tools, Winter Ann. ASME mtg., MD, *Amer. Mach.,* **16,** 19 (1990).

75. W.S. Ricci, Machining metal matrix composites, AMTL, SME MR-87-827, presented at Composites in Manufacturing Conf., Long Beach, CA, 28 p, Dec. 1987.

76. M.M. Schwartz, *Handbook of Structural Ceramics,* McGraw-Hill Book Co., 524 p, 1992.

77. P. O'Reilly, J. Monaghan, D.M.R. Taplin, The machining of metal-matrix composites – An overview, in D. Taylor, ed., Proc. Irish Fracture and Durability Conference, Trinity College, Dublin, pp 81-93, Sept. 1989.

78. V.P. McConnell, Metal-matrix composites: Materials in transition Part II, *Advanced Composites,* pp 63-67, July/Aug 1990.

79. Flexible assembly subsystem in cost-effective production, USAF Manufacturing Tech. Program Status Report, pp 4-5, July 1992.

80. E.R. More, Manufacturing methods for composite fan blades, Hamilton Standard Div., United Technologies Corp, AFML-TR-76-138, Final Rep., April 1974 - Feb. 1976, Contr. F33615-74-C-5135.

81. S.M. Boszor, R.S. Ecklund, Manufacturing technology for advanced propulsion materials, Phase V Composite airfoil manufacturing scale-up for fan applications, United Technologies P&W, WL-TR-91-8009, June 1991, FR 21372, Final Tech. Rept., Sept. 1985 - Sept. 1990, Cont # F33615-85-C-5152, 87+ p.

82. R.K. Miller, Waterjet cutting: Technology and industrial applications, Fairmont Press Inc., 154 p, 1991.

83. D.F. Wightman, Waterjet cutting and hydrobrasive machining of aerospace components, MR90-672, SME, 13 p, Aug. 1990.

84. M. Hashish, Optimization factors in abrasive waterjet machining, *ASME Transactions,* **113** (1) 29 (1991).

85. M. Hashish, Current capabilities for precision machining with abrasive waterjets: Nontraditional machining, SME, MR91-520, 12 p, Nov. 1991.

86. M. Hashish, State-of-the-art of abrasive waterjet machining operations for composites, Machining of Composite Materials Symposium, Chicago, IL, ASM Inter., pp 65-73, Nov. 1992.

87. W. Kulischenko, Abrasive jet machining, SME, MR 76694, in D. Richerson (ed.), Bullen Ultrasonics, Advanced Ceramics Conf., Ceramic Applications in Manufacturing, Dearborn, MI, pp 176-84, 1986.

88. Advanced composites for aerospace require waterjet technology, MAN, pp 40-3, Sept. 1992.

89. Waterjet cutting expands opportunities for West Coast fabricator, *Welding Design & Fabrication,* p 18, April 1993.

90. D.C. Davis, Robotic abrasive water jet cutting of composite aerostructure components, SME Robotic Systems in Aerospace Manufacturing Conference, Sep 1987, ed. by T.J. Drozda, Composites Applications The Future is Now, pp 382-96, 1989.

91. D. Stover, Out on the cutting edge, *Advanced Composites,* pp 46-53, Sept/Oct 1992.

92. M. Ramulu, D. Arola, Water jet and abrasive water jet cutting of unidirectional graphite/epoxy composite, *Composites,* **24** (4) 299 (1993).

93. E.S. Geskin, L. Tisminetski, D. Verbitsky, et.al., Investigation c waterjet machining of composites, Proc. of the Machining of Composite Materials Symposium, ASM Materials Week, Chicago, IL, pp 81-7, Nov. 1992.

94. M. Hashish, Comparative evaluation of abrasive liquid jet machining systems, ASME Transactions, **115** (1) 44 (1993).

95. G. Chryssolouris, P. Sheng, N. Anastasia, Laser grooving of composite materials with the aid of a water jet, ibid, pp 62-72.

96. G. Street, C. Ferlito, S. Riches, Trends in laser cutting of advanced materials, *TWI Bulletin 5*, pp 108 10, Sept/Oct 1992.

97. D. Doyle, J. Kokosa, Laser processing of kevlar: hazardous chemical byproducts, *Proc. LAMP 87*, Osaka, Japan, pp 285-89, May 1987.

98. G.L. Viegelahn, S. Kawall, R.J. Scheuerman, et.al., LASER cutting of fiberglass/ polyester resin composites, *Advanced Composite Materials: New Developments and Applications Conference Proceedings*, Detroit, MI, pp 143-47, Sept 30-Oct 3, 1991.

99. W.S. Lau, T.M. Yue, C.Y. Jiang, et.al., Pulsed Nd:YAG laser cutting of SiC/Al-Li metal matrix composite, *Proceedings of the Machining of Composite Materials Symposium*, ASM Materials Week, Chicago, IL, pp 29-33, Nov. 1992.

100. K.C. Meinert, Jr., R.P Martukanitz, R.B. Bhagat, Laser processing of discontinuously reinforced aluminum composites, *Proceedings of American Soc. for Composites, 7th Technical Conference*, pp 168-77, Oct. 1992.

101. Technological and economic assessment of advanced ceramic materials, Volume 6: A case study of ceramic cutting tools, Charles River Associates Inc., PB85-113132, prepared for NIST, NBS GCR 84-470-6,65 p, Aug. 1984.

102. E.P. Rothman, G.B. Kenney, H.K. Bowen, Potential of ceramic materials to replace cobalt, chromium, manganese, and platinum in critical applications, MIT Industrial Liaison Program Report 9-11-85, Materials Processing Center, MIT, Cambridge MA, Cont #333-6530.2.

103. R. Barrett, T.F. Page, The interactions of an Al_2O_3-SiC-whisker-reinforced composite ceramic with liquid metals, *Wear*, **138**, 225-37.

104. A.R. Thangaraj, K.J. Weinmann, On the wear mechanisms a cutting performance of silicon carbide whisker-reinforced alum ASME Transactions, **114** (3) 301 (1992).

105. H. Xiao, X. Ai, H.S. Yang, Effect of whisker orientation on toughening behaviour and cutting performance of SiC_w-Al_2O_3 composite, *Mater. Sci. and Tech.*, **9**, 21 (1993).

106. K. Shintani, T. Oiji, Y. Fujimura, et.al., Cutting performance of toughened ceramic tools, Inter. Conf. on Evolution of Advanced Materials, Associazione Italiana Di Metallurgia Italy, 638p, pp 267-72, 1989.

107. K.H. Smith, The application of whisker reinforced and phase transformation toughened materials in machining of hardened steels and nickel-based alloys, ed. by G. Schneider, Jr., ASM Inte and SCTE Mtg., San Diego, CA, pp 81-88, Nov. 1989.

108. L.D. Maloney, Make way for "Engineered Ceramics," *Design News*, pp 64-74, March 1989.

109. J.C. Withers, Ceramic composite cutting tool insert for increased productivity, ultra highspeed machining, REPT 0717891, NSF/ISI-89014, NTIS Alert, **92**, (29) 89 (1989).

110. Ceramic Matrix Composites, BCC, Norwalk, CT, GB 11 0R, Nov. 1990.

111. T.J. Drozda, Ceramic tools find new applications, Manuf. Eng., pp 110-15, May 1985.

112. T.B. Troczynski, D. Ghosh, S. Das Gupta, et.al., Advanced ceramic materials for metal cutting, Advanced Structural Materials, ed. D.S. Wilkinson, Montreal, Canada, Pergamon Press, New York, NY, pp 157-68, Aug. 1988.

113. T.L. Ashley, The application of sialon and silicon nitrides in metal cutting, Manufacturing '92, Chicago, IL, MR-92-355-8, 8 p, Sept. 1992.

114. T. Ekstrom, M. Nygren, SiAlON ceramics, *J. Am. Ceram. Soc.,* **75** (2) 259 (1992).

115. W.W. Gruss, Ceramic tools improve cutting performance, *Ceram. Bull.,* **67,** 993 (1988).

116. E.D. Whitney, P.N. Vaidyanathan, Engineered ceramics for high-speed machining, in J.A. Swartley-Loush (Ed.), *Proc. ASM and SCTE Conf. on Advanced Tool Materials for Use in High Speed Machining,* Scottsdale, AZ, pp 77-82, Feb. 1987.

117. D. Agranov, presented at the CIRP Conf., Israel, Aug. 1986.

118. A.J. Klein, *AM&P,* **9,** 26 (1986).

119. J.D. Christopher, N. Zlatin, New cutting tool materials, SME Tech Paper MR 74 101, 1974; and in D.W. Richerson (Ed.), *Ceramics Applications in Manufacturing,* SME, pp 105-09, 1988.

120. E.R. Billman, et.al., Machining with Al_2O_3-SiC-whisker cutting tools, *Ceram. Bull.,* **67** (6) 1,016 (1988).

121. J.J. Schuldies, J.A. Branch, Ceramic composites Emerging manufacturing processes and applications, SME Using Advanced Ceramics in Manufacturing Applications Conference, SME EM 91-250, 17 p, June 1991.

122. D.A. Clarke, Industrial applications and markets for ceramic matrix composites, Rept. No. PNR 90753, Rolls Royce Reprint, 31 p, March 1990.

BIBLIOGRAPHY – CHAPTER IV

C. Martin, B. Calés, P. Vivier, et.al., Electrical discharge machinable ceramic composites, European Materials Research Society, *Mater. Sci and Engr.,* **A109,** 351 (1989).

G.L. Gunderson, J.A. Lute, The use of preformed holes for increased strength and damage tolerance of advanced composites, *Proc. of the Amer. Soc. for Composites, 7th Tech. Conf.,* Univ. Park, PA, pp 460-64, Oct. 1992.

P. Niskanen, W.R. Mohn, Versatile metal-matrix composites, *AM &P,* pp 39-41, March 1988.

C. Lane, The effect of different reinforcement on PCD tool life for aluminum matrix composites, ASM Inter. Annual Mtg., Chicago, Nov. 1992, Session I-IB.

G. Michel, A new approach to machinability testing: Application to machining of metal matrix composites, ibid.

S. Sullivan, Machining, trimming and drilling of metal matrix composites for structural applications, ibid.

A. Jawaid, Turning and drilling of metal matrix composites, ibid.

Z. Yuanjie, Tools and their design for machining of composite materials, ibid.

Flowing severance, *Aerospace Composites and Materials,* **4** (1) 10 (1992).

R&D Tackles Ceramics Machining, *Manufacturing Engr.,* **109** (4) 20 (1992).

Laser Cutting and Machining; Metals; (June 1986 - Jan. 1990);, Citations from the Compendex Database, NTIS, NZ-9096, 51 p, Jan. 1990.

B. Wielage, J. Drozak, Crack free laser cutting of ceramic and metal-ceramic composites, Dortmund Univ., Lehrstuhl fuer werlstofftechnologie, 38 p, 1990.

M.N. Alias, R. Brown, Damage to composites from electrochemical processes, Corrosion, **48** (5) 273 (1992).

G. Santhanakrishnan, R. Krishnamurthy, S.K. Malhotra, Investigation into the machining of carbon-fibre-reinforced plastics with cemented carbides, *J. Mater. Processing Technol.,* **30** (3) 263 (1992).

D. Gardner, Making ceramics work for you, *Design News,* pp 93-7, March 1990.

M. Mehta, A.A. Soni, Investigation of optimum drilling conditions and hole quality in PMR-15 polyimide/graphite composite laminates, Structural Composites, Proc. of the 6th Annual ASM/ESD Advanced Composites Conf., Detroit, MI, pp 633-5, Oct. 1990.

H.D. Pierson, Handbook of Chemical Vapor Deposition (CVD); Principles, Technology and Applications, Mater. and Sci. and Process Technol Series, Noyes Data Corp., pp 347-67, 1992.

D. Bhattacharyya, K. Xia, J.L. Mihelich, et.al., Machinability study on a ceramic micro-sphere-reinforced aluminum matrix composite, SME, Dearborn, MI, EM92-254, 7 p, Sept. 1992.

P.M. Stephan, Diamond films enhance machining with ceramics, *Ceram. Soc. Bull.,* **71** (11) 1,623 (1992).

A. Jawaid, S. Barnes, S.R. Ghadimzadeh, Drilling of particulate aluminum silicon carbide metal matrix composites, *Proc. of the Machining of Composites Materials Symposium, ASM,* Chicago. pp 35-47, Nov. 1992.

Oak Ridge National Laboratory, Adv. Materials Devel. Program, ceramic technology for advanced heat engines program plan, 1983-1993, ORNL/ M-1248, July 1990.

S.J. Burden, J. Hong, J.W. Rue, et.al., Comparison of hot-isostatically-pressed and uniaxially hot-pressed alumina-titanium-carbide cutting tools, *Ceram. Bull.,* **67** (6) 1,003 (1988).

Office of Technology Assessment, Advanced Materials by Design, U. S. Congress, OTA-E351, U.S. Govt. Printing Office, Washington, D.C., 1988.

H. Iwanek, G. Grathwol, R. Hamminger, et.al., Machining of ceramics by different methods, Sym. on Ceramic Materials and Components for Engines, Verlag Deutsche Keramische Gesellschaft, pp 417-23, 1986.

L.M. Sheppard, Machining of advanced ceramics, *AM&P,* **132** (6) 40 (1987).

M.J. McGinty, C.W. Preuss, Machining ceramic fiber metal matrix composite, *ASM Conf. Proceeding of High Productivity Machining . Materials and Process,* ed. by V.K. Sariw, ASM Pub., pp 231-44, 1985.

W. Hoover, Metal matrix composite-processing, microstructures and properties, 12th Riso Inter. Symposium, ed. by N. Hansen et.al., Roskilde, Denmark: Riso, pp 387-92, 1991.

S. Brusethaug, O. Reiso, ibid, pp 247-55.

H. Ho-Cheng, C.K.H. Dharan, Delamination during drilling in composite laminates, *ASME J. of Engr. for Industry,* **112,** 236 (1990).

B.A. Mackey, A practical solution for the machining of accurate, clean, burr free holes in Kevlar composites, *Proc.37th Conf. of the Plastics Industry,* pp 1-5 (24D), 1982.

H. Schulz, Machining composite materials with Syndite PCD, Ind. Diamond Rev., pp 263-65.

I.I. Opisova, E.L. Shvedkov, Cutting ceramics, *Soviet Powder Metallurgy and Metal Ceramics,* **31** (9) 751 (1992).

R. Komanduri, Advanced ceramic tool materials for machining, *Int. J. Refract. Metals Hard Mater.,* **8** (2) 125 (1989).

M. Furukawa, M. Shiroyama, V. Takano, et.al., SiC whisker reinforced complex ceramic cutting tool "Whiskal," Nip. Tunsgt. Rev., **22,** 34 (1989).

K. Upadhya, editor, Developments in Ceramic and Metal-Matrix Composites, *Proceedings of symposium 1992 TMS Annual Mtg.,* San Diego, CA, March 1992.

R. McFaul, Water treatment process for waterjet cutting, SME, MS91-499, Los Angeles, CA, 15 p, Sept. 1991.

W. Koenig, F.J. Trasser, Laser beam cutting of fiber reinforced polymers, Fraunhofer Inst. fuer Produktions technologie, Aachen, Germany, F.R., 145 p, June 1990.

S.L. Gunderson, J.A. Lute, The use of preformed holes for increased strength and damage tolerance of advanced composites, *J. of Reinf. Plastics and Composites,* **12** (5) 559 (1993).

D. Evans, SFPM & RPM - High Speed Machining, SME, MR92-194, Oakbrook, IL, 10 p, April 1992.

P. Newnham, S. Abrate, Finite element analysis of heat transfer anisotropic solids: application to manufacturing problems, Proc the Amer. Soc. for Composites, 7th Tech. Conf., Univ. Park, PA, pp 125-31, Oct. 1992.

T.C. Ramaraj, R. Janakiram, Analytical investigation of the ultrasonic machining to metal matrix composites, SME, EM89-822, Oct. 1989.

F. Zawistowski, Investigation into operation of new special drills for drilling holes in thin components made of composite materials, Technische Hogeschool, Delft, Netherlands, LR519, 1987.

R.A. Weiss, Portable air feed peck drilling of graphite composite titanium and other materials in dissimilar combinations, SME, AD89-642, Oct. 1989.

N. Tamari, et.al., Electric discharge machining of composite ceramics of Si_3N_4-SiC whiskers, *J. Ceram. Ind. Assoc.,* **94,** 1,231 (1986).

H.K. Tonshoff, C. Emmelmann, Laser cutting of advanced ceramics, *Ann. CIRP,* **38,** 219, (1989).

K.J.A. Brookes, Guide to the world's advanced cutting tool materials, *AM,* pp 63-8, March 1990.

M.A. Moreland, Ultrasonic advantages revealed in "The Hole Story," in J D. Richerson, ed., Bullen Ultrasonics, Advanced Ceramics, Conf., Ceramic Applications in Manufacturing, SME, Dearborn, MI, pp 156-62, 1986.

K. Isomura, T. Fukuda, K. Ogasahara, et.al., Machinable Si_3N_4-BN composite ceramics with high thermal shock resistance, high errosion resistance, Unitecr '89, Proc. Unified Int. Tech. Conf. on Refractories, vol 1, ACS Inc., ed. by L.J. Trostel, Anaheim, CA, pp 624-34, Nov. 1989.

M. Mehta, T.J. Reinhart, A.H. Soni, Effect of fastener hole drilling anomolies on structural integrity of PMR-15/Gr composite laminates, Proc. of the Machining of Composites Materials Symposium, ASM, Chicago, IL, pp 113-26, Nov. 1992.

H. Hocheng, H.P. Puw, K.C. Yao, Experimental aspects of drilling of some fiber-reinforced plastics, ibid, pp 127-38.

P.T. Eubank, B. Bozkurt, Recent developments in understanding the fundamentals of spark erosion for composite materials, ibid, pp 159-66.

R. Gilmore, Ultrasonic machining of composite materials, ibid, pp 189-96.

A. Di Ilio, A. Paoletti, V. Tagliaferri, et.al., Progress in drilling of composite materials, ibid, pp 199-203.

P.F. Becher, G.C. Wei, Toughening behaviour in SiC whisker reinforced aluminum, *J. Am. Ceram. Soc.,* **67,** p C259 (1984).

X. Ai, H. Xiao, Application of ceramic cutting tools in engineering industry, Beijing, Publishing House of the Machinery Ministry of China, p 260, 1988.

S.G. Howarth, A.B. Strong, Edge effects with waterjet and laser beam cutting of advanced composite materials, 35th Inter. SAMPE Sym., Anaheim, CA, pp 1,684-97 April 1990.

R.M. Fairhurst, et.al., "DIAJET" A new abrasive waterjet cutting technique, Proceedings of the 8th Inter. Sym. on Jet Cutting Technology, BHRA, Durham, England, pp 395-402, 1986.

M. Hashish, Prediction of depth of cut with abrasive waterjets, Sym. on Modeling of Materials Processing, A.A. Tseng, ed., ASME, New York, MD **3,** 65 (1987c).

M. Hashish, Turning, milling, and drilling with abrasive waterjets, *Proceedings of the 9th Inter. Sym. on Jet Cutting Technology,* BHRA, Japan, 1988a.

J.V. Owen, Cutting it with lasers, *Manufacturing Engineers,* **108** (6) 35 (1992).

D.G. Lee, K.S. Kim, Y.K. Kwak, Manufacturing of a SCARA type direct-drive robot with graphite fiber epoxy composite material, *Robotica,* **9,** 219 (1991).

K.S. Kim, D.G. Lee, Y.K. Kwak, Cutting (milling) characteristics of carbon fiber/epoxy composites, *Trans. Korean Soc. Mech. Eng.,* **14** (1) 237 (1990).

T. Kaneeda, CFRP cutting mechanism, *Proc. NAMRC XV,* Columbus, OH, SME, pp 216-21, 1989.

G. Chryssolouris, P. Sheng, Aspects of surface quality in laser machining of composite materials, *Proceedings of the NAMRC XVIII,* pp 250-55, May 1990.

G. Chryssolouris, P. Sheng, W.C. Choi, Three-dimensional laser machining of composite materials, *ASME J. of Engr. Mater. and Technol.,* **112,** 387 (1990).

J. Powell, C. Wykes, A comparison between CO_2 laser cutting and competitive techniques, *Proc. 6th Int. Conf. Lasers in Manufacturing,* pp 135-53, May 1989.

M. Flaum, Cutting of fibre-reinforced polymers with a cw CO_2 laser, SPIE 801, High Power Lasers, pp 142-49, 1987.

K. Riech, CO_2 laser cutting of fibre-reinforced materials, Proc. ECLAT 90, pp 777-88, Sept. 1990.

A. Di Iorio, et.al., Laser cutting of aramid FRP, Proc. LAMP 87, Osaka Japan, pp 291-96, May 1987.

GLOSSARY

ABL bottle: An internal-pressure-test vessel about 18 in (457 mm) in diameter and 24 in (610 mm) long, used to determine the quality and properties of the filament-wound material in the vessel.

Abhesive: A film or coating applied to one solid to prevent (or greatly decrease) the adhesion to another solid with which it is to be placed in intimate contact, e.g., a parting or mold-release agent.

Ablative plastic: A material which absorbs heat (while part of it is being consumed by heat) through a decomposition process (pyrolysis) taking place near the surface exposed to the heat.

ABS: Acrylonitrile-butadiene-styrene.

Accelerator: A material mixed with a catalyzed resin to speed up the chemical reaction between the catalyst and resin; used in polymerizing resins and vulcanizing rubbers; also known as promoter or curing agent.

Acicular powder: Needle shaped particles.

Activated sintering: An enhanced sintering process involving a treatment which reduces the activation energy for atomic motion. The increase in sintering rate allows faster sintering, lower sintering temperatures, or improved properties.

Activation: A process that increases the surface area of a material such as charcoal or alumina. In the case of charcoal, the surface of the material is oxidized and minute cavities are created which are capable of absorbing gas atoms or molecules.

Activator: An additive used to promote the curing of matrix resins and reduce curing time. (See also Accelerator.)

Addition: A polymerization reaction in which no by-products are formed.

Addition polymerization: Polymerization in which monomers are joined together without the splitting off of water or other simple molecules.

Additive: Any substance added to another, usually to improve properties.

Adherend: A body held to another body by an adhesive.

Adhesion: The state in which two surfaces are held together at an interface by forces or interlocking action or both.

Adhesion, mechanical: Adhesion between surfaces in which the adhesive holds the parts together by interlocking action.

Adhesive, contact: See Contact adhesive.

Adhesive failure: A rupture of adhesive bond that appears to be a separation at the adhesive/adherend interface.

Adhesive film: A synthetic resin adhesive, usually of the thermosetting type, in the form of a thin dry film of resin, used under heat or pressure as an interleaf in the production of laminated materials.

Adhesiveness: The property defined by the adhesion stress $A = F/S$, where F = perpendicular force to glue line and S = surface.

Adsorption: The formation of a layer of gas on the surface of a solid (or occasionally a liquid). The two types of adsorption are (a) chemisorption where the bond between the surface and the attached atoms, ions, or molecules is chemical, and (b) physisorption where the bond is due to van der Waals' forces.

Agglomeration: A tendency for fine particles to stick together and appear as larger particles.

Aggregate: A hard fragmented material used with an epoxy binder, as in epoxy tools.

Aging: The process or the effect on materials of exposure to an environment for an interval of time.

AIMS: Advanced integrated-manufacturing system.

Air-bubble void: Noninterconnected spherical air entrapment within and between the plies of reinforcement.

Air locks: Surface depressions on a molded part, caused by trapped air between the mold surface and the plastic.

Air vent: Small outlet to prevent entrapment of gases.

Alloy, alloying: An alloy is formed by the mutual solution within a single phase or crystal structure of two or more elements. Even in alloys where several phases coexist, all the phases present are likely to be alloyed, i.e. contain two or more constituent elements.

Alloy powder: A powder in which each particle is composed of the same alloy of two or more metals.

Ambient: The surrounding environmental conditions, e.g., pressure or temperature.

Amorphous: Material with a random or non-symmetrical atomic arrangement. In metals and ceramics it means a non-crystalline state; in polymers it means a random arrangement of the atomic chains of the polymer. For alloys, amorphous structures are obtained by rapidly quenching molten droplets or mechanical alloying.

Amorphous plastic: A plastic that has no crystalline component. There is no order or pattern to the distribution of the molecules.

Anistropic: Exhibiting different properties in response to stresses applied along axes in different directions.

Anistropic laminate: One in which the strength properties are different in different directions.

Anisotropy of laminates: The difference of the properties along the directions parallel to the length or width into the lamination planes or parallel to the thickness into the planes perpendicular to the lamination.

Antistatic compounds: Compounds intermediate between insulators (high resistivity) and conductive compounds (low resistivity).

APIs: Addition-reaction polyimides.

Aramid: Aromatic polyamide fibers characterized by excellent high-temperature, flame-resistance, and electrical properties. Aramid fibers are used to achieve high-strength, high-modulus reinforcement in plastic composites. More usually found as polyaramid-a synthetic fiber (trade name Kevlar).

Arc resistance: The total time in seconds that an intermittent arc can play across a plastic surface without rendering the surface conductive.

Areal weight: The weight of fiber per unit area (width times length) of tape or fabric.

Aromatic: Unsaturated hydrocarbon with one or more benzene ring structure in the molecule.

Ash content: The solid residue remaining after a reinforcing substance has been incinerated or strongly heated.

Aspect ratio: The ratio of length to diameter of a fiber.

A stage: An early stage in the polymerization reaction of certain thermosetting resins (especially phenolic) in which the material, after application to the reinforcement, is still soluble in certain liquids and is fusible; sometimes referred to as resole. (See also B stage, C stage.)

ATLAS: Automated tape lay-up system.

Atomization: The dispersion of molten metal into droplets by a rapidly moving stream of gas or liquid, or by centrifugal force.

Atomized metal powder: Metal powder produced by the disintegration and subsequent solidification of a molten metal stream.

Attenuation: The process of making thin and slender, as applied to the formation of fiber from molten glass. Reduction of force or intensity as in the decrease of an electrical signal.

Autoclave: A closed vessel for conducting a chemical reaction or other operation under pressure and heat.

Autoclave molding: After lay-up, the entire assembly is placed in steam autoclave at 50 to 100 lb/in^2 (23.4 to 47.6 Pa); additional pressure achieves higher reinforcement loadings and improved removal of air.

Automatic mold: A mold for injection or compression molding that repeatedly goes through the entire cycle, including ejection, without human assistance.

Automatic press: A hydraulic press for compression molding or an injection machine which operates continuously, being controlled mechanically, electrically, hydraulically, or by a combination of these methods.

Axial winding: In filament-wound reinforced plastics, a winding with the filaments parallel to the axis.

Back draft: An area of interference in an otherwise smooth-drafted encasement; an obstruction in the taper which would interfere with the withdrawal of the model from the mold.

Backpressure: Resistance of a material, because of its viscosity, to continued flow when a mold is closing.

Bag molding: A technique in which the consolidation of the material in the mold is effected by the application of fluid pressure through a flexible membrane.

Balanced design: In filament-wound reinforced plastics, a winding pattern so designed that the stresses in all filaments are equal.

Balanced-in-plane contour: In a filament-wound part, a head contour in which the filaments are oriented within a plane and the radii of curvature are adjusted to balance the stresses along the filament with the pressure loading.

Balanced laminate: All lamina except those at 0°/90° are placed in plus/minus pairs (not necessarily adjacent) symmetrically about the lay-up centerline.

Balanced twist: An arrangement of twist in a plied yarn or cord which will not cause twisting on itself when the yarn or cord is held in the form of an open loop.

Barcol hardness: A hardness value obtained by measuring the resistance to penetration of a sharp steel point under a spring load. The instrument, the Barcol Impressor, gives a direct reading on a scale of 0 to 100. The hardness value is often used as a measure of the degree of cure of a plastic.

Bare glass: Glass (yarns, roving, or fabrics) from which the sizing or finish has been removed or before it has been applied.

Base: The reinforcing material (glass fiber, paper, cotton, asbestos, etc.) which is impregnated with resin in the forming of laminates.

Batch: A measured mix of various materials. (See also Lot.)

Batt: Felted fabrics; structures built by the interlocking action of fibers themselves without spinning, weaving, or knitting. (See also Felt.)

Bearing area: The diameter of the hole times the thickness of the material.

Bearing strength: The bearing stress at that point on the stress-strain curve where the tangent is equal to the bearing stress divided by n% of the bearing-hole diameter.

Bearing stress: The applied load in pounds divided by the bearing area. (Maximum bearing stress is the maximum load in pounds sustained by the specimen during the test divided by the original bearing area.)

Bias fabric: A fabric in which warp and fill fibers are at an angle to the length.

Biaxial load: (1) A loading condition in which a laminate is stressed in at least two different directions in the plane of the laminate. (2) A loading condition of a pressure vessel under internal pressure and with unrestrained ends.

Biaxial winding: In filament winding, a type of winding in which the helical band is laid in sequence, side by side, with no crossover of fibers.

Bidirectional laminate: A reinforced plastic laminate with the fibers oriented in various directions in the plane of the laminate; a cross laminate. (See also Unidirectional laminate.)

Binder: The resin or cementing constituent of plastic compound which holds the other components together; the agent applied to glass mat or preforms to bond the fibers before laminating or molding, or a material added to a powder for the specific purpose of cementing together powder particles which would not otherwise sinter into a strong body.

Bioglass: This is a fairly complex ceramic, based on silica glass, developed for its strength, compatibility, stability and function in the form of implant devices in the human body.

Biomaterial: Synthetic material used for implants in the human (and animal) body.

Bismaleimide: A type of polyimide that cures by an addition reaction, avoiding formation of volatiles, and has temperature capabilities between those of epoxy and polyimide.

Blank: A pressed, presintered or fully sintered compact, usually in the unfinished condition, requiring cutting, machining or some other operation to give a final shape.

Blanket: Plies which have been laid up in a complete assembly and placed on or in the mold all at one time (flexible-bag process); also the form of bag in which the edges are sealed against the mold.

Bleeder cloth: A layer of woven or nonwoven material, not a part of the composite, that allows excess gas and resin to escape during cure.

Bleedout: In filament winding, the excess liquid resin that migrates to the surface of the winding.

Blend: A polymer material containing two or more constituents within a given phase. Blends are also referred to as polymer alloys.

Blister: Undesirable rounded elevation of the surface of a plastic with boundaries that are more or less sharply defined, resembling in shape a blister on the human skin; the blister may burst and become flattened.

Block coploymer: An essentially linear copolymer in which there are repeated sequences of polymer segments of different chemical structure-some of which may be crystalline in nature, others may be amorphous.

Blocking: An unintentional adhesion between plastic films or between a film and another surface usually corrected by antistatic additives.

BMC: Bulk-molding compound.

Bond strength: The amount of adhesion between bonded surfaces; a measure of the stress required to separate a layer of material from the base to which it is bonded. (See also Peel strength.)

Boron fiber: A fiber usually of a tungsten-filament core with elemental boron vapor deposited on it to impart strength and stiffness.

Boss: Protuberance on a plastic part designed to add strength, to facilitate alignment during assembly, to provide for fastenings, etc.

Bottom plate: A steel plate fixed to the lower section of a mold, often used to join the lower section of the mold to the platen of the press.

Braiding: Weaving fibers into a tubular shape.

Breather: A loosely woven material that does not come in contact with the resin but serves as a continuous vacuum path over a part in production.

Breathing: (1) Opening and closing a mold to allow gases to escape early in the molding cycle (also see degassing). (2) Permeability to air of plastic sheeting.

Bridging: A region of a contoured part which has cured without being properly compacted against the mold.

Broad goods: Woven glass, synthetic fiber, or combinations thereof over 18 in (457 mm) wide. Fibers are woven into fabrics that may or may not be impregnated with resin, usually furnished in rolls.

B stage; An intermediate stage in the reaction of certain thermosetting resins in which the material swells when in contact with certain liquids and softens when heated but may not dissolve or fuse entirely; sometimes referred to as resistol. The resin in an uncured prepreg or premix is usually in this stage. (See also A stage, C stage.)

Bubble: A spherical internal void; globule of air or other gas trapped in a plastic.

Buckling: Crimping of fibers in a composite material, often occurring in glass-reinforced thermoset due to resin shrinkage during cure.

Bulk density: The density of a molding material in loose form (granular, nodular, etc.), expressed as a ratio of weight to volume.

Burn-off: The removal of an additive by preheating prior to sintering.

Burst strength: Hydraulic pressure required to burst a vessel of given thickness; commonly used in testing filament-wound composite structures.

Butt joint: See joint.

Butt wrap: Tape wrapped around an object in an edge-to-edge fashion.

CAD: Computer-aided design.

CAM: Computer-aided manufacturing.

Carbon-carbon: A composite of carbon fiber in a carbon matrix.

Carbon fiber: An important reinforcing fiber known for its light weight, high strength, and high stiffness that is produced by pyrolysis of an organic precursor fiber such as rayon, polyacrylonitrile (PAN), and pitch, in an inert environment at temperatures above 1,800°F (982°C). The material may also be graphitized by heat treating above 3,000°F (1,649°C).The term is often used interchangeably with the term graphite; however, carbon fibers and graphite fibers differ. The basic differences lie in the temperature at which the fibers are made and heat treated and in the structures of the resulting fibers. Graphite fibers have a more "graphitic" structure than carbon fibers.

Carbon/graphite fibers: Usually prepared from mesophase pitch or polacrylonitrile (PAN). Used as reinforcements in composites.

Carbonization: In carbon fiber manufacturing, it is the process step in which the heteroatoms are pyrolyzed in an inert environment at high temperature to leave a very high carbon content material (about 95 percent carbon and higher). The carbonization temperature can range up to 2,000°C. Temperatures above 2,000°C (up to 3,000°C) are often referred to as graphitization temperatures. The range of temperature employed is influenced by precursor, individual manufacturing process, and properties desired.

Casting: An object or finished shape obtained by solidifaction of a substance in a mold.

Catalyst: A substance which changes the rate of a chemical reaction without itself undergoing permanent change in its composition; a substance which markedly speeds up the cure of a compound when added in small quantity compared with the amounts of primary reactants. (See also Accelerator, Curing agent, Hardener, Inhibitor.)

Catenary: A measure of the difference in length of the strands in a specified length of roving as a result of unequal tension; the tendency of some strands in a taut horizontal roving to sag lower than the others.

Caul: A sheet the size of the platens used in hot pressing.

Cavity: (1) Depression in mold. (2) The space inside a mold into which a resin is poured. (3) The female portion of a mold. (4) That portion of the mold which encloses the molded article. (5) That portion which forms the outer surface of molded article (often referred to as the die). (6) The space between matched molds; depending on the number of such depressions, molds are designated as single or multiple-cavity.

CC: Carbonaceous heat-shield composites (C-C is carbon-carbon).

Cemented carbide: A solid composite consisting of a metal carbide and a binder phase, usually cobalt or nickel aluminide. The composite is formed by liquid phase sintering a mixture of the carbide and binder metal powders.

Centerless grinding: A technique for machining parts having a circular cross section, consisting of grinding the rod which is fed without mounting it on centers. Grinding is accomplished by working the material between wheels rotating at different speeds; the faster, abrasive wheel cuts the stock. Variations of the basic principle can be used to grind internal surfaces.

Centrifugal atomization: The formation of spherical particles by combining a melt with a centrifugal force such that the melt is thrown off into droplets which spheroidize prior to solidification.

Cermet: A body consisting of ceramic particles bonded with a metal.

Centrifugal casting: A high-production technique for cylindrical composites, such as pipe, in which chopped strand mat is positioned inside a hollow mandrel designed to be heated and rotated as resin is added and cured.

Ceramic-matrix composites: Materials consisting of a ceramic or carbon fiber surrounded by a ceramic matrix, usually silicon carbide.

CFG Iron: Compact-flake-graphite iron.

Charge: The measurement or weight of material (liquid, preformed, or powder) used to load a mold at one time or during one cycle.

Chase: (1) The main body of the mold, which contains the molding cavity or cavities, or cores, the mold pins, the guide pins or the bushings, etc. (2) An enclosure of any shape used to shrink-fit parts of a mold cavity in place to prevent spreading or distortion in hobbing or to enclose an assembly of two or more parts of a split cavity block.

Chemical vapor deposition (CVD): A process in which desired reinforcement material is deposited from vapor phase onto a continuous core; boron on tungsten, for example.

Chill: (1) To cool a mold by circulating water through it. (2) To cool a molding with an air blast or immersing it in water.

Circuit: In filament winding (1) one complete traverse of the fiber-feed mechanism of a winding machine; (2) one complete traverse of a winding band from one arbitrary point along the winding path to another point on a plane through the starting point and perpendicular to the axis.

Circumferential ("circ") winding: In filament-wound reinforced plastics a winding with the filaments essentially perpendicular to the axis.

Clamping plate: A mold plate fitted to the mold and used to fasten the mold to the machine.

Clamping pressure: In injection molding and transfer molding the pressure applied to the mold to keep it closed, in opposition to the fluid pressure of the compressed molding material.

Coagulation bath: A liquid bath that serves to harden viscous polymer strands into solid fibers after extrusion through a spinnerette. Used in wet spinning processes such as rayon or acrylic fiber manufacture.

Coalescence: The merging of two objects into a larger object, as can be observed in particles during fabrication, and pores and grains during sintering.

Co-curing: Simultaneous bonding and curing of components.

Coefficient of elasticity: The reciprocal of Young's modulus in a tension test.

Coefficient of expansion: The fractional change in dimension of material for a unit change in temperature. Also called coefficient of thermal expansion (CTE).

Coefficient of friction: A measure of the resistance to sliding of one surface in contact with another surface.

Coefficient of thermal expansion α: The change in length per unit length produced by a unit rise in temperature.

Cohesion: (1) The propensity of a single substance to adhere to itself. (2) The internal attraction of molecular particles toward each other. (3) The ability to resist partition from the mass. (4) Internal adhesion. (5) The force holding a single substance together.

Cold isostatic pressing: The compaction of a powder under isostatic pressure conditions at room temperature using a flexible mold and a high hydrostatic pressure in a hydraulic pressure chamber (CIP).

Cold pressing: Forming a compact at room temperature low enough to avoid sintering - usually at room temperature.

Cold-setting adhesive: A synthetic resin adhesive capable of hardening at normal room temperature in the presence of a hardener.

Commingled yarn: A hybrid yarn made with two types of materials intermingled in a single yarn, for example, thermoplastic filaments intermingled with carbon filaments to form a single yarn.

Compact: An object produced by the compression of metal powders.

Compaction: The shaping, deformation, and densification of a powder by the application of pressure through a tool material.

Compatibility: The ability of two or more substances combined with each other to form a homogeneous composition with useful plastic properties.

Composite: A homogeneous material created by the synthetic assembly of two or more materials (a selected filler or reinforcing elements and compatible matrix binder) to obtain specific characteristics and properties. Composites are subdivided into the following classes on the basis of the form of the structural constituents: fibrous: The dispersed phase consists of fibers; flake: the dispersed phase consists of flat flakes; laminar: composed of layer of laminar constituents; particulate: dispersed phase consists of small particles, skeletal: composed of a continuous skeletal matrix filled by a second material.

Compression mold: A mold which is open when the material is introduced and which shapes the material by heat and by pressure of closing.

Compression molding: A technique for molding thermoset plastics in which a part is shaped by placing the fiber and resin into an open mold cavity, closing the mold, and applying heat and pressure until the material has cured or achieved its final form.

Compression molding pressure: The unit pressure applied to the molding material in the mold.

Compressive modulus E_c: Ratio of compressive stress to compressive strain below the proportional limit. Theoretically equal to Young's modulus determined from tensile experiments.

Compressive strength: (1) The ability of a material to resist a force that intends to crush. (2) The crushing load at the failure of a specimen divided by the original sectional area of the specimen.

Compressive stress: The compressive load per unit area of the original cross section carried by the specimen during the compression test.

Condensation: A polymerization reaction in which simple by-products (for example, water) are formed.

Conductivity: (1) Reciprocal of volume resistivity. (2) The conductance of a unit cube of any material.

Consolidation: A processing step that compresses fiber and matrix to reduce voids and achieve a desired density.

Contact adhesive: An adhesive which for satisfactory bonding requires the surfaces to be joined to be no father apart than about 0.004 in (0.1 mm).

Contact molding: A process for molding reinforced plastics in which reinforcement and resin are placed on a mold, cure is at room temperature using a catalyst-promoter system or by heat in an oven, and no additional pressure is used.

Contact-pressure resins: Liquid resins which thicken or polymerize on heating and require little or no pressure when used for bonding laminates.

Continuous filament: An individual flexible rod of glass of small diameter of great or indefinite length.

Continuous-filament yarn: Yarn formed by twisting two or more continuous filaments into a single continuous strand.

Continuous roving: Parallel filaments coating with sizing, gathered together into a single or multiple strands, and wound into a cylindrical package. It may be used to provide continuous reinforcement in woven roving, filament winding, pultrusion, prepregs, or high strength molding compounds, or it may be used chopped.

Cooling fixture: A fixture used to maintain the shape or dimensional accuracy of a molding or casting after it is removed from the mold and until the material is cool enough to hold its shape..

Co-polymer: A composite of two or more polymer types.

Core: (1) The central member of a sandwich construction to which the faces of the sandwich are attached. (2) A channel in a mold for circulation of heat-transfer media.

Corrosion: This is a phenomenon in which certain environments, e.g. salt water, attack and (through an electrochemical reaction) break down the thin oxide films that normally protect metals in air. Corrosion resistant metals such as stainless steel contain sufficient chromium that a hard tough film of Cr_2O_3 quickly forms at the surface, thereby protecting the metal under normal wear and tear.

Count: (1) For fabric the number of warp and filling yarns per inch in woven cloth. (2) For yarn the size based on relation of length and weight. Basic unit is a tex.

Coupling agent: Any chemical substance designed with two different types of fibers in individual yarns, for example, thermoplastic fibers woven side by side with carbon fibers.

CP: (1) Cross-ply. (2) Resinous heat-shield composites.

CPIs: Condensation-reaction polyimides.

Crack: An actual separation of molding material visible on opposite surfaces of the part and extending through the thickness; a fracture.

Crazing: Fine cracks which may extend in a network on or under the surface of plastic material.

Creel: A device for holding the required number of roving balls or supply packages in the desired position for unwinding onto the next processing step.

Creep: The change in dimension of plastic under load over a period of time, not including the initial instantaneous elastic deformation; at room temperature it is called cold flow.

Crimp: The waviness of a fiber; it determines the capacity of fibers to cohere under light pressure; measured either by the number of crimps or waves per unit length or by the percent increase in extent of the fiber on removal of the crimp.

Critical length: The minimum length of a fiber necessary for matrix shear loading to develop fiber ultimate strength by a matrix.

Critical loading: The maximum volume fraction of solid particles which can be incorporated in a polymer binder without forming pores.

Critical strain: The strain at the yield point.

Critical longitudinal stress (fibers): The longitudinal stress necessary to cause internal slippage and separation of a spun yarn; the stress necessary to overcome the interfiber friction developed as a result of twist.

Cross-laminated: Laminated so that some of the layers of material are oriented at right angles to the remaining layers with respect to the grain or strongest direction in tension. Balanced construction above the centerline of the thickness of the laminate is normally assumed. (See also Parallel-laminated.)

Crosswise direction: Refers to cutting specimens and to application of load. For rods and tubes, crosswise is the direction perpendicular to the long axis. For other shapes or materials that are stronger in one direction than another, crosswise is the direction that is weaker. For materials that are equally strong in both directions, crosswise is an arbitrarily designed direction at right angles to the length.

Crystalline orientation: The term crystalline applies to section of all chemical fibers, which consist of alternate crystalline and amorphous (noncrystalline) regions. These regions are influenced by manufacturing conditions and can be controlled to some extent. Crystalline orientation implies the extent to which these crystalline regions can be aligned parallel to the fiber axis. Crystalline orientation is an important factor in determining the mechanical properties of carbon fibers.

Crystallinity: The quality of having a molecular structure with atoms arranged in an orderly, three-dimensional pattern.

C stage: The final stage in the reaction of certain thermosetting resins in which the material is relatively insoluble and infusible; sometimes referred to as resite. The resin in a fully cured thermoset molding is in this stage. (See also A stage, B stage.)

Cull: Material remaining in a transfer chamber after the mold has been filled. (Unless there is a slight excess in the charge, the operator cannot be sure the cavity will be filled.)

Cure: To change the properties of a resin by chemical reaction, which may be condensation or addition; usually accomplished by the action of heat or catalyst, or both, and with or without pressure.

Curing Agent: Hardener, a catalytic or reactive agent added to a resin to cause polymerization. Curing agents participate in the polymerization process. They may be latent-curable only at elevated temperatures-or they may be activated at room temperature (25°C).

Curing temperature: Temperature at which a cast, molded, or extruded product, a resin-impregnated reinforcement, an adhesive, etc., is subjected to curing.

Curing time: The length of time a part is subjected to heat or pressure, or both, to cure the resin; interval of time between the instant relative movement between the moving parts of a mold ceases and the instant pressure is released. (Further cure may take place after removal of the assembly from the conditions of heat or pressure.)

Cycle: The complete, repeating sequence of operations in a process or part of a process. In molding, the cycle time is the elapsed time between a certain point in one cycle and the same point in the next.

D glass: A high-boron-content glass made especially for laminates requiring a precisely controlled dielectric constant.

Damage tolerance: A measure of the ability of structures to retain load-carrying capability after exposure to sudden loads (for example, ballistic impact).

Damping (mechanical): Mechanical damping gives the amount of energy dissipated as heat during the deformation of material. Perfectly elastic materials have no mechanical damping. Damping also diminishes the intensity of vibrations.

Daylight: The distance in the open position between the moving and fixed tables (platens) of a hydraulic press. For the multidaylight press, daylight is the distance adjacent platens.

Debinding: A step between molding and sintering where the majority of the binder used in molding is extracted by heat, solvent, or other techniques.

Debond: An unplanned nonadhered or unbonded region in an assembly.

Deep-draw mold: A mold having a core which is long in relation to the wall thickness.

Deflection temperature under load: The temperature at which a simple beam has deflected a given amount under load (formerly called heat-distortion temperature).

Deformation under load: The dimensional change of a material under load for a specific time following the instantaneous elastic deformation caused by the initial application of the load; also called cold flow or creep.

Degassing: See Breathing.

Delaminate: To split a laminated plastic material along the plane of its layers. (See also Laminate.)

Delamination: Physical separation or loss of bond between laminate plies.

Denier: A yarn and filament numbering system in which the yarn number is equal numerically to the weight in grams of 30,000 ft (9,144 m) (used for continuous filaments). The lower the denier the finer the yarn.

Design allowable: A limiting value for a material property that can be used to design a structural or mechanical system to a specified level of success with 95% statistical confidence. B-basis allowable: material property exceeds the design allowable 90 times out of 100. A-basis allowable: material property exceeds the design allowable 99 times out of 100.

Dew point: A temperature which provides a relative measure of the moisture content in an atmosphere.

Die: The part or parts making up the confining form in which a powder is pressed.

Dielectric: A nonconductor of electricity.

Dielectric constant ε: (1) The ratio of the capacity of a capacitor having a dielectric material between the plates to that of the same capacitor when the dielectric is replaced by a vacuum. (2) A measure of the electrical charge stored per unit volume at unit potential.

Dielectric curing: Curing a synthetic thermosetting resin passing an electric charge from a high-frequency generator through the resin.

Dielectric loss: The energy eventually converted into heat in a dielectric placed in a varying electric field.

Dielectric strength: The ability of material to resist the flow of an electrical current.

Dimensional stability: Ability of a plastic part to retain the precise shape to which it was molded, cast or otherwise fabricated.

Dislocation: A linear discontinuity in the perfection of a crystal. A sufficiently large external force acting on the crystal may cause the dislocation, or groups of dislocations, to move, and this is the basis of plastic deformation.

Displacement angle: In filament winding the distance of advance of the winding ribbon on the equator after one complete circuit.

Doctor roll: A device for regulating the amount of liquid material on the rollers of a spreader; also called doctor bar.

Doily: In filament winding the planar reinforcement applied to a local area between windings to provide extra strength in an area where a cutout is to be made, e.g. port openings.

Dome: In filament winding the portion of a cylindrical container that forms the integral ends of the container.

Dopant, doping: A very small amount of an alloying addition which, when dissolved by a material, may modify its properties. An example is the doping of silicon by phosphorus or boron to make it semiconducting.

Doubler: Localized area of extra layers of reinforcement, usually to provide stiffness or strength for fastening or other abrupt load transfers.

Draft: The taper or slope of the vertical surfaces of a mold designed to facilitate removal of molded parts.

Draft angle: The angle between the tangent to the surface at that point and the direction of ejection.

Drape: The ability of preimpregnated broad goods to conform to an irregular shape; textile conformity.

Dry spot: (1) Of a laminate the area of incomplete surface film on laminated plastics. (2) In laminated glass an area over which the interlayer and the glass have not become bonded. (See also Resin-starved area.)

Dry winding: Filament winding using preimpregnated roving, as differentiated from wet winding. (See also Wet winding.)

Dry lay-up: Construction of a laminate by layering preimpregnated reinforcement (partly cured resin) in a female or male mold, usually followed by bag molding or autoclave molding.

DS: Directionally solidified.

DSC: Differential scanning calorimeter. Instrumentation for measuring chemical reactions by observing exotherm or endothermic (heat rise or heat input) reactions of materials- usually over a programmed temperature cycle.

Dwell: (1) A pause in the application of pressure to a mold, made just before the mold is completely closed, to allow gas to escape from the molding machine. (2) In filament winding the time the traverse mechanism is stationary while the mandrel continues to rotate to the appropriate point for the traverse to begin a new pass.

Dynamic compaction: Explosive or gas gun compaction of powder at shockwave velocities.

Edgewise: Refers to cutting specimens and to the application of load. The load applied edgewise when it is applied to the edge of the original sheet or specimen. For compression-molded specimens of square cross section the edge is the surface parallel to the direction of motion of the molding plunger. For injection-molded specimens of square cross section this surface is selected arbitrarily; for laminates the edge is the surface perpendicular to the laminae. (See also Flatwise.)

E glass: A borosilicate glass; the type most used for glass fibers for reinforced plastics; suitable for electrical laminates because of its high resistivity. (Also called electric glass.)

Ejection: Removal of a molding from the mold impression by mechanical means, by hand, or by using compressed air.

Ejection ram: A small hydraulic ram fitted to a press to operate the ejection pins.

Elastic deformation: The part of the total strain in a stressed body which disappears upon removal of the stress.

Elasticity: The property of plastics materials by virtue of which they tend to recover their original size and shape after deformation.

Elastic limit: The greatest stress which a material is capable of sustaining without permanent strain remaining upon the complete release of the stress. A material is said to have passed its elastic limit when the load is sufficient to initiate plastic (nonrecoverable) deformation.

Elastic recovery: The fraction of a given deformation that behaves elastically.

$$\text{Elastic recovery} = \frac{\text{elastic extension}}{\text{total extension}}$$

$$\text{Elastic recovery} = \begin{cases} 1 & \text{for perfectly elastic material} \\ 0 & \text{for perfectly plastic material} \end{cases}$$

Electroformed molds: A mold made by electroplating metal on the reverse pattern on the cavity.

Electrolytic powder: Powder produced by electrolytic deposition and subsequent pulverization. The resulting powder has a dendritic or sponge shape.

Elemental powder: Powder of a single chemical species like iron, nickel, titanium, copper or cobalt, with no alloying ingredients.

Elongation: Deformation caused by stretching; the fractional increase in length of a material stressed in tension. (When expressed as percentage of the original gage length, it is called percentage elongation.)

Elutriation: Classification of powder particles by size using a rising stream of gas or liquid acting against the setting due to gravity, similar to air classification.

EMC: Elastomeric-molding tooling compound.

End: A strand of roving consisting of a given number of filaments gathered together (the group of filaments is considered an end or strand before twisting and; a yarn after twist has been applied); an individual warp yarn, thread, fiber, or roving.

End count: An exact number of ends supplied on a ball or roving.

Endothermic atmosphere: A reducing gas atmosphere used in sintering. It is produced by the reaction of hydrocarbon fuel gas and air over a catalyst with the aid of an external heat source. The resulting atmosphere is low in carbon dioxide and water vapor with relatively large percentages of hydrogen and carbon monoxide.

Endurance limit: See Fatigue limit.

Epitaxial films: A term identified with semiconductors, involving the build-up, layer by layer, of thin films with molecular dimensions. Applicable to polymers, ceramics, and metals.

Epoxy plastics: Plastics based on resins made by the reaction of epoxides or oxiranes with other materials such as amines, alcohols, phenols, carboxylic acids, acid anhydrides, and unsaturated compounds.

Equator: In filament winding the line in a pressure vessel described by the junction of the cylindrical portion and the end dome.

Equiaxed powder: Particles with approximately the same size in all three (perpendicular) dimensions.

Equilibrium: The stable state of a phase or structure. These stable states are illustrated for most binary and many ternary alloys in the form of equilibrium diagrams.

ERM: Elastic-reservoir molding.

Erosion: The dissolution of the metal compact at the surface where a liquid infiltrant flows into the part.

Eutectic: An alloy having the composition indicated by the eutectic point on an equilibrium phase diagram.

Even tension: The process whereby each end of roving is kept in the same degree of tension as the other ends making up that ball of roving. (See also Catenary.)

Exotherm: The liberation or evolution of heat during curing of a plastic product.

Exothermic atmosphere: A reducing gas atmosphere used in sintering, produced by partial or complete combustion of a hydrocarbon fuel gas and air. The maximum combustible content is 25%.

Expansion: The increase in dimensions of a compact due to unbalanced chemical reactions, pore formation, or pore growth during sintering.

Extrusion: The conversion of a billet into lengths of uniform cross-section by forcing metal powder, or a metal powder and plastic mixture, through a die orifice of desired cross-section.

Exude: The expulsion of a low melting temperature alloying ingredient from the pore structure due to poor wetting.

Fabric: A material constructed of interlaced yarns, fibers, or filaments, usually planar.

Fabric, nonwoven: A material formed from fibers or yarns without interlacing.

Fabric, woven: A material constructed of interlacing yarns, fibers, or filaments.

Fabricating, fabrication: The manufacture of plastic products from molded parts, rods, tubes, sheeting, extrusions, or other form of appropriate operations such as punching, cutting, drilling, and tapping. Fabrication includes fastening plastic parts together or to other parts by mechanical devices, adhesives, heat sealing, or other means.

Fan: In glass-fiber forming the fan shape that is made by the filaments between the bushing and the shoe.

Fatigue: The failure or decay of mechanical properties after repeated applications of stress. (Fatigue tests give information on the ability of a material to resist the development of cracks, which eventually bring about failure as a result of a large number of cycles.)

Fatigue life: The number of cycles of deformation required to bring about failure of the test specimens under given set of oscillating conditions.

Fatigue limit: The stress below which a material can be stressed cyclically for an infinite number of times without failure.

Fatigue strength: (1) The maximum cyclic stress a material can withstand for a given number of cycles before failure occurs. (2) The residual strength after being subjected to fatigue.

Felt: A fibrous material made from interlocked fibers by mechanical or chemical action, moisture, or heat; made from asbestos, cotton, glass, etc. (See also Batt.)

Fiber: Relatively short lengths of very small cross section of various materials made by chopping filaments (converting); also called filament, thread, or bristle. (See also Staple fibers.)

Fiber-composite material: A material consisting of two or more discrete physical phases, in which a fibrous phase is dispersed in a continuous matrix phase. The fibrous phase may be macro-, micro-, or submicroscopic, but it must retain its physical identity so that it could conceivably be removed from the matrix intact.

Fiber glass: An individual filament made by attenuating molten glass. (See also Continuous filament, Staple fibers.)

Fiber diameter: The measurement of the diameter of individual filaments.

Fiber-matrix interface: The region separating the fiber and matrix phases, which differs from them chemically, physically, and mechanically. In most composite materials, the interface has a finite thickness (nanometers to thousands of nanometers) because of diffusion or chemical reactions between the fiber and matrix. Thus, the interface can be more properly described by the terms interphase or interfacial zone. When coatings are applied to the fibers or several chemical phases have well-defined microscopic thickness, the interfacial zone may consist of several interfaces. In this book "interface" is used to mean both "interphase" and "interfacial zone".

Fiber orientation: Fiber alignment in a nonwoven or a mat laminate where the majority of fibers are in the same direction, resulting in a higher strength in that direction.

Fiber pattern: (1) Visible fibers on the surface of laminates or moldings. (2) The thread size and weave of glass cloth.

Fiber placement: A continuous process for fabricating composite shapes with complex contours and/or cutouts by means of a device that lays preimpregnated fibers (in tow form) onto a nonuniform mandrel or tool. It differs from filament winding (below) in several ways: There is no limit on fiber angles; compaction takes place online via heat, pressure, or both; and fibers can be added and dropped as necessary. The process produces more complex shapes and permits a faster putdown rate than filament winding.

Filament: Any fiber whose aspect ratio (length to effective diameter) is for all practical purposes infinity, i.e., a continuous fiber. For a noncircular cross section, the effective diameter is that of a circle which has the same (numerical) area as the filament cross section. The smallest unit of a fibrous material. The basic units formed during drawing and spinning, which are gathered into strands of fiber (tows) for use in composites. Filaments usually are of extreme length and very small diameter, usually less than 25 microns (a mil). Normally filaments are not used individually.

Filaments: Individual glass fibers of indefinite length, usually as pulled from the stream of molten glass flowing through an orifice of the bushing. In the operation, a number of fibers are gathered together to make a strand or end of roving or yarn.

Filament weight ratio: In a composite material, the ratio of filament weight to the total weight of the composite.

Filament winding: A process for fabricating a composite structure in which continuous reinforcements (filament wire, yarn, tape, or other) impregnated with a matrix material either previously or during the winding are placed over a rotating removable form or mandrel in a prescribed way to meet certain stress conditions. Generally the shape is a surface of revolution, which may or may not include end closures. When the right number of layers has been applied, the wound form is cured and the mandrel removed.

Fill: Yarn oriented at right angles to the warp in a woven fabric.

Filler: A relatively inert material added to a plastic mixture to reduce cost, modify mechanical properties, serve as base for color effects, or improve the surface texture. (See also Binder, Reinforced plastic.)

Fillet: A rounded filling for the internal angle between two surfaces of a plastic molding.

Filling yarn: The transverse threads or fibers in a woven fabric, i.e., fibers running perpendicular to the warp; also called weft.

Film adhesive: A synthetic resin adhesive usually of the thermosetting type in the form of a thin dry film of resin with or without a paper carrier.

Fillout: See Lack of fillout.

Fine-grained microstructure: Most crystalline materials are polycrystalline, comprising of many small crystals separated by crystal or grain boundaries. A fine-grained microstructure is a polycrystalline material with a fine grain size.

Finish: A material applied to the surface of fibers in a fabric used to reinforce plastics, and intended to improve the physical properties of the reinforced plastics over those obtained using reinforcement without finish.

Flake powder: A flat or scale-like powder which is relatively thin with a large aspect ratio.

Flame resistance: Ability of a material to extinguish flame once the source of heat is removed. (See also Self-extinguishing resin.)

Flame retardants: Chemicals used to reduce or eliminate the tendency of a resin to burn. (For polyethylene and similar resins, chemicals such as antimony trioxide and chlorinated paraffins are useful.)

Flame-retarded resin: A resin compounded with certain chemicals to reduce or eliminate its tendency to burn.

Flame spraying: Method of applying a plastic coating in which finely powdered fragments of the plastic, together with suitable fluxes, are projected through a cone of flame onto a surface.

Flammability: Measure of the extent to which material will support combustion.

Flash: The portion of the charge that flows or is extruded from the mold cavity during molding; extra plastic attached to a molding along the parting line, which may be removed before the part is considered finished.

Flash mold: A mold designed to permit the escape of excess molding material; such a mold relies upon the backpressure to seal the mold and put the piece under pressure.

Flat lay: (1) The property of nonwarping in laminating adhesives. (2) An adhesive material with good noncurling and nondistention characteristics.

Flatwise: Refers to cutting specimens and the application of load. The load is applied flatwise when it is applied to the face of the original sheet or specimen.

Flexural modulus: The ratio, within the elastic limit, of the applied stress on a test specimen in flexure to the corresponding strain in the outermost fibers of the specimen.

Flexural rigidity: (1) For fibers this is a measure of the rigidity of individual strands or fibers; the force couple required to bend a specimen to unit radius of curvature. (2) For plates the measure of rigidity is $D = EI$, where E is the modulus of elasticity and I is the moment of inertia, or

$$d = \frac{Eh^2}{12(1-v)} \; in/lb$$

where E = modulus of elasticity; h = thickness of plate; and v = Poisson's ratio.

Flexural strength: (1) The resistance of a material to breakage by bending stresses. (2) the strength of a material in bending expressed as the tensile stress of the outermost fibers of a bent test sample at the instant of failure. For plastics this value is usually higher than the straight tensile strength. (3) the unit resistance to the maximum load failure by bending, usually in kips per square inch (megapascals).

Flow: The movement of resin under pressure, allowing it to fill all parts of the mold; flow or creep is the gradual but continuous distortion of a material under continued load, usually at high temperatures.

Foamed plastics: Resins in sponge form; may be flexible or rigid; cells may be closed or interconnected and density anywhere from that of the solid parent resin to 2 lb/ft^3 (32 kg/m^3).

Foam-in-place: Foam deposition requiring the foaming machine to be brought to the work (as opposed to bringing the work to the foaming machine).

FOD: Foreign-object damage.

Force: (1) The male half of the mold, which enters the cavity, exerting pressure on the resin and causing it to flow (also called punch). (2) Either part of a compression mold (top force and bottom force).

Forging: The plastic deformation of a metal into a desired shape. The deformation is usually performed hot at high strain rates and may be done in constraining dies.

FP: Polycrystalline alumina fiber.

Fracture: Rupture of the surface without complete separation of laminate.

Fracture toughness: A measure of the damage tolerance of a material containing initial flaws or cracks.

FRAT: Fiber-reinforced advanced titanium.

FRP: Fibrous-glass-reinforced plastic, any type of plastic-reinforced cloth, mat, strands, or any other form of fibrous glass.

Gage length: Length over which deformation is measured.

Gap: In filament winding the space between successive windings, which are usually intended to lie next to each other.

Gas atomized powder: A rounded or spherical powder formed by the disintegration of a melt stream by the gas expansion nozzle. The particles solidify during free flight after atomization.

Gel: A semisolid system consisting of a network of solid aggregates in which liquid is held; the initial jelly like solid phase that develops during formation of a resin from a liquid.

Gelation time: For synthetic thermosetting resins the interval of time between introduction of a catalyst into a liquid adhesive system and gel formation.

Gel coat: A resin applied to the surface of a mold and gelled before lay-up. (The gel coat becomes an integral part of the finished laminate and is usually used to improve surface appearance, etc.)

Gel point: The stage at which a liquid begins to exhibit pseudo-elastic properties, also conveniently observed from the inflection point on a viscosity-time plot. Also called gel time.

Geodesic: The shortest distance between two points on a surface.

Geodesic isotensoid: Constant-stress level in any given filament at all points in its path.

Geodesic-isotensoid contour: In filament-wound reinforced plastic pressure vessels a dome contour in which the filaments are placed on geodesic paths so that the filaments will exhibit uniform tension throughout their lengths under pressure loading.

Geodesic ovaloid: A contour for end domes, the fibers forming a geodesic line. The forces exerted by the filaments are proportioned to meet hoop and meridional stresses at any point.

Glass: An inorganic product of fusion which has cooled to a rigid condition without crystallizing. Glass is typically hard and relatively brittle and has a conchoidal fracture.

Glass fiber: A glass filament that has been cut to a measurable length. Staple fibers of relatively short length are suitable for spinning into yarn.

Glass filament: A form of glass that has been drawn to a small diameter and extreme length. Most filaments are less than 0.005 in (0.13 mm) in diameter.

Glass-filament bushing: The unit through which molten glass is drawn in making glass filaments.

Glass finish: A material applied to the surface of a glass reinforcement to improve its effect upon the physical properties of the reinforced plastic; also called bonding agent.

Glass flake: Thin, irregularly shaped flakes of glass typically made by shattering a continuous thin-walled tube of glass.

Glass former: An oxide which forms a glass easily, also one that contributes to the network of silica glass when added to it.

Glass stress: In a filament-wound part, usually a pressure vessel, the stress calculated using only the load and the cross-sectional area of the reinforcement.

Glass transition: The reversible change in an amorphous polymer or in amorphous regions of a partially crystalline polymer from (or to) a viscous or rubbery condition to (or from) a hard and relatively brittle state. The glass transition generally occurs over a relatively narrow temperature region and is similar to the solidifaction of a liquid to a glassy state; it is not a phase transition. The glass transition temperature is the approximate midpoint of the temperature range over which glass transition takes place.

Glass-transition temperature T_g: The approximate temperature at which increased molecular mobility results in significant changes in properties of a cured resin. The measured value of T_g can vary, depending on the test method.

Glass volume percent: The product of the specific gravity of a laminate and the percent glass by weight divided by the specific gravity of the glass.

Grain: An individual crystal within polycrystalline (multi-grain) material, in which adjacent grains have substantially different orientations.

Grain boundary: Each grain in a polycrystalline material is separated by a grain boundary about two atoms wide.

Green density: The powder density after compaction.

Green strength: The strength of the as-pressed powder compact.

Greige: Fabric before finishing; yarn or fiber before bleaching or dyeing. Also called gray goods, greige goods, greige gray.

Growth: The increase in dimensions of a powder compact which may occur during sintering; the opposite dimensional change from shrinkage.

Guide pin: A pin which guides mold halves into alignment on closing.

Guide-pin bushing: The bushing through which the guide pin moves when the mold is closed.

Gusset: A piece used to give added size or strength in a particular location of an object; the folded-in portion of a flattened tubular film.

Hand: The softness of a piece of fabric, as determined by the touch (individual judgment).

Hand lay-up: The process of placing (and working) successive plies of reinforcing material or resin-impregnated reinforcement in position on a mold by hand, then cured to the formed shape.

Hardener: (1) A substance or mixture added to a plastic composition to promote or control the curing action by taking part in it. (2) A substance added to control the degree of hardness of the cured film. (See also Catalyst.)

Hard metal: Another name for the cemented carbides which reflects their high hardness after sintering.

Hardness: The resistance to surface indentation, usually measured by the depth of penetration (or arbitrary units related to depth of penetration) of a blunt point under a given load using a particular instrument according to a prescribed procedure. (See also Barcol hardness, Rockwell hardness number.)

Hat: A member in the shape of a hat.

Heat affected zone: When metal is fusion welded, the areas near the hot part of the weld metal will be heated. This heating cycle can cause substantial changes in microstructure (e.g. grain size) and properties.

Heat build-up: The temperature rise in a part resulting from the dissipation of applied strain energy as heat.

Heat-convertible resin: A thermosetting resin convertible by heat to an infusible and insoluble mass.

Heat-distortion temperature: Temperature at which a test bar deflects a certain amount under specified temperature and a stated load.

Heat resistance: The property or ability of plastics and elastomers to resist the deteriorating effects of elevated temperatures.

Heavy alloy: A range of high density, machinable alloys nominally based on tungsten with small concentrations of alloying additions such as nickel, iron, manganese, or copper. The alloys are liquid phase sintered from mixed elemental powders to create a composite material.

High modulus organic fibers: High-modulus organic fibers (HMOFs) are a new generation of organic fibers possessing unusually high tensile strength and modulus but generally poor compressive and transverse properties. Unlike conventional fibers that have a chain-folded structure, all HMOF fibers have an oriented chain-extended structure. One approach to producing such a polymer is to make a structurally modified, highly oriented version of a "conventional' polymer such as high-modulus polyethylene. A second approach is to synthesize polymers that have inherently ordered rigid molecular chains in an extended conformation. Examples of the latter are liquid crystalline aramids and liquid crystalline polyesters (LCPs).

High-pressure laminates: Laminates molded and cured at pressures not lower than 1 kip/in^2 (7 MPa) and more commonly at 1.2 to 2 kips/in^2 (8.3 to 13.8 MPa).

High-pressure molding: A molding process in which the pressure used is greater than 1 kip/in^2 (7 MPa).

HM: High-modulus.

HMC: High-strength molding compound.

HME: High-vinyl-modified epoxy.

Honeycomb: Manufactured product of resin-impregnated sheet material (paper, glass, fabric, etc.) or sheet metal formed into hexagonal-shaped cells; used as a core material in sandwich construction.

Hoop stress: The circumferential stress in a material of cylindrical form subjected to internal and external pressure.

Hot isostatic pressing: A process for densifying ceramic or metallic powders combining temperature and high pressure gas to densify a material into a net-shape component using a pressure-tight outer envelope.

Hot pressing: The high pressure, low strain rate forming of a compact at a temperature high enough to induce sintering and creep processes.

Hot working: The plastic deformation of a metal at a temperature and strain rate that minimizes work hardening.

Hybrid: The result of attaching a composite body to another material such as aluminum, steel, etc.,on two reinforcing agents in the matrix such as graphite and glass.

Hybrid composite: A composite laminate consisting of laminae of two or more composite material systems. A combination of two or more different fibers, such as carbon and glass or carbon and aramid, into structure. Tapes, fabrics, and other forms may be combined; usually only the fibers differ.

Hydraulic press: A press in which the molding force is created by the pressure exerted on a fluid.

Hydrophilic: Capable of absorbing or absorbing water.

Hydrophobic: Capable of repelling water.

Hydroscopic: Capable of absorbing and retaining atmospheric moisture.

IFAC: Integrated-flexible automation center.

ILC: Integrated-laminating center.

IM: Intermediate-modulus.

Impact strength: The ability of a material to withstand shock loading; the work done in fracturing a test specimen in a specified manner under shock loading.

Impact toughness: Toughness refers to a material's resistance to crack growth. Impact toughness is its crack growth resistance under conditions of impact or rapid loading.

Impregnate: In reinforced plastics to saturate the reinforcement with a resin.

Impregnated fabric: A fabric impregnated with a synthetic resin. (See also Prepreg.)

Induction furnace: In the carbon fiber process, a carbonization furnace that utilizes induction heating to eliminate the heteroatoms from the carbon fiber structure. Heat distribution is obtained by a combination of induced heat to a furnace muffle/susceptor within the induction coil and radiation from the interior surfaces of the furnace chamber in temperatures in excess of 3,000°C. Sometimes an induction furnace is used in combination with a resistance furnace to achieve the desired pyrolysis temperature (i.e., the resistance furnace may be used to expose the fiber to a temperature as high as 2,000°C, and if higher temperatures are required, as may be the case in graphitization, the fiber could then be passed through an induction furnace).

Inert filler: A material added to a plastic to alter the end-item properties through physical rather than chemical means.

Infrared: The part of the electromagnetic spectrum between the visible-light range and the radar range; radian heat in this range, and infrared heaters are much used in sheet thermosetting.

Inhibitor: A substance which retards a chemical reaction; used in certain types of monomers and resins to prolong storage life.

Initial modulus: See Modulus of elasticity.

Injection molding: A technique developed for processing thermoplastics which are heated and forced under pressure into closed molds and cooled.

Inorganic: Designating or pertaining to the chemistry of all elements and compounds not classified as organic; matter other than animal or vegetable, such as earthy or mineral matter. (See also Organic.)

Insert: An integral part of a plastic molding consisting of metal or other material which may be molded into position or pressed into the molding after the molding is complete.

Insert pin: A pin which keeps an inserted part (insert) inside the mold by screwing or friction; it is removed when the object is being withdrawn from the mold.

Instron: An instrument used to determine the tensile and compressive properties of materials.

Insulating resistance: The electric resistance between two conductors or systems of conductors separated only by insulating material.

Insulator: (1) A material of such low electric conductivity that the flow of current through it can usually be neglected: (2) A material of low thermal conductivity.

Interface: The junction point or surface between two different media; on glass fibers, the contact area between glass and sizing or finish; in a laminate, the contact area between the reinforcement and the laminating resin.

Interlaminar: Existing or occurring between two or more adjacent laminae.

Interlaminar shear: The shearing force tending to produce displacement between two laminae along the plane of their interface; usually the weakest element of a composite.

Interlaminar shear strength: The maximum shear stress existing between layers of a laminated material.

Internal stress: Stress created within an adhesive layer by the movement of the adherends at differential rates or by the contraction or expansion of the adhesive layer. (See also Stress.)

Interstitial: A non-metallic atom in an alloy or compound small enough to occupy sites in the lattice between the metal atoms. In a dilute solution of interstitials in a metal, such as carbon in iron, interstitial atoms can migrate (diffuse) without the need for vacancies.

Ion implantation: A sophisticated process in which ions are impacted into the surface of samples causing modifications in properties of the surface layers. Since the process occurs at low temperatures the implanted ions cause extreme distortion in the lattice structure of the substrate material, greatly affecting its properties.

Ionic Bond: Atomic bonding due to electrostatic attraction between charged particles (ions) resulting from the transfer of electrons.

IPS: Integrated process system.

Irreversible: (1) Not capable of redissolving or remelting. (2) Descriptive of chemical reactions which proceed in a single direction and are not capable of reversal (as applied to thermosetting resins).

Isocyanate plastics: Plastics based on resins made by the condensation of organic isocyanates with other compounds . (See also Urethane plastics.)

Isostatic pressing: The compaction of a powder by subjecting it to equal pressure from all directions; hydrostatic compaction.

Isotropic laminate: One in which the strength properties are equal in all directions.

Izod impact test: A destructive test designed to determine the resistance of a plastic to the impact of a suddenly applied force.

Joint: The location at which two adherends are held together with a layer of adhesive; the general area of contact for a bonded structure. Butt joint: the edge faces of the two adherends are at right angles to the other faces of the adherends. Scarf joint: a joint made by cutting away similar angular segments of two adherends and bonding them with the cut areas fitted together. Lap joint: a joint made by placing one adherend partly over another and bonding together the overlapping portions.

Lack of fillout: Characteristic of an area, occurring usually at the edge of a laminated plastic, where the reinforcement has not been wetted with resin.

Lacquer: Solution or natural or synthetic resins in readily evaporating solvents, used as a protective coating.

Laminate: (1) To unite sheets of material by a bonding material usually with pressure and heat (normally used with reference to flat sheets). (2) A product made by so bonding. (See also Bidirectional laminate, Unidirectional laminate.) A composite material made up of a single ply or layer or series of layers with each layer consisting of a reinforcing fiber imbedded in a matrix.

Each ply or layer oriented in a predetermined manner in order to maximize the properties of the laminate.

Laminated molding: A molded plastic article produced by bonding together, under heat and pressure in a mold, layers of resin-impregnated laminating reinforcement; also called laminated plastics.

Laminate ply: One layer of a product which is evolved by bonding together two or more layers of materials.

Land: The portion of a mold which provides the separation or cutoff of the flash from the molded article.

Lap: In filament winding the amount of overlay between successive windings, usually intended to minimize gapping.

Lap joint: See Joint.

Lattice: The geometrically perfect network structure on which crystalline materials are based.

Lay: (1) In glass fiber the spacing of the roving bands on the roving package expressed in the number of bands per inch. (2) In filament winding the orientation of the ribbon with some reference, usually the axis of rotation.

Lay-flat: See Flat lay.

Lay-up: (1) As used in reinforced plastics, the reinforcing material placed in position in the mold. (2) The process of placing the reinforcing material in position in the mold. (3) The resin-impregnated reinforcement. (4) The component materials, geometry, etc., of a laminate.

Lengthwise direction: Refers to cutting specimens and the application of loads. For rods and tubes, lengthwise is the direction of the long axis. For other shapes of materials that are stronger in one direction than in the other lengthwise is the direction that is stronger. For materials that are equally strong in both directions lengthwise is an arbitrarily designated direction that may be with the grain, direction of flow in manufacture, longer direction, etc. (See also Crosswise direction.)

LHS: Low-cost high-strength.

LIM: Liquid injection molding.

Linear expansion: The increase of a planar dimension, measured by the linear elongation of a sample in the form of a beam which is exposed to two given temperatures.

Liner: In a filament-wound pressure vessel the contiguous, usually flexible, coating on the inside surface of the vessel used to protect the laminate from chemical attack or to prevent leakage under stress.

Liquid crystal: Liquid crystal materials are usually made up of rigid, rod-like molecules. They can become ordered in either solution or melt phase, which means that the molecules aggregate under certain conditions so that the materials are anisotropic. It is very easy to achieve high orientation with these materials during either liquid or melt extrusion, so the resulting extrudate (e.g.,fiber or film) has a excellent tensile properties.

Liquid-crystal polymers: A newer type of thermoplastic, melt processible, with high orientation in molding, improved tensile strength, and high-temperature capability.

Liquid phase sintering: Sintering at room temperature where a liquid and solid coexist due to chemical reactions, partial melting, or eutectic liquid formation.

Liquidus: The maximum temperature at which equilibrium exists between the molten glass and its primary crystalline phase.

LMC: Low-pressure molding compound.

Load-deflection curve: A curve in which the increasing flexural loads are plotted on the ordinate and the deflections caused by those loads are plotted on the abscissa.

Longos: Low-angle helical or longitudinal windings.

Loop tenacity: The strength value obtained by pulling two loops, like two links in a chain, against each other in order to test whether a fibrous material will cut or crush itself; also called loop strength.

Loss on ignition: Weight loss, usually expressed as percent of total, after burning off an organic sizing from glass fibers or an organic resin from a glass-fiber laminate.

Lot: A specific amount of material produced at one time and offered for sale as a unit quantity.

Low-pressure laminates: In general, laminates molded and cured at pressures from 0.4 kips/in^2 (2.8 MPa) down to and including pressure obtained by the mere contact of plies.

Low-pressure molding: The distribution of relatively uniform low pressure [0.2 kip/in^2 (1.4 MPa) or less] over a resin-bearing fibrous assembly of cellulose, glass, asbestos, or other material, with or without application of heat from an external source, to form a structure possessing definite physical properties.

LST: Large Space Telescope.

LWV: Lightweight vehicle.

Macerate: (1) To chop or shred fabric for use as a filler for a molding resin. (2) The molding compound obtained when so filled.

Mandrel: (1) The core around which paper-, fabric-, or resin-impregnated glass is wound to form pipes, tubes, or vessels. (2) In extrusion the central finger of a pipe or tubing die.

Mat: A fibrous material for reinforced plastic consisting of randomly oriented chopped filaments or swirled filaments with a binder, available in blankets of various widths, weights, and lengths.

Mat binder: Resin applied to glass fiber and cured during the manufacture of mat, used to hold the fibers in place and maintain the shape of the mat.

Matched metal molding: A reinforced-plastic manufacturing process in which matching male and female metal molds are used (similar to compression molding) to form the part, as opposed to low-pressure laminating or spray-up.

Matrix: The dominating lattice structure of a given phase in a material. The essentially homogeneous resin or polymer material in which the fiber system of a composite is imbedded. Both thermoplastic and thermoset resins may be used, as well as metals, ceramics, and glasses.

Matrix metal: The continuous phase in a polyphase alloy or mechanical mixture; the physically continuous metallic constituent in which separate particles of another constituent are embedded.

Matte: A nonspecular surface having diffused reflective powers.

Mechanical adhesion: Adhesion between surfaces in which the adhesive holds the parts together by interlocking action.

Mechanical alloying (MA): The formation of an alloy powder by milling elemental powders for a prolonged time; frequently used to create amorphous or dispersion strengthened alloy powders.

Melamine plastics: Plastics based on melamine resins.

Mesh: The screen number representing the number of wires per inch in a square grid. The higher the mesh number, the smaller the opening size.

Mesophase: An intermediate phase in the formation of carbon fiber from a pitch precursor. This is a liquid crystal phase in the form of microspheres, which, upon prolonged heating above 400°C, coalesce, solidify, and form regions of extended order. Heating above 2,000°C leads to the formation of graphite-like structure.

Metallic Bond: The interatomic chemical bond in metals characterized by delocalized electrons in the energy bands. The atoms are considered to be ionized with the positive ions occupying the lattice positions. The valence electrons are free to move. The bonding force is the electrostatic attraction between ions and electrons.

Metallic fiber: Manufactured fiber composed of metal, plastic-coated metal, metal-coated plastic, or a core completely covered by metal.

M glass: A high-beryllia-content glass designed especially for high modulus of elasticity.

Micron: A unit of length replaced by the micrometer (μm); 1 μm = 10^{-6} m = 10^{-3} mm = 0.00003937 in = 39.4 μin.

Microstructure: A structure with heterogeneities that can be seen through a microscope.

Mil: The unit used in measuring the diameter of glass-fiber strands, wire, etc. (1 mil = 0.001 in).

Milled fibers: Continuous glass strands hammer-milled into small modules of filamentized glass. Useful as anticrazing reinforcing fillers for adhesives.

MMC: Metal-matrix composite; material in which continuous carbon, silicon carbide, or ceramic fibers are embedded in a metallic matrix.

Modulus: A number which expresses a measure of some property of a material, e.g., modulus of elasticity, shear modulus, etc.; a coefficient of numerical measurement of a property. Using "modulus' alone without modifying terms is confusing and should be discouraged. For example, Young's modulus (E) refers to the constant of elasticity, or stiffness of a material.

Modulus in compression: See Compressive modulus.

Modulus in flexure: See Flexural modulus.

Modulus, initial, or Young's modulus: See Modulus of elasticity.

Modulus in shear: See Shear modulus.

Modulus in tension: See Tensile modulus.

Modulus of elasticity: The ratio of the stress or applied load to the strain or deformation produced in a material that is elastically deformed. If a tensile strength of 2 kips/in^2 (14 MPa) results in an elongation of 1%, the modulus of elasticity is 2/0.01 = 200 kips/in^2 (1,379 MPa); also called Young's modulus.

Modulus of elasticity in torsion: The ratio of the torsion stress to the strain in the material over the range for which this value is constant.

Modulus or rigidity: See Flexural rigidity.

Modulus of rupture: See Flexural strength.

Mohs hardness: A measure of the scratch resistance of a material; the higher the number the greater the scratch resistance (diamond is 10).

Moisture absorption: The pickup of water vapor from air by a material; it relates only to vapor withdrawn from the air by a material and must be distinguished from water absorption. (See also Water absorption.)

Mold: (1) The cavity or matrix in or on which the plastic composition is placed and from which it takes form. (2) To shape plastic parts or finished articles by heat or pressure. (3) The assembly of all parts that function collectively in the molding process.

Mold-release agent: A liquid or powder used to prevent sticking of molded articles in a cavity.

Mold seam: Line on a molded or laminated piece, differing in color or appearance from the general surface, caused by the parting line of the mold.

Mold shrinkage: (1) The immediate shrinkage which a molded part undergoes when it is removed from a mold and cooled to room temperature. (2) The difference in dimensions, expressed in inches per inch (millimeters per millimeter) between a molding and the mold cavity in which it was molded (at normal temperature measurement). (3) The incremental difference between the dimensions of the molding and the mold from which it was made, expressed as a percentage of the dimension of the mold.

Molding: The shaping of a plastic composition in or on a mold, normally accomplished under heat and pressure; sometimes used to denote the finished part.

Molding compounds: Plastics in a wide range of forms (especially granules or pellets) to meet specific processing requirements.

Molding cycle: (1) The time required for the complete sequence of operations on a molding press to produce one set of moldings. (2) The operations necessary to produce a set of moldings without reference to time.

Molding press: A press used to form compacts from powder.

Molding pressure: The pressure applied to the ram of an injection machine or press to force the softened plastic to fill the mold cavities completely. (See also Compression molding pressure.)

Molding, pressure-bag: See Pressure-bag molding.

Molding pressure, compression: See Compression molding pressure.

Monofilament: (1) A single fiber or filament of indefinite length generally produced by extrusion. (2) A continuous fiber sufficiently large to serve as yarn in normal textile operations; also called monofil.

Monomer: (1) A simple molecule capable of reacting with like or unlike molecules to form a polymer. (2) The smallest repeating structure of a polymer, also called a mer.

Morphology: The overall form of a polymer structure, that is crystallinity, branching, molecular weight, etc. Also, the study of the fine structure of a fiber or other material, such as basal plane orientation across a carbon/graphite fiber.

Multiple-cavity mold: A mold with two or more mold impressions; i.e., a mold that produces more than one molding per molding cycle.

Multicircuit winding: In filament winding a winding that requires more than one circuit before the band repeats by lying adjacent to the first band.

Multifilament yarn: A multitude of fine, continuous filaments (often 5 to 100), usually with some twist in the yarn to facilitate handling. Sizes range from 5 to 10 denier up to a few hundred denier. Individual filaments in a multifilament yarn are usually about 1 to 5 denier.

Nanoscale: Powders or microstructures with sizes that can be measured in nanometers. Typically the powders are less than 100 nm in size.

Near-net shape (NNS): A compact which has the general shape of the final product, but is oversized and requires final machining.

Needles: Elongated rod-like particles.

Nesting: In reinforced plastics placing plies of fabric so that the yarns of one ply lie in the valleys between the yarns of the adjacent ply (nested cloth).

Net shape: A compact manufactured to the final density and dimensions without the need for machining.

Netting analysis: The analysis of filament-wound structures which assumes that the stresses induced in the structure are carried entirely by the filaments, the strength of the resin being neglected, and that the filaments possess no bending or shearing stiffness, carrying only the axial tensile loads.

Nol ring: A parallel filament-wound test specimen used for measuring various mechanical-strength properties of the material by testing the entire ring or segments of it.

Nondestructive inspection (NDI): A process or procedure for determining material or part characteristics without permanently altering the test subject. Nondestructive testing (NDT) is broadly considered synonymous with NDI.

Nonhygroscopic: Not absorbing or retaining an appreciable quantity of moisture from the air (water vapor).

Nonrigid plastic: A plastic which has a stiffness or apparent modulus of elasticity not over 10 kips/in^2 (69 MPa) at 73.4°F (23°C).

Nonwoven fabric: A planar structure produced by loosely bonding together yarns, roving etc. (See also Fabric.)

Notch sensitivity: The extent to which the sensitivity of a material to fracture is increased by the presence of a surface inhomogeneity such as a notch, a sudden change in section, a crack, or a scratch. Low notch sensitivity is usually associated with ductile materials and high notch sensitivity with brittle materials.

Novolac: A phenolic-aldehyde resin which remains permanently thermoplastic unless a source of methylene groups is added; a linear thermoplastic B-staged phenolic resin. (See also Thermoplastic.)

Offset yield strength: The stress at which the strain exceeds by a specific amount (the offset) an extension of the initial proportional portion of the stress-strain curve.

Open-cell-foamed plastic: A cellular plastic in which there is a predominance of interconnected cells.

Orange peel: An uneven surface resembling that of an orange peel; said of injection moldings with unintentionally rugged surfaces.

Order: This usually refers to the structure of an alloy of stoichiometric composition. For example, compounds such as aluminum oxide (Al_2O_3) or titanium nitride (TiN) are perfectly ordered, each component species forming its own lattice structure, and with the combination forming a superlattice structure. It should be noted, however, that in the present text order is sometimes used to describe the symmetrical crystalline state of a metal, as opposed to an unordered amorphous state.

Organic: Designating or composed of matter originating in plant or animal life or composed of chemicals of hydrocarbon origin, natural or synthetic.

Oriented materials: Materials, particularly amorphous polymers and composites, whose molecules and/or macroconstituents are aligned in a specific way. Oriented materials are anisotropic. Orientation is generally uniaxial or biaxial.

Orthotropic: Having three mutually perpendicular planes of elastic symmetry.

Out-life: The period of time a prepreg material remains in a handleable form and with properties intact outside of the specified storage environment; for example, out of the freezer in the case of thermoset prepregs.

Overcuring: The beginning of thermal decomposition resulting from too high a temperature or too long a molding time.

Overflow groove: Small groove used in molds to allow material to flow freely to prevent weld lines and low density and to dispose of excess material.

Overlap: A simple adhesive joint, in which the surface of one adherend extends past the leading edge of another.

Overlay sheet: A nonwoven fibrous mat (in glass, synthetic fiber, etc.) used as the top layer in a cloth or mat lay-up, to provide a smoother finish or minimize the appearance of the fibrous pattern.

Package: The method of supplying the roving or yarn.

PAN: Polyacrylonitrile.

Parallel-laminated: Laminated so that all the layers of material are oriented approximately parallel with respect to the grain or strongest direction in tension. (See also Cross laminated.)

Parameter: An arbitrary constant, as distinguished from a fixed or absolute constant. Any desired numerical value may be given as a parameter.

Particle size: The controlling linear dimension of an individual particle, as determined by analysis with screens or other instruments.

Particle size analyzer: An automated device for determination of the particle size distribution.

Parting agent: See Release agent.

Parting line: The linear mark on a compact where two separate tool or die pieces mated during shaping. In injection molding it is where the two halves of the die joined together.

PBBI: Polybutadiene bisimide.

PBT: Polybutylene terephtalate; polybenzothiazole.

Peel ply: The outside layer of a laminate which is removed or sacrificed to achieve improved bonding of additional plies and leaves a clean resin-rich surface ready for bonding.

Peel strength: Bond strength, in pounds per inch of width, obtained by peeling the layer. (See also Bond strength.)

Permanence: The property of a plastic which describes its resistance to appreciate change in characteristics with time and environment.

Permanent set: The deformation remaining after a specimen has been stressed in tension a prescribed amount for a definite period and released for a definite period.

PES: Polyethersulfone.

PET: Polyethylene terephthalate.

Phase: A microstructural constituent of one given lattice structure or type. A material from one to several coexisting phases.

Phenolic, phenolic resin: A synthetic resin produced by the condensation of an aromatic alcohol with an aldehyde, particularly of phenol with formaldehyde. (See also A stage, B stage, C stage, Novolac.)

Physical vapor deposition: A process for depositing extremely thin layers of material onto a substrate through a vapor or gaseous phase passing over it.

PI: Polyimide.

Pick: (1) An individual filling yarn, running the width of a woven fabric at right angles to the warp, also called fill, woof, weft. (2) To experience tack. (3) To transfer unevenly from an adhesive applicator mechanism due to high surface tack.

Pinch-off: In blow molding a raised edge around the cavity in the mold which seals off the part and separates the excess material as the mold closes around the parison.

Pinhole: A tiny hole in the surface of, or through, a plastic material, usually not occurring alone.

Pin, insert: See Insert pin.

Pit: Small regular or irregular crater in the surface of a plastic, usually with width about the same order of magnitude as the depth.

Pitch: A high molecular weight material left as a residue from the destructive distillation of coal and petroleum products. Pitches are used as base materials for the manufacture of certain high-modulus carbon fibers and as matrix precursors for carbon-carbon composites.

Planar helix winding: A winding in which the filament path on each dome lies on a plane which intersects the dome while a helical path over the cylindrical section is connected to the dome paths.

Planar winding: A winding in which the filament path lies on a plane intersecting the winding surface.

Plasma spraying: A process for spraying materials onto a substrate or surface in which the materials, in powder form, are heated in a hot plasma created by an electric arc. The technique is particularly useful for laying down surface coatings of ceramic materials.

Plastic: A material that contains as an essential ingredient an organic substance of high molecular weight, is solid in its finished state, and at some stage in its manufacture or processing into finished articles can be shaped by flow; made of plastic. A rigid plastic is one with a stiffness or apparent modulus of elasticity greater than 100 kips/in^2 (690 MPa) at 73.4°F (23°C). A semirigid plastic has a stiffness or apparent modulus of elasticity between 10 and 100 kips/in^2 (69 and 690 MPa) at 73.4°F (23°C).

Plasticate: To soften by heating or kneading.

Plastic deformation: Change in dimensions of an object under load that is not recovered when the load is removed; opposite of elastic deformation. (See also Elastic recovery.)

Plastic flow: Deformation under the action of a sustained force; flow of semisolids in molding plastics.

Plasticize: To make a material moldable by softening it with heat or a plasticizer.

Plastic tooling: Tools constructed of plastics, generally laminates or casting materials.

Platens: The mounting plates of a press, to which the entire mold assembly is bolted.

Plates: Powder particles with a flat shape and considerable thickness in contrast to flakes which are thin with respect to the width.

Plied yarn: A yarn formed by twisting together two or more single yarns in one operation.

Ply: The number of single yarns twisted together to form a plied yarn; one of the layers that make up a stack or laminate.

PMR: Polyimides. Polymerization-of-monomer-reactant polyimides.

Poisson's ratio ν: A constant relating change in cross-sectional area to change in length when a material is stretched;

$$\nu = \begin{cases} \frac{1}{2} & \text{for rubbery materials} \\ \frac{1}{4} - \frac{1}{2} & \text{for crystals and glasses} \end{cases}$$

Polar winding: A winding in which the filament path passes tangent to the polar opening at one end of the chamber and tangent to the opposite side of the polar opening at the other end. A one-circuit pattern is inherent in the system.

Polyacrylonitrile (PAN): Used as a base material or precursor in the manufacture of certain carbon fibers. The fiber-forming acrylic polymers are high in molecular weight and are produced commercially either by solution polymerization or suspension polymerization. Both techniques utilize free-radical-initiated addition polymerization of acrylonitrile and small percentages of other monomers. Commercial precursor fibers are more than 90 percent acrylonitrile based.

Polyamide: A polymer in which the structural units are linked by amide or thioamide groupings; many polyamides are fiber-forming.

Polyamideimide: A polymer containing both amide and imide (as in polyamide) groups; its properties combine the benefits and disadvantages of both.

Polycrystalline: This refers to a crystalline material containing several (usually very many) grains.

Polyesters: Thermosetting resins produced by dissolving unsaturated, generally linear alkyd resins in a vinyl active monomer, e.g., styrene, methyl styrene, or diallyl phthalate. The two important commercial types are (1) liquid resins that are cross-linked with styrene and used either as impregnates for glass or carbon fiber reinforcements in laminates, filament-wound structures, and other built-up constructions, or as binders for chopped-fiber reinforcements in molding compounds; and (2) liquid or solid resins cross-linked with other esters in chopped-fiber and mineral-filled molding compounds.

Polyetheretherketone (PEEK): A linear aromatic crystalline thermoplastic. A composite with a PEEK matrix may have a continuous- use temperature as high as 250°C.

Polyimide: A polymer produced by heating polyamic acid; a highly heat-resistant resin [>600°F (> 316°C)] suitable for use as a binder or an adhesive. Similar to a polyamide, differing only in the number of hydrogen molecules contained in the groupings. May be either thermoplastic or thermosetting.

Polymer: A high-molecular-weight organic compound, natural or synthetic, whose structure can be represented by a repeated small unit (mer), e.g., polyethylene, rubber, cellulose. Synthetic polymers are formed by addition or condensation polymerization of monomers. Some polymers are elastomers, some plastics. When two or more monomers are involved, the product is called a copolymer.

Polymerization: A chemical reaction in which the molecules of a monomer are linked together to form large molecules whose molecular weight is a multiple of that of the original substance. When two or more monomers are involved, the process is called copolymerization or hetero-polymerization.

Polymerize: To unite molecules of the same kind into a compound having the elements in the same proportion but possessing much higher molecular weight and different physical properties.

Poly(phenylene): A high carbon content (94.7 percent) polymer that has the monomeric repeat unit - (C_6H_4).

Polyphenylene sulfide (PPS): A high-temperature thermoplastic useful primarily as a molding compound. Known for chemical resistance.

Polysulfone (PS): A high-temperature resistant thermoplastic polymer with the sulfone linkage, with a Tg of 190°C.

Polyvinyl Chloride (PVC): A vinyl-type thermoplastic resin formed by the addition reaction of vinyl chloride monomer.

Porosity: The ratio of the volume of air or void contained within the boundaries of a material to the total volume (solid material plus air or void), expressed as a percentage.

Positive mold: A mold designed to apply pressure to a piece being molded with no escape of material.

Postcure: Additional elevated-temperature cure, usually without pressure, to improve final properties and/or complete the cure. Complete cure and ultimate mechanical properties of certain resins are attained only by exposure of the cured resin to higher temperatures than those of curing.

Postforming: The forming, bending, or shaping of fully cured, C-stage thermoset laminates that have been heated to make them flexible. On cooling, the formed laminate retains the contours and shape of the mold over which it has been formed.

Pot life: The length of time a catalyzed resin system retain viscosity low enough to be used in processing; also called working life.

Powder: Particles of matter characterized by a small size, less than 1 mm in size.

Powder metallurgy: The art and science of producing metal powders and of utilizing metal powders for the production of massive materials and shaped objects.

Prealloyed powder: Each particle contains an intimate mixture of two or more elements in a prescribed ratio to form an alloy; examples include brass, bronze, steel, and stainless steel.

Precipitate: A minor phase or particle that has precipitated by the movement and coalescence of individual atoms from solid solution during heat treatment.

Precipitation-hardened: This refers to a material in which the precipitation reaction has been so finely dispersed that the material becomes hardened. The small particles provide a barrier to dislocation movement during plastic deformation.

Precure: The full or partial setting of a synthetic resin or adhesive in a joint before the clamping operation is complete or before pressure is applied.

Precursor: For carbon fibers, the rayon, PAN, or pitch fibers from which carbon fibers are made.

Preform: (1) A preshaped fibrous reinforcement formed by the distribution of chopped fibers by air, water flotation, or vacuum over the surface of a perforated screen to the approximate contour and the thickness desired in the finished part. (2) A preshaped fibrous reinforcement of mat or cloth formed to the desired shape on a mandrel or mock-up before being placed in a mold press. (3) A compact "pill" formed by compressing premixed material to facilitate handling and control of uniformity of charges for mold loading.

Preform binder: A resin applied to the chopped strands of a preform, usually during its formation, and cured so that the preform will retain its shape and be handleable.

Preimpregnation: The practice of mixing resin and reinforcement and effecting partial cure before use or shipment to the user. (See also Prepreg.)

Premix: A molding compound prepared prior to, and apart from, the molding operations and containing all components required for molding, i.e., resin, reinforcement, fillers, catalysts, release agents, and other compounds.

Prepreg: Ready-to-mold material in sheet form, which may be cloth, mat, or paper impregnated with resin and stored for use. The resin is partially cured to a B stage and supplied to the fabricator, who lays up the finished shape and completes the cure with heat and pressure.

Presintering: The heating of a compact to a temperature lower than the normal sintering temperature to gain strength for subsequent handling, including machining.

Pressure: Force measured per unit are. Absolute pressure is measured with respect to zero. Gage pressure is measured with respect to atmospheric pressure.

Pressure-assisted sintering: Sintering with the application of an external pressure. It is often performed by initially sintering in vacuum and subsequently pressurizing the furnace to densify any remaining closed pores.

Pressure-bag molding: A process for molding reinforced plastics, in which a tailored flexible bag is placed over the contact lay-up on the mold, sealed, and clamped in place. Fluid pressure, usually compressed air, is exerted on the bag, and the part is cured.

Primary structure: One critical to flight safety.

Primer: A coating applied to a surface before the application of an adhesive or lacquer, enamel, or the like to improve the performance of the bond.

Promoter: See Accelerator.

Proportional limit: The greatest stress which a material is capable of sustaining without deviation from proportionality of stress or strain (Hooke's law); it is expressed in force per unit area, usually in kips per square inch (megapascals).

Prototype: A model suitable for use in complete evaluation of form, design, and performance.

Pultrusion: Reversed extrusion of resin-impregnated roving in the manufacture of rods, tubes, and structural shapes of a permanent cross section. After passing through the resin dip tank the roving is drawn through a die to form the desired cross section.

Quasi-isotropic: Approximating isotropy by orientation of plies in several directions.

Quenching: This is a term, commonly used in the heat treatment of metals, in which a certain microstructure or property is obtained by rapidly cooling the sample from some high temperature condition. Blacksmiths of old have used this technique, for example, when hardening steel by quenching it from glowing-red temperature to oil or salt water, and forming martensite (a solution of carbon and iron).

Ramping: A gradual, programmed increase/decrease in temperature or pressure to control the cure or cooling of composite parts.

Random pattern: A winding with no fixed pattern. If a large number of circuits are required for the pattern to repeat, a random pattern is approached; a winding in which the filaments do not lie in an even pattern.

Rapid solidification (RS): The extraction of heat from molten material at cooling rates of 10^4°C/s or higher, resulting in new microstructures, compositions, or phases.

Rayon: A synthetic fiber made up primarily of regenerated cellulose. In the process, cellulose derived from wood pulp, cotton linters, or other vegetable matter is dissolved into a viscose spinning solution. The solution is extruded into an acid-salt coagulating bath and drawn into continuous filaments. Rayon fibers were one of the first precursor materials to be used in the manufacture of carbon fibers. They have now been almost completely replaced by PAN and pitch fibers as starting materials for carbon fiber manufacture due to low yields, high processing costs, and limited physical property formation.

Reactive sintering: A novel sintering process where an exothermic reaction is initiated in a mixture of dissimilar (elemental) powders. The reaction produces a compound (carbide, boride, aluminide, or other compound) and the heat from the reaction is used to simultaneously sinter the product phase.

Recycle: This is a function of growing importance in today's energy and ecologically conscious world. It is much cheaper to remelt scrap metal than produce new metal from the ore. Recycling of materials today, however, is beginning to involve most materials, including all metals, paper, plastic and glass.

Refractory: A metal or ceramic having a high melting temperature, usually over 1,700°C. Example refractory metals are tungsten, molybdenum, rhenium, and zirconium.

Reinforced molding compound: Compound supplied by raw material producer in the form of ready-to-use materials, as distinguished from premix. (See also Premix.)

Reinforced plastic: A plastic with strength properties greatly superior to those of the base resin, resulting from the presence of reinforcements embedded in the composition.

Reinforcement: A strong inert material bonded into a plastic to improve its strength, stiffness, and impact resistance. Reinforcements are usually long fibers of glass, asbestos, sisal, cotton, etc., in woven or nonwoven form. To be effective, the reinforcing material must form a strong adhesive bond with the resin. ("Reinforcement" is not synonymous with "filler".)

Release agent: A material which is applied in a thin film to the surface of a mold to keep the resin from bonding to it.

Release film: An impermeable film layer that does not bond to the composite during cure.

Resilience: (1) The ratio of energy returned on recovery from deformation to the work input required to produce the deformation (usually expressed as a percentage). (2) The ability to regain an original shape quickly after being strained or distorted.

Resin: A solid, semisolid, or pseudo-solid organic material which has an indefinite (often high) molecular weight, exhibits a tendency to flow when subjected to stress, usually has a softening or melting range, and usually fractures conchoidally. Most resins are polymers. In reinforced plastics the material used to bind together the reinforcement material, the matrix. (See also Polymer.)

Resin applicator: In filament winding the device which deposits the liquid resin onto the reinforcement band.

Resin content: The amount of resin in a laminate expressed as a percent of total weight or total volume.

Resin, liquid: An organic polymeric liquid which becomes a solid when converted into its final state for use.

Resin-rich area: Space which is filled with resin and lacking reinforcing material.

Resin-starved area: Area of insufficient resin, usually identified by low gloss, dry spots, or fiber show.

Resistance furnace: In the carbon fiber process, a carbonization furnace that utilizes resistance heating to eliminate the heteroatoms from the carbon fiber structure. In this furnace, heat distribution is obtained by a combination of direct radiation from the resistors and reradiation from the interior surfaces of the furnace chamber to temperatures in excess of 1,000°C.

Resistivity: The ability of a material to resist passage of electric current through its bulk or on a surface.

Retarder: See Inhibitor.

Reverse helical winding: As the fiber-delivery arm traverses one circuit, a continuous helix is laid down, reversing direction at the polar ends; contrasted to biaxial, compact, or sequential winding in that the fibers cross each other at definite equators, the number depending on the helix. The minimum crossover would be 3.

Rib: A reinforcing member of a fabricated or molded part.

Ribbon: A fiber having essentially a rectangular cross section, where the width-to-thickness ratio is at least 4:1.

Rigid plastic: See Plastic.

RIM: Reaction-injection molding.

Rockwell hardness number: A value derived from the increase in depth of an impression as the load on an indenter is increased from a fixed minimum value to a higher value and then returned to the minimum value.

Room-temperature-curing adhesives: Adhesives that set (to handling strength) within 1 h at 68 to 86°F (20 to 30°C) and later reach full strength without heating.

Roving: In filament winding a collection of bundles of continuous filaments either as untwisted strands or as twisted yarns. Rovings may be lightly twisted, but for filament winding they are generally wound as bands or tapes with as little twist as possible. Glass rovings are predominantly used in filament winding.

Roving cloth: A textile fabric, coarse in nature, woven from rovings.

RPP: Reinforced pyrolzed plastic.

RRIM: Reinforced reaction-injection molding.

RTM: Resin-transfer molding. A molding process in which catalyzed resin is transferred into an enclosed mold into which the fiber reinforcement has been placed; cure normally is accomplished without external heat. RTM combines relatively low tooling and equipment costs with the ability to mold large structural parts.

S glass: A magnesia-alumina-silicate glass, especially designed to provide filaments with very high tensile strength.

Sandwich constructions: Panels composed of a lightweight core material (honeycomb, foamed plastic, etc.) to which two relatively thin, dense, high-strength faces or skins are adhered.

Sandwich heating: A method of heating a thermoplastic sheet before forming by heating both sides of the sheet simultaneously.

SAP: Sintered-aluminum powder.

Satin: A plastic finish having a satin or velvety appearance.

Scarf joint: See Joint.

Scratch: Shallow mark, groove, furrow, or channel normally caused by improper handling or storage.

Scrim: A low-cost, nonwoven open-weave reinforcing fabric made from continuous-filament yarn in an open-mesh construction.

Secondary structure: One not critical to flight safety.

Segregation: In this process alloying elements diffuse out of solid solution and collect at defects in the material such as grain boundaries and nonuniform distribution of ingredients, such as powder separation by size, shape, or density, or chemical separation in the microstructure of a solidified material.

Self-extinguishing resin: A resin formulated which will burn in the presence of a flame but which will extinguish itself within a specified time after the flame is removed.

Self-propagating high-temperature synthesis: An exothermic reaction of mixed dissimilar powders which forms a compound and liberates heat. The heat further induces a reaction between unreacted particles to give a continual reaction wave once ignited.

Selvage: The edge of a woven fabric finished off so as to prevent the yarns from raveling.

Semirigid plastic: See Plastic.

Sequential winding: See Biaxial winding.

Set: (1) To convert into a fixed or hardened state by chemical or physical action, such as condensation, polymerization, oxidation, vulcanization, gelation, hydration, or evaporation of volatiles. (2) The irrecoverable deformation or creep usually measured by a prescribed test procedure and expressed as a percentage of original dimension.

Set up: To harden, as in curing.

Shear: An action or stress resulting from applied forces and tending to cause two contiguous parts of a body to slide relative to each other in a direction parallel to their plane of contact.

Shear edge: The cutoff edge of the mold.

Shear modulus G: The ratio of shearing stress τ to shearing strain γ within the proportional limit of a material.

Shelf life: The length of time a material can be stored under specified conditions without harmful changes in its properties, also called storage life.

Shoe: A device for gathering filaments into a strand in glass-fiber forming. (See Chase.)

Short-beam shear strength: The interlaminar shear strength of a parallel-fiber-reinforced plastic material as determined by three-point flexural loading of a short segment cut from a ring specimen.

Shrinkage: The relative change in dimension between the length measured on the mold when it is cold and the length on the molded object 24 h after it has been taken out of the mold.

Sialon: This is a commercially successful ceramic alloy, comprising of silicon (Si), aluminum (Al), oxygen (O) and nitrogen (N). In its alloy form it thus becomes: (Si)(Al)ON.

Silicon carbide fiber: A reinforcing fiber with high strength and modulus: density is equal to that of aluminum. It is used in organic- and metal-matrix composites.

Silicone plastics: Plastics based on resins in which the main polymer chain consists of alternating silicon and oxygen atoms with carbon-containing side groups; derived from silica (sand) and methyl chloride.

Silicones: Resinous materials derived from organosiloxane polymers, furnished in different molecular weights including liquids and solid resins and elastomers.

Single-circuit winding: A winding in which the filament path makes a complete traverse of the chamber, after which the following traverse lies immediately adjacent to the previous one.

Sintering: A thermal process (fusion of metallic or ceramic powders) which increases the strength of a powder mass by bonding adjacent particles via diffusion or related atomic level events. Most of the properties of a powder compact are improved and frequently density increases with sintering. The process occurs in the solid state by interparticle diffusion.

Size: Any treatment consisting of starch, gelatin, oil, wax, or other suitable ingredient applied to yarn or fibers at the time of formation to protect the surface and facilitate handling and fabrication or to control the fiber characteristics. The treatment contains ingredients which provide surface lubricity and binding action but, unlike a finish, no coupling agent. Before final fabrication into a composite, the size is usually removed by heat-cleaning and a finish is applied.

Sizing: (1) Applying a material on a surface in order to fill pores and thus reduce the absorption of the subsequently applied adhesive or coating. (2) To modify the surface properties of the substrate to improve adhesion. (3) The material used for this purpose, also called size. A coating put on the fiber, usually at the time of manufacture, to protect the surface and aid the process of handling and fabrication or to control the fiber characteristics. Most standard sizes used for aerospace-grade carbon fibers are epoxy based.

Sizing content: The percent of the total strand weight made up by the sizing, usually determined by burning off the organic sizing ("loss on ignition").

Skein: A continuous filament, strand, yarn, roving, etc., wound up to some measurable length and generally used to measure various physical properties.

Skin: The relatively dense material that may form the surface of the cellular plastic or sandwich.

Slip casting: A forming process applicable to small powders where the powder is dispersed in a low viscosity fluid to allow shaping by casting the slurry into a porous mold which extracts the fluid.

SMC: Sheet-molding compound.

S-N curve: Stress per number of cycles to failure. (See also Stress-strain.)

Soft flow: The behavior of a material which flows freely under conventional conditions of molding and which, under such conditions, will fill all the interstices of a deep mold where a considerable distance of flow can be demanded.

Solid solution: Very few practical materials are pure in form or consist of only one element. If one element or more are dissolved completely in a base material in the solid state, then the whole is referred to as a solid solution.

Specification: A detailed description of the characteristics of a product and of the criteria which must be used to determine whether the product is in conformity with the description.

Specific gravity: The ratio of the weight of any volume of a substance to the weight of an equal volume of another substance taken as standard at a constant or stated temperature.

Specific heat: The quantity of heat required to raise the temperature of a unit mass of a substance 1 degree under specified conditions.

Specimen: An individual piece or portion of a sample used to make a specific test; of specific shape and dimensions.

Spectroscopy: When a focused beam of X-rays or electrons interacts with a material, the emitted X-rays or electrons become highly characteristic of the material through which they have passed. These emissions produce well defined spectra and provide a "finger print" of the material being studied.

SPF/DB: Superplastic-forming diffusion bonding.

Spherical powder: Powder with a uniform spherical shape and a size which can be characterized by a diameter. Gas atomized powders are often spherical.

Spinnerette: A metal disc containing numerous minute holes used in yarn extrusion. The spinning solution or molten polymer is forced through the holes to form the yarn filaments.

Spiral: In glass-fiber forming the device that is used to make the strand traverse back and forth across the forming tube.

Splice: To join two ends of glass-fiber yarn or strand, usually by means of an air-drying glue.

Spline: (1) To prepare a surface to its desired contour by working a paste material with a flat-edged tool; the procedure is similar to screeding concrete. (2) The tool itself.

Split-cavity blocks: Blocks which, when assembled, contain a cavity for molding articles having undercuts.

Split mold: A mold in which the cavity is formed of two or more components, known as splits, held together by an outer chase.

Split-ring mold: A mold in which a split-cavity block is assembled in a chase to permit forming undercuts in a molded piece. The parts are ejected from the mold and then separated from the piece.

SPMC: Solid polyester molding compound.

Spray: A complete set of moldings from a multi-impression injection mold, together with the associated molded material.

Spray-up: Techniques in which a spray gun is used as the processing tool. In reinforced plastics, for example, fibrous glass and resin can be simultaneously deposited in a mold. In essence, roving is fed through a chopper and ejected into a resin stream, which is directed at the mold by either of two spray systems. In foamed plastics, very fast-reacting urethane foams or epoxy foams are fed in liquid streams to the gun and sprayed onto the surface. On contact, the liquid starts to foam.

Sprayed-metal molds: Molds made by spraying molten metal onto a master until a shell of predetermined thickness is achieved. The shell is then removed and backed up with plaster, cement, casting resin, or other suitable material. Used primarily as a mold in sheet-forming process.

Spun roving: A heavy, low-cost glass-fiber strand consisting of filaments that are continuous but doubled back on each other.

Stabilization: In carbon fiber manufacture the process used to render the carbon fiber precursor infusible prior to carbonization.

Staple fibers: Fibers of spinnable length manufactured directly or by cutting continuous filaments to relatively short lengths [generally less than 17 in (432 mm)].

Starved area: An area in a plastic part which has an insufficient amount of resin to wet out the reinforcement completely. This condition may be due to improper wetting or impregnation or excessive molding pressure.

Starved joint: An adhesive joint which has been deprived of the proper film thickness of adhesive due to insufficient adhesive spreading or application of excessive pressure during lamination.

Static fatigue: Failure of a part under continued static load; analogous to creep-rupture failure in metals testing but often the result of aging accelerated by stress.

Static modulus: The ratio of stress to strain under static conditions, calculated from static stress-strain tests, in shear, compression, or tension.

Stiffness: The relationship of load and deformation; a term often used when the relationship of stress to strain does not conform to the definition of Young's modulus. (See Stress-strain.)

Storage life: See Shelf life.

Strain ε: Although strain has several definitions, which depend upon the system being considered, for small deformations, engineering strain is applicable and is most common definition of strain.

Strain relaxation: See Creep.

Strands: A primary bundle of continuous filaments (or slivers) combined in a single compact unit without twist. These filaments (usually 51,102 or 51,204) are gathered together in the forming operations.

Strand count: The number of strands in a plied yarn or in a roving.

Strand integrity: The degree to which the individual filaments making up the strand or end are held together by the sizing applied.

Strength, flexural: See Flexural strength.

Stress σ: Most commonly defined as engineering stress, the ratio of the applied load P to the original cross-sectional area Ao.

Stress concentration: Magnification of the level of an applied stress in the region of a notch, void, or inclusion.

Stress corrosion: Preferential attack of areas under stress in a corrosive environment, where this factor alone would not have caused corrosion.

Stress crack: External or internal cracks in a plastic caused by tensile stresses less than that of its short-time mechanical strength. The stresses which cause cracking may be present internally or externally or may be combinations of these stresses. (See also Crazing.)

Stress relaxation: The decrease in stress under sustained constant strain, also called stress decay.

Stress-strain: Stiffness, expressed in kips per square inch (megapascals), at a given strain.

Stress-strain curve: Simultaneous readings of load and deformation, converted into stress and strain, plotted as ordinates and abscissas, respectively, to obtain a stress-strain diagram.

Structural adhesive: An adhesive used for transferring loads between adherends.

Structural bond: A bond that joins basic load-bearing parts of an assembly; the load may be either static or dynamic.

Substitutional: This refers to atoms of an alloying addition to a metal which occupy, or substitute, positions or atoms of the matrix material.

Superalloys: Multiphase alloys, usually cobalt- or nickel-based, which are very heat resistant.

Superplastic: Most metals are plastic in the sense that they can be plastically deformed up to a reduction in area at fracture of some 20%. Superplastic metals can be deformed several 100% before fracturing. The mechanism by which this occurs does not involve dislocations or glide planes, but occurs instead by a rotation of grains by diffusion at grain boundaries.

Surfacing mat: A very thin mat, usually 0.007 to 0.020 in (0.18 to 0.51 mm) thick, of highly filamentized fiber glass used primarily to produce a smooth surface on a reinforced plastic laminate. (See Overlay sheet.)

Surface resistance (electric): The surface resistance between two electrodes in contact with a material is the ratio of the voltage applied to the electrodes to that portion of the current between them which flows through the surface layers.

Surface resistivity (electric): The ratio of potential gradient parallel to the current along the surface of a material to the current per unit width of surface.

Surface treatment: A material applied to fibrous glass during the forming operation or in subsequent processes, i.e., size or finish. In carbon fiber manufacturing, surface treatment is the process step whereby the surface of the carbon fiber is oxidized in order to promote wettability and adhesion with the matrix resin in the composite.

Syntactic foam: A cellular plastic which is put together by incorporating preformed cells (hollow spheres or microballoons) in a resin matrix; opposite of foamed plastic, in which the cells are formed by gas bubbles released in the liquid phase by chemical or mechanical action.

Synthetic resin: A complex, substantially amorphous, organic semisolid or solid material (usually a mixture) built up by chemical reaction of comparatively simple compounds, approximating the natural resins in luster, fracture, comparative brittleness, insolubility in water, fusibility or plasticity, and some degree of rubberlike extensibility but commonly deviating widely from natural resins in chemical constitution and behavior with reagents.

Tack: Stickiness of an adhesive or filament-reinforced resin prepreg material.

Tack range: The length of time an adhesive will remain in the tacky-dry condition after application to the adherend and under specified conditions of temperature and humidity.

Tack stage: The length of time deposited adhesive film exhibits stickiness or tack or resists removal or deformation of the cast adhesive.

Tangent line: In a filament-wound bottle, any diameter at the equator.

Tape: A composite ribbon consisting of continuous or discontinuous fibers that are aligned along the tape axis parallel to each other and bonded together by a continuous matrix phase.

Tape laying: A fabrication process in which prepreg tape is laid side by side or overlapped to form a structure.

TD: Thoria-dispersed.

Tenacity: The strength of a yarn or of a filament of a given size; equals breaking strength divided by denier.

Tensile bar: A compression- or injection-molded specimen of specified dimensions used to determine the tensile properties of a material.

Tensile modulus: The ratio of the tension stress to the strain in the material over the range for which this value is constant.

Tensile strength or stress: The maximum tensile load per unit area of original cross section, within the gage boundaries, sustained by the specimen during a tension test. It is expressed as kips per square inch (megapascals). Tensile load is interpreted to mean the maximum tensile load sustained by the specimen during the test, whether this coincides with the tensile load at the moment of rupture or not.

TFRS: Tungsten-fiber-reinforced superalloy.

Theoretical density: The true crystal density for a material corresponding to the limit attainable through full density products without pores.

Thermal conductivity: Ability of a material to conduct heat; the physical constant for quantity of heat that passes through a unit cube of a substance in unit time when the difference in temperature of two faces is 1 degree.

Thermal expansion, Coefficient of: See Coefficient of thermal expansion.

Thermal oxidative stability: The resistance of a fiber, resin, or composite material to degradation upon exposure to elevated temperature in an oxidizing atmosphere. It is often measured as percent weight loss after exposure to a specified temperature for a set period of time. It may also be measured as the percentage of retained properties after elevated-temperature exposure.

Thermoplastic: Capable of being repeatedly softened by increase of temperature and hardened by decrease in temperature; applicable to those materials whose change upon heating is substantially physical rather than chemical and can be shaped by flow into articles by molding and extrusion. Many natural resins may be described as thermoplastic. Polyester and PVC are examples.

Thermoset: A plastic which changes into a substantially infusible and insoluble material when it is cured by application of heat or by chemical means. Prior to becoming infusible, thermo-setting polymers such as phenolic and melamine formaldehyde possess thermoplastic qualities which permit processing. Epoxy is an example.

Thixotropic: Gel-like at rest but fluid when agitated; having high static shear strength and low dynamic shear strength at the same time.

Thread count: The number of yarns (threads) per inch (millimeter) in either lengthwise (warp) or crosswise (fill) direction of woven fabrics.

TLM: Tape-laying machine.

TMC: Thick molding compound.

Toggle action: A mechanism which exerts pressure developed by the application of force on a knee joint, used as a method of closing presses and applying pressure at the same time.

Tolerance: The guaranteed maximum deviation from the specified nominal value of a component characteristic at standard or stated environmental conditions.

Tooling resins: Plastic resins, chiefly epoxy and silicone, that are used as tooling aids.

Torsional rigidity (fibers): The resistance of a fiber to twisting.

Toughness: The energy required to break a material, equal to the area under the stress-strain curve.

Tow: A large bundle of continuous filaments, generally 10,000 or more, not twisted, usually designated by a number followed by "K", indicating multiplication by 1,000; for example, 12 K tow has 12,000 filaments.

Transfer molding: Method of molding thermosetting materials in which the plastic is first softened by heating and pressure in a transfer chamber and then forced by high pressure through suitable sprues, runners, and gates into the closed mold for final curing.

Transfer pot: (1) A heating cylinder. (2) Transfer chamber in a transfer mold.

Transformation: The transition from one crystalline phase to another. A phase transformation can be brought about either by a diffusion controlled reaction or by a diffusionless (martensitic) shear process. The transformation is brought about essentially by chemical changes in energy of atomic bonding that occur on changing the temperature.

Transient liquid phase sintering: A sintering cycle characterized by the formation and disappearance of a liquid phase during heating. The initial compact has at least two differing chemistries and the first liquid must be soluble in the remaining solid.

Transition temperature: The temperature at which the properties of a material change.

Transverse rupture strength: A three point fracture test applied to brittle materials or green compacts to assess relative strength.

Twist: The turns about its axis per unit of length in a yarn or other textile strand. (See also Balanced twist.)

TZM: Trade name of molybedenum alloy wire.

UDC: Unidirectional composites.

UHM: Ultrahigh-modulus.

Ultimate elongation: The elongation at rupture.

Ultimate tensile strength: The ultimate or final stress sustained by a specimen in a tension test; the stress at moment of rupture.

Ultraviolet: Zone of invisible radiations beyond the violet end of the spectrum of visible radiations. Since ultraviolet wavelengths are shorter than the visible, their photons have more energy, enough to initiate some chemical reactions and to degrade most plastics.

UMC: Unidirectional molding compound.

Unbond: Area of a bonded surface in which bonding of adherends has failed to occur, or where two prepreg layers of a composite fail to adhere to each other; also denotes areas where bonding is deliberately prevented so as to simulate a defective bond.

Undercut: Having a protuberance or indentation that impedes withdrawal from a two-piece, rigid mold; any such protuberance or indentation, depending on the design of the mold (tilting a model in designing its mold may eliminate an apparent undercut).

Unidirectional: Refers to fibers that are oriented in the same direction, such as unidirectional fabric, tape, or laminate, often called UD.

Unidirectional laminate: A reinforced plastic laminate in which substantially all the fibers are oriented in the same direction.

Urethane plastics: Plastics generally reacted with polyols, e.g., poly esters or poly ethers, when reactants are joined by formation of a urethane linkage.

Vacancy: A "point defect" in a crystal, an atom missing from the lattice structure. It has the important function of aiding the process of diffusion.

Vacuum-bag molding: A process for molding reinforced plastic in which a sheet of flexible transparent material is placed over the lay-up on the mold and sealed. A vacuum is applied between the sheet and the lay-up. The entrapped air is mechanically worked out of the lay-up and removed by the vacuum, and the part is cured; also called bag molding.

Van der Waals' Forces: Interatomic and intermolecular forces of electrostatic origin. These forces arise due to the small instantaneous dipole moments of the atoms. They are much weaker than valence-bond forces and inversely proportional to the seventh power of the distance between the particles (atoms or molecules).

Vapor: A gas at a temperature below the critical temperature so that it can be liquefied by compression without lowering the temperature.

Veil: An ultrathin mat similar to a surface mat, often composed of organic fibers as well as glass fibers.

VIM: Vibrational microlamination.

Virgin filament: An individual filament which has not been in contact with any other filament or any other hard material.

Viscosity: The property of resistance to flow exhibited within the body of a material expressed in terms of relationship between applied shearing stress and resulting rate of strain in shear.

Void content: The percentage of voids in a laminate can be calculated by the use of the formula void % = 100 - x, where x is usually a weight percent.

Voids: Gaseous pockets trapped and cured into a laminate; unfilled spaces in a cellular plastic substantially larger than the characteristic individual cells.

Volatile content: The percent of volatiles drive off as a vapor from a plastic or an impregnated reinforcement.

Volatile loss: Weight loss by vaporization.

Volatiles: Materials in a sizing or a resin formulation capable of being driven off as a vapor at room temperature or slightly above.

Warp: (1) The yarn running lengthwise in a woven fabric; a group of yarns in long lengths and approximately parallel, put on beams or warp reels for further textile processing, including weaving. (2) A change in dimensions of a cured laminate from its original molded shape.

Warpage: Distortion in a compact which occurs during sintering or heat treatment typically due to nonuniform heating or density gradients in the initial compact.

Water absorption: Ratio of the weight of water absorbed by a material upon immersion to the weight of the dry material. (See also Moisture absorption.)

Water jet: A high-pressure stream of water used for cutting organic composites.

Weathering: The exposure of plastics outdoors. In artificial weathering plastics are exposed to cyclic laboratory conditions of high and low temperatures, high and low relative humidities, and ultraviolet radiant energy, with or without direct water spray, in an attempt to produce changes in their properties similar to those observed on long continuous exposure outdoors. Laboratory exposure conditions are usually intensified beyond those in actual outdoor exposure to achieve an accelerated effect.

Weave: The particular manner in which a fabric is formed by interlacing yarns and usually assigned a style number. In plain weave, the warp and fill fibers alternate to make both fabric faces identical; in satin weave, the pattern produces a satin appearance, with the warp tow over several fill tows and under the next one (for example, eight-harness satin would have warp tow over seven fill tows and under the eighth).

Web: A textile fabric, paper, or a thin-metal sheet of continuous length handled in roll form, as contrasted with the same material cut into sheets.

Weft: The transverse threads of fibers in a woven fabric; fibers running perpendicular to the warp; also called filler, filler yarn, woof.

Weldability: A material has good weldability when it can be joined by fusion welding without problems of cracking or excessive porosity occurring.

Wet flexural strength (WFS): The flexural strength after water immersion, usually after boiling the test specimen for 2 h in water.

Wet lay-up: The reinforced plastic which has liquid resin applied as the reinforcement is laid up; the opposite of dry lay-up or prepreg. (See also Dry lay-up, Prepreg.)

Wet-out: The condition of an impregnated roving or yarn wherein substantially all voids between the sized strands and filaments are filled with resin.

Wet-out rate: The time required for a plastic to fill the interstices of a reinforcement material and wet the surface of the reinforcement fibers; usually determined by optical or light-transmission means.

Wetting agent: A surface-active agent that promotes wetting by decreasing the cohesion within a liquid.

Wetting angle: The equilibrium angle formed by a liquid-solid-vapor combination; it provides a relative measure of spreading and capillary forces.

Wet winding: In filament winding the process of winding glass on a mandrel where the strand is impregnated with resin just before contact with the mandrel. (See also Dry winding.)

Whisker: A very short fiber form of reinforcement, usually crystalline and have almost no crystalline defects. Numerous materials, including metals, oxides, carbides, halides, and organic compounds, have been prepared in the form of whiskers. They are often used to reinforce resin- and metallic-matrix composites.

Winding, biaxial: See Biaxial winding.

Winding pattern: (1) The total number of individual circuits required for a winding path to begin repeating by laying down immediately adjacent to the initial circuit. (2) A regularly recurring pattern of the filament path after a certain number of mandrel revolutions, leading to the eventual complete coverage of the mandrel.

Winding tension: In filament winding the amount of tension on the reinforcement as it makes contact with the mandrel.

Wire: A metallic filament.

Working life: The period of time during which a liquid resin or adhesive, after mixing with catalyst, solvent, or other compounding ingredients, remains usable. (See also Gelation time, Pot life.)

Woven fabrics: Fabrics produced by interlacing strands at more or less right angles.

Woven roving: A heavy glass-fiber fabric made by weaving roving.

Wrinkle: A surface imperfection in laminated plastics that has the appearance of a crease in one or more outer sheets of the paper, fabric, or other base which has been pressed in.

Wrought: A material fabricated by conventional fusion metallurgy techniques, with a final stage of plastic deformation to improve mechanical properties.

X axis: The axis in the plane of the laminate used as 0° reference; the y axis is the axis in the plane of the laminate perpendicular to the x axis; the z axis is the reference axis normal to the laminate plane in composite laminates.

XMC: Directionally reinforced molding compound.

Yarn: An assemblage of twisted fibers or strands, natural or manufactured, to form a continuous yarn suitable for use in weaving or otherwise interweaving into textile materials. (See also Continuous filament.)

Yield point: The first stress in a material, less than the maximum attainable stress, at which an increase in strain occurs without an increase in stress. Only materials that exhibit this unique phenomenon of yielding have a yield point.

Yield strength: The stress at which a material exhibits a specified limiting deviation from the proportionality of stress to strain; the lowest stress at which a material undergoes plastic deformation. Below this stress, the material is elastic; above it, viscous.

Young's Modulus: The ratio of the applied load per unit area of cross section to the increase in length of a body obeying Hooke's law. (See Modulus of elasticity.)